# Introducing and Implementing Autodesk® Revit®

D1158551

# Introducing and Implementing Autodesk® Revit®

**LAY CHRISTOPHER FOX**
**JAMES J. BALDING, AIA**

**autodesk®**
press

THOMSON
DELMAR LEARNING

Australia • Canada • Mexico • Singapore • Spain • United Kingdom • United States

# Introducing and Implementing Autodesk® Revit®

Lay Christopher Fox / James J. Balding, AIA

**Vice President, Technology and Trades SBU:**
Alar Elken

**Editorial Director:**
Sandy Clark

**Senior Acquisitions Editor:**
James DeVoe

**Senior Development Editor:**
John Fisher

**Marketing Director:**
Cyndi Eichelman

**Channel Manager:**
Fair Huntoon

**Marketing Coordinator:**
Casey Bruno

**Production Director:**
Mary Ellen Black

**Production Editor:**
Thomas Stover

**Production Manager:**
Andrew Crouth

**Art & Design Specialist:**
Mary Beth Vought ·

**Editorial Assistant:**
Katherine Bevington

COPYRIGHT © 2005 by Delmar Learning, a division of Thomson Learning, Inc. Thomson Learning™ is a trademark used herein under license.

Printed in Canada
2 3 4 5 XXX 06 05 04

For more information contact Delmar Learning, Executive Woods 5 Maxwell Drive, PO Box 8007, Clifton Park, NY 12065-8007

Or find us on the World Wide Web at www.delmarlearning.com

**ALL RIGHTS RESERVED.**
No part of this work covered by the copyright hereon may be reproduced in any form or by any means—graphic, electronic, or mechanical, including photocopying, recording, taping, Web distribution, or information storage and retrieval systems—without the written permission of the publisher.

For permission to use material from the text or product, contact us by

Tel. (800) 730-2214
Fax (800) 730-2215
www.thomsonrights.com

**Library of Congress Cataloging-in-Publication Data:**
Fox, Lay Christopher.
   Introducing and implementing Autodesk Revit / Lay Christopher Fox, James J. Balding.-- 1st ed.
        p. cm.
   Includes bibliographical references and index.
   ISBN 1-4018-5049-9 (pbk.)
   1. Architectural design--Data processing. 2. Architectural drawing--Data processing. 3. Computer-aided design. I. Balding, James J. II. Title.

NA2728.F69 2005
729'.0285--dc22

2004010371

## NOTICE TO THE READER

Publisher does not warrant or guarantee any of the products described herein or perform any independent analysis in connection with any of the product information contained herein. Publisher does not assume, and expressly disclaims, any obligation to obtain and include information other than that provided to it by the manufacturer.

The reader is expressly warned to consider and adopt all safety precautions that might be indicated by the activities herein and to avoid all potential hazards. By following the instructions contained herein, the reader willingly assumes all risks in connection with such instructions.

The publisher makes no representation or warranties of any kind, including but not limited to, the warranties of fitness for particular purpose or merchantability, nor are any such representations implied with respect to the material set forth herein, and the publisher takes no responsibility with respect to such material. The publisher shall not be liable for any special, consequential, or exemplary damages resulting, in whole or part, from the readers' use of, or reliance upon, this material.

# CONTENTS

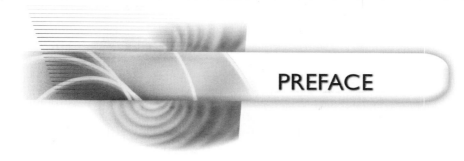

# PREFACE

## WELCOME

"Get the right tool for the job."

**—Mr. Natural**

The above words entered the popular lexicon in the free-wheeling 1960s, spoken by a metaphoric cartoon character who combined confident disregard of hidebound conventional thinking—a common commodity in those days—with a wise, practical outlook, rare enough at any time.

In today's design world the tools of the trade are computerized, and new, "different" applications or releases appear every couple of years. How to recognize the right tool amidst today's clamor of constant change? What marks a real advance in technology that will benefit designers and the firms that employ them? How do we take advantage of changes in tools while we're constantly involved in the rigors of meeting expectations and deadlines? Are there definitive answers to these questions?

This book has a double purpose:

- to introduce Autodesk Revit as the next generation in architectural design, a fully 3-dimensional building modeler of exceptional power, depth, and ease of use, that functions well at all phases of the design process, from concept sketches all the way through construction documentation

- to prepare individuals and firms for the process of implementing this new tool into their current work flow and processes. Successful implementation will require planning, training, and a transition phase from current Computer Aided Design applications. We will use AutoCAD, the worldwide standard, as a reference.

## CAD AND REVIT: WHAT ARE THE DIFFERENCES?

CAD stands for Computer Aided Design (or Drafting). This acronym explains succinctly the approach that mid-priced design software (what individuals and small firms can afford) has embodied until the last few years: the computer has been used as an aid to an age-old design process that hasn't changed substantially even as the tools became electronic. CAD software still requires the user to design line-by-line for the most part, and lines representing a wall in a plan do not create or appear in elevation, section, or 3D views of that wall.

Revit is a robust, affordable, and efficient software application created specifically for architectural designers. It is not based on the drafting language of lines, arcs, and circles. Revit is a building modeler database, with drafting tools included for those inevitable cases where simple linework is the most efficient way to illustrate a design concept or detail. Revit differs fundamentally from CAD; it is not a drafting application that has been expanded to attempt limited 3D content and views. Revit models consist of objects that are completely definable in size, structure, appearance, data content, and relation to other model components. All possible views of the model update with each command, including sheets and schedules. Much of current design process routine, especially propagating changes throughout a set of drawings, disappears as a result. Designers can spend the bulk of their time designing, not transcribing.

## WHAT DOES PARAMETRIC DESIGN MEAN?

Good design of any kind is composed of harmonious relationships, which are based on rules, or parameters. Revit objects are placed in relation to all other objects in a model, and parameters govern the relations. Designers can now establish relationships among building components that will hold through design revisions unless specifically removed. Suppose that an array of windows in a wall works best if the windows are equidistant from one another, the end windows are six feet from the corners, and the sills are two feet from the floor. Revit will recognize and apply those constraints, and adjust the positions of the windows if the wall length or floor height changes. The potential use of constraints flows throughout a model.

Revit carries parametric specifications much further than the previous simple example of windows in a wall. All types of components can carry information (job address, object size or placement specifications, supplier, cost, department, electrical requirements, etc.) according to supplied or user-defined parameters. This information can be collected in schedules or diagrams, and exported to other applications.

Revit's use of parameters underlies every aspect of the user interface. There is no drafting-oriented organization such as layers to control the appearance of objects; much of the learning curve for CAD applications therefore does not pertain to new users.

## WHO SHOULD READ THIS BOOK?

If you currently work in architectural design and your firm is considering or moving towards Revit, this book will explain the basics of the Revit tools that you, your co-workers, or your employees will be using every day. An important part of implementing Revit in an active design company's workflow is blending this powerful new tool with the often vast store of valuable information in existing drawing files. The first chapters of the text deal with importing AutoCAD information into Revit and exporting DWG files from Revit, so that the new technology can be transparent to clients, collaborators, or other departments or offices in your firm. We make mention throughout the text (see particularly Appendix A) of the effects Revit will have on workflow, process, personnel, and attitudes in organizations of nearly any size making this particular technology transfer.

If you are a student of architectural design, this book will show you the very latest software tool in the marketplace. You will be among the first to design buildings parametrically, entirely in three dimensions, without having had to learn the mechanics of representational lines, arcs, and circles beforehand. A few words of caution that your professors will no doubt echo: you will never become a good designer if you don't understand and practice good drafting techniques. Architects who can't successfully create clear sketches or readable detail pages do not win prizes, or even commissions.

## FEATURES OF THIS TEXT

### WHAT YOU WILL FIND IN THIS BOOK

Each chapter in the book covers specific, logically-sequential skill development using Revit in a planned progression from basic modeling through more complicated refinements, data extraction, and illustration. All work is based on tutorial exercises, with conceptual explanations, tips, notes, and cautions where appropriate. Each chapter ends with review questions suitable for quizzes.

Chapter 1 provides an overview of Revit's emerging position in the market of architectural design software, and contains a tutorial exercise showing how to create, revise, and document a simple building model. Chapter 2 covers template files, their uses, and pre-loaded content; this chapter also shows how Revit can import information from AutoCAD during the creation of a custom titleblock. These chapters should be of particular interest to company principals, supervisors, or CAD managers.

Chapters 3 through 10 use a hypothetical multi-building, multi-phase project on a college campus as the framework for their exercises. Chapter 3 covers the creation of a site plan with building shell, using information imported from outside Revit, and also shows how to export information from a Revit file to AutoCAD. Chapter 4 covers file linking, sketching, and massing components. Chapter 5 develops modeling techniques in the context of multi-user design teams, and introduces project phasing. Many of Revit's most powerful modes and tools are covered in these exercises.

Chapters 6, 7, and 8 concentrate on various aspects of annotation, within the context of the hypothetical project workflow. Chapter 6 covers customization and management of the appearance of annotation, from tags to titleblocks. Chapter 7 works through Revit's extensive and powerful scheduling capabilities. Chapter 8 moves from schedules into area and room plans, with customizable color fills. These chapters show—in broad strokes, concentrating on design development—how to create, manage, and enhance the "information" side of Revit.

Chapter 9 covers graphic output: perspective views, rendering, and animations. Chapter 10 wraps up with Revit's tools for organizing and managing design options, and a look at Revit family creation, itself a topic worthy of its own book. Some of the best features of the latest release appear here.

## WHAT YOU WON'T FIND IN THIS BOOK

This text is not a Revit help manual or command dictionary. Its scope covers all aspects of design in Revit—sketching, mass modeling, conceptual design, design development, construction documentation, data export, and illustration. Space limitations prevent us from covering every tool available in the software, or all the possible control options for the tools and methods we do illustrate. Where appropriate the text shows or mentions alternate ways of accessing tools or completing tasks.

## STYLE CONVENTIONS

Text formatting and style conventions used in this book are as follows:

| Text Element | Example |
| --- | --- |
| Step-by-Step Tutorials | 1.1 Perform these steps |
| Menu Selections | Select File>Import/Link>RVT… |
| Keys you press are in SMALL CAPS | Press CTRL or ESC |
| User input is in **bold** | Change the Eye Elevation value to **55' 6"**. |
| Files and Paths are shown in *italic* | *Content\Imperial Library\Doors* |

## HOW TO USE THIS BOOK

The order of chapters has been planned to follow normal design firm flow and process. Chapters 1 and 2 are introductions of particular value for instructors, firm principals, and CAD managers, although the exercises presume no extensive design experience. If you are new to Revit, we recommend that you complete the exercises in order. If certain chapters do not pertain to the work you or your firm does, feel free to skip topics as you see fit.

## FILES ON THE CD

Certain tutorial exercises use startup files. These files are on the CD enclosed with this book. Users can load files from the CD into convenient locations on their own systems, or as specified by their instructor(s). Revit files are on the CD in unzipped rvt format. Additional supplement files are also in native format (doc, etc.).

## WE WANT TO HEAR FROM YOU

We welcome your suggestions and comments regarding *Introducing and Implementing Autodesk® Revit®*. Please send your correspondence to:

The REVIT Team
c/o Autodesk Press
Executive Woods
5 Maxwell Drive
P.O. Box 15015
Clifton Park, NY 12065

Web sites:   www.autodeskpress.com
www.archimagecad.com

## REVIT RELEASES AND SUBSCRIPTIONS

Autodesk has announced plans to make substantive new releases of Revit roughly once a year, with an interim (.x) release halfway through the cycle. The latest build of Revit will be available as a download from the Autodesk website, or you may request a free CD. Without a license, Revit works for a 60-day demonstration period. After that time you will need to purchase a subscription in order to save work or print.

The exercises in this book were tested with Release 6.0. Release 6.1 appeared as the text was going to press. While the authors have tried to coordinate exercises and illustrations with the latest available information, there may be inconsistencies between this book and the behavior of the software you have installed. Exercise files on the CD are in 6.1 format. Check the CD or websites for additional information not available at press time.

## ABOUT THE AUTHORS

Lay Christopher Fox is an independent architectural drafter/illustrator, author, and educator. He has co-authored two books on Architectural Desktop® and one on Revit® prior to this text. He has written columns on AutoCAD® and Revit for *AUGIWORLD Magazine,* the bi-monthly newsletter of Autodesk User Group International (AUGI®). Chris is president of the Rochester (NY) Area AutoCAD Users Group, an adjunct professor at Rochester Institute of Technology (RIT)—teaching AutoCAD to engineering students—and was an instructor of AutoCAD, Architectural Desktop, and Inventor® at the former Autodesk® Training Center on the RIT campus. He has been a speaker at Autodesk University (2003). He received his undergraduate degree in literature at Harvard, and has pursued technical training in Rochester, NY, and Adelaide, South Australia.

Jim Balding is a licensed architect with more than 17 years of experience integrating technology into the architectural field. He is currently employed with Wimberly Allison Tong & Goo (WATG) in Newport Beach, CA. Jim earned his Bachelor of Environmental Science degree from the University of Colorado, Boulder. He has been a member of the Autodesk® Revit® Client Advisory board since its inaugural meeting and is currently serving as the Revit Product Chair for AUGI®. He is also currently serving as the South Coast Revit Users Group (SCRUG) president. Jim has spoken at several technology conferences and is one of the top-rated speakers at Autodesk University. He has developed a successful Autodesk Revit implementation strategy and is currently bringing the seven offices of WATG up to speed in its use.

## ACKNOWLEDGEMENTS

The authors would like to thank all of those who have contributed to this project and to our personal and professional development over these many years.

First, thanks to the editors at Delmar/Thomson Learning/Autodesk Press: Jim Devoe, John Fisher, Sharon Popson, MaryBeth Vought, MaryEllen Martino, and Katherine Bevington. Particular thanks to John Fisher for his patience and understanding.

Special thanks to those who reviewed the text and exercises:

Michael Gatzke—Des Moines Area Community College, Ankeny, Iowa
John Knapp—Metropolitan Community College, Omaha, Nebraska
Robert Mencarini—Autodesk

The authors would also like to acknowledge the help of others who took the time to review the book in its formative stages and assist with feedback. You went above and beyond the call of duty:

Scott Davis, Greg Cashen, Justin Kelly, Carl Walls

The international Revit community of users, while small in numbers compared to users of CAD products, is lively and committed to mutual growth. Recognition should go to Chris Zoog for starting a thriving web discussion group where users new and experienced share comments, concerns, and content. By now this forum may have moved to the site of Autodesk Users Group International (www.augi.com). Along with the ZoogDesign Discussion forums the Revit Users Group International (www.rugi.org) will be joining the augi.com site. Many thanks to Candy Cox, Jerry Cox and Robert Cox for their inspiration, commitment and dedication to founding and maintaining such a fine site. Revitcity is another source for shared content, forums and general information regarding Revit. Revit/Autodesk has long supported a busy newsgroup, alt.cad.revit. Many discussion groups exist at Autodesk's discussion group site, news.autodesk.com.

Jim Balding not only wrote a significant part of the book text, but he worked as Technical Editor on the rest. Little did he know what he was getting into—many thanks from Chris.

Thanks to John Shanley of Phoenix Creative Graphics for the book's composition, and to Rachel Pearce Anderson for copy editing. Your efforts keep the world from knowing how rough a rough draft can be.

Chris would like to make particular acknowledgement to John Clauson and David Harrington of AUGI, who encouraged me to keep writing newsletter articles for years, and Elise Moss (also of AUGI), who generously and graciously let me assist her on a number of book projects.

Jim would like to thank the many people that have had an influence on his career as an architect and leader in architectural technology. Many thanks to: Chuck Cutforth, Ed Conway, M. G. Barr, Gerald Cross, Curt Kamps, Charlie Wyse, Jim Grady, Perry Brown, and last but certainly not least, Larry "Figurehead" Rocha. Tom (Buck) and Collette, thank you very much for all of your love, support, and understanding while I was under your wings. I love you both very much.

Finally, we would both like to recognize the Revit Team—Leonid Raiz, Irwin Jungreis, Marty Rozmanith, Steve Burri, Dave Heaton, Rick Rundell, David Conant, the development and support staff, and the hard workers too numerous to name at the Factory. Your pride and inspiration shine through, and do you honor.

## DEDICATIONS

This book is dedicated to my wife, Diane, and sons, Judson and Luke. Without their understanding and patience there would be no book at all. Thank you for saying, "that's ok, we can do it another time," when there was too much work and too little time to do everything. I love you.

—Jim

This book is dedicated to my wife, Sally Goers Fox, a bright-eyed, intrepid bird of passage, with all my love and strength forever. Coming in on a wing and a prayer…

—Chris

# Executive Summary

## INTRODUCTION

This chapter addresses the needs of owners, principals, or upper-level executives of architectural design firms that currently produce drawing files and construction documents using AutoCAD or AutoCAD-based software. It will also be of use to program committees and chairpersons for technical schools or university architectural departments who seek to meet the current and future needs of the architectural design marketplace.

Students in drafting or design courses will find the "quick tour" exercise in this chapter a helpful introduction to the basic modeling and documentation tools in Revit, in preparation for more advanced concepts in later exercises.

## OBJECTIVES

- Understand Revit's emerging position as an architectural design tool
- Prepare for successful implementation of Revit in your current work process
- Experience an overview of the software

## WHAT'S A REVIT?

Autodesk, Inc. is the world's premier producer of mid-price design software. Its general-purpose AutoCAD drafting package is the world standard. It has sold millions of copies worldwide, and there is an entire industry of third-party developers dedicated to creating task-specific or industry-specific design applications based on the AutoCAD engine. Autodesk has developed and successfully marketed "Desktop" vertical applications to push AutoCAD into mechanical, architectural, civil, and land-development specialties.

Architectural Desktop is now in its sixth major release, Architectural Desktop 2005, with an approximate 250,000 licenses worldwide. In April of 2002, Autodesk acquired Revit Technology Corporation (formerly Charles River Software), who's sole product was Revit, a parametric building modeler package. "Revit" is taken from architect's slang for "revise it," a nearly constant refrain in building design. Autodesk has established

Revit as its non-AutoCAD based building design and documentation system and its strategic platform for software serving the building industry. Why the change, and what does it mean for architectural design firms using AutoCAD and Architectural Desktop?

Autodesk's move away from the DWG format (AutoCAD's output file) for architectural models is consistent with a previous transition the company made in the mechanical design field. Autodesk developed Inventor, a parametric modeler of 3D solids that uses file formats incompatible with the DWG, and has used it to replace the previously successful and widely accepted Mechanical Desktop. With Revit, Autodesk has simply purchased a corresponding non-AutoCAD technology for the building industry rather than build it in-house. Here again, Autodesk has been consistent. Architectural Desktop itself is an outgrowth of a purchase—Autodesk bought out Softdesk software in the mid 1990s. Softdesk had developed AutoArchitect, an add-on package to AutoCAD for architectural documentation. Autodesk reworked AutoArchitect and released it as Architectural Desktop; other packages Softdesk had developed turned into Autodesk's Building Systems.

Autodesk executives have recognized the limitations inherent in the DWG format, whether for coordinating large mechanical assemblies or the complexities of building design and construction documentation. The results of their aggressive search for the next phase of design software have been the development of Inventor and the purchase of Revit.

Firms using AutoCAD or Architectural Desktop, that may have a huge inventory of DWG files and an ongoing commitment to developing/purchasing/applying peripheral applications to enhance parts of their work process such as schedules or space planning, will undeniably face a transition period while they come to grips with this change in the tools of the trade. This book is designed to show what changes actually lie in store for firms and institutions that decide to embrace the new shape of architectural design, how to manage and make best use of the parametric capabilities of Revit, and what job skills, office practices, and client benefits are likely to develop in answer to the new conditions.

As the decision-makers of your respective firms or schools, whether large or small, your responsibilities include keeping your company or students equipped with the best tools, training, and techniques you can afford in order to attract, service, and satisfy clients. You already comprehend the amount of effort involved in maintaining a competitive edge, the need for constant awareness of possibilities for improvement, and the tendency for company culture to slide from satisfaction to complacency when success comes—or from eagerness to despair when the well seems to run dry. Whether you are a one-person firm or someone responsible for offices across the globe, your role means periodic reassessment of whether or how to bring change to your firm's work methods.

If you are reading this book because you have come to the conclusion that maintaining your company's current investment in AutoCAD is not as important as learning to make effective use of the next generation of tools, then you understand the costs of doing business and are willing to pay the price of keeping ahead of the pack rather than the price of falling behind.

## REVIT'S STRENGTHS

### SIMPLICITY MEANS SPEED

One of the biggest differences between Revit and any other CAD architectural design software currently in widespread use lies well under its surface, but shows up in the entire interface and organization of its information handling. Revit is a complete architectural design software application with drafting components, not drafting software that has been modified or expanded to make architectural design possible. This means a number of major differences from CAD packages, AutoCAD in particular.

- Every object created in Revit is classified according to its function (walls, doors, structural members, components such as furniture, etc.); the relations between objects are also based on architectural function. Walls "host" doors and windows and the necessary openings, for example. Geographic positioning, size, and other properties of objects can be explicitly controlled or made subject to parameters, so that relationships can be established which will survive changes. A door can always center on a given wall, and windows on either side can be equidistant from one another, no matter how long the wall or the size of the windows. This eliminates many steps checking and rechecking dimensions during design iteration (as shown in Figures 1–1 and 1–2).

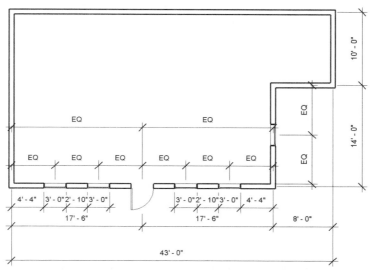

**Figure 1–1** *Doors and windows spaced equally in walls of certain lengths*

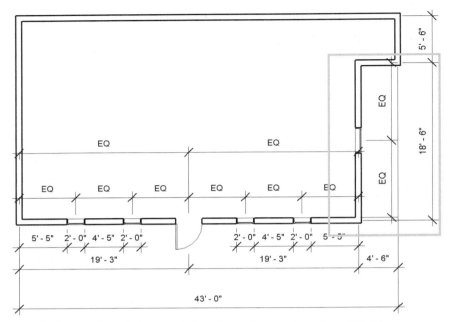

**Figure 1–2** *Drag walls and change window size—the equal spacing still holds*

- The visibility of objects is controlled in views (plan, elevation, section, 3D) according to properties of the views and the objects. There are no layers or their equivalent used, which means vast simplification of the interface. Revit's appearance is largely black lines on a white background, but this can be changed to as many colors or patterns for as many different purposes as the user may desire. The default visual settings will suffice for the great majority of users, and the visible lineweights correspond to printed lineweights. The function and detail level setting of each view control the detail visible—no material symbols or section hatching in walls at coarse scale or in reflected ceiling plans, but hatch visible in floor plans at fine scale (as shown in Figures 1–3 and 1–4). This eliminates a large amount of setup and correlation between the screen view and the plotted results—no plot configuration files or layer standards are necessary.

| Visibility | Edit... |
|---|---|
| Model Graphics Style | Hidden Line |
| Detail Level | Fine |
| Discipline | Coarse |
| Phase Filter | Medium |
| Phase | Fine |

**Figure 1–3** *The view detail level control*

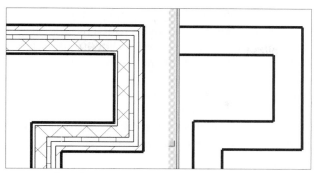

**Figure 1–4** *The same walls—fine detail level in the left view, coarse detail right*

- The views in Revit are extremely powerful—a fundamental organization principle. All come from the existing model, so changes in the model are automatically propagated throughout all possible views of it (see Figure 1–5). An elevation will always contain exactly the number of windows in the exact location that the plan shows, and a wall section will always show the correct components of the wall no matter how many times the wall type has been changed, with no manual updating procedure required of the user. This saves significant amounts of design time over the course of a project.

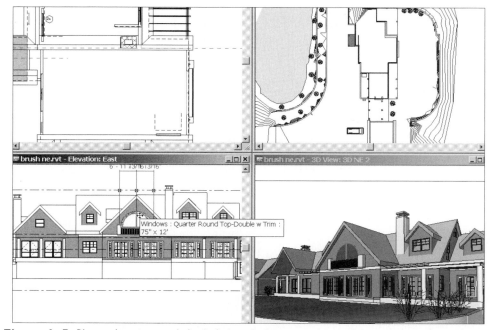

**Figure 1–5** *Plans, elevation, and shaded view. A change to the model in one view (note the highlighted window) shows immediately in the other views.*

- Revit includes "traditional" line-arc-circle drafting tools and editing commands, which are used, as one would expect, for sketching custom components or 2D detailing. There is no visible coordinate system in Revit, but a system of work planes on which one drafts or sketches. These work planes utilize reference planes that can be governed by parameters when creating components, so part of a column can always be twice as wide (or as long) as another part, or the column will always run from its base level to the one above (plus or minus an offset value if desired) even as floor heights change. Design intent can be specified to carry through the iterative process that is architectural design.

Revit also separates itself from other available architectural applications in a number of ways:

- Plotting sheets and annotations are smart—Figure 1–6 shows how a plan tag will renumber itself according to the sheet and detail number of its reference as the sheet package develops, eliminating a source of much error and time spent checking annotations in the latter stages of documentation.

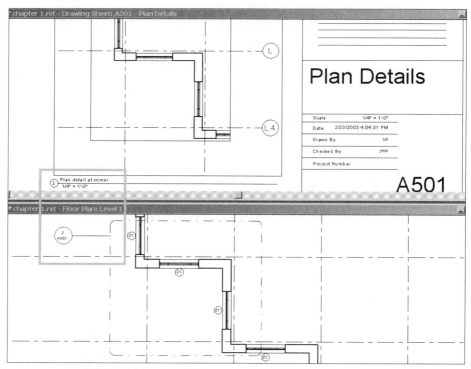

**Figure 1–6** *A callout in the plan view knows its referenced sheet and detail number. Move the detail to another sheet, and the plan symbol updates wherever it is visible.*

- Nearly any object with parametric properties can be scheduled, and nearly all objects are or can easily be made parametric. Schedules (which are views) are live and reflect changes in quantities or properties.

- Room and area tags are easy to create and apply, and feed directly into space planning calculations, schedules, and colored views.

- Revit includes a rendering engine (currently AccuRender) that makes lit, colored views, with material maps applied, possible at any time in the design process, for any part of the model. Early visualizations of conceptual models are so easy some Revit users have resorted to subterfuge so as not to respond "too quickly" to client requests for renderings.

All this boils down to greater ease of use for more users than design systems that are based on drafting methods rather than design categories. There have been totally object-based architectural design packages available before, but they were not as easy to use, nor did their interfaces provide the ready visual understanding and ease of plotting that Revit does, and those applications are no longer viable in the market.

## QUICK PRODUCTIVITY

Revit's entire system has been carefully put together to provide a design environment that architects will readily understand and find easy to use—all software strives for the elusive goal of providing an intuitive experience for the user, and Revit, in its basics, comes very close to the ideal.

As a result, designers who may not be CAD-literate can expect to create models, edit them, change their appearance, and create specifications in considerably less time than it takes to learn a drafting package and use that drafting package to create a building design. Revit does not require existing CAD skills, and much productive time can ride on that simple difference.

Experienced CAD designers who are not thrown for a loop by not having layers to turn on and off, or by not having X-Y-Z coordinates to enter when defining the location of an object, can find themselves productive in a surprisingly short training cycle. Competent architectural drafters who are not simply CAD jockeys will find their drafting skills as much in demand as ever. There is a world of 3D content yet to create for this young program.

## SIZE AND STRENGTH MATTER

Revit is a robust application, capable of generating complex and complete building models of considerable (though not enormous) size in its current form. Its hardware requirements are substantial but fully in line with those of current CAD workstations.

Revit is designed to hold a single building model in each file. It contains a mechanism (Worksets) that allows teams to parcel out and work on parts of a model and correlate their updates to a central file. File linkages (corresponding to external references in AutoCAD-based products) allow for assemblies of models or campus layouts combined in a single file location.

Since Revit contains both costing and rendering modules, architects can produce (at least nominal) estimates and detailed presentation images without leaving the application. The controls of the AccuRender module within Revit are easy to learn, production and capture of useful images or animations is quick, and no updating of the model or file linking is necessary. For firms with a separate graphics department apart from building design teams, Autodesk has white-paper information on exporting Revit models into VIZ, and the company is working on providing connections between Revit and VIZ on the order of the current links between Architectural Desktop models and VIZ.

For those worried that Revit might represent a "flavor of the month" that will soon disappear and leave early adopters worse off than before, Autodesk's position in the global market as a premier supplier of design and visualization applications provides a stable development and support environment for Revit.

Architecture is a notoriously conservative industry. As late as the mid-1990s nearly half of the architectural firms in the U.S. had next-to-no investment in CAD technology in their offices. Before the development of Architectural Desktop, when AutoCAD was Autodesk's flagship design product, each new release of the software occasioned hand-wringing and consternation as firms agonized over the decision whether to keep CAD stations and standards current or not. Now that the software decision before today's firms involves a new file format, different design techniques, changing office roles, and fluid client relationships, the basic question is actually simpler than before.

Autodesk stated from the moment it acquired Revit that this application is its building modeling solution of the future. The company obviously does not expect wholesale rejection of its initiative from the architecture industry. Nor does Autodesk expect quick or eager acceptance of this announced move away from the only CAD standard many architectural firms have ever known, and the myriad third-party applications developed by an entire industry devoted to extending AutoCAD and the DWG.

Autodesk has substantial resources to help firms making a commitment to Revit understand and manage the transition. Neither the product nor the company is going to evaporate any time soon. Autodesk, while aggressive when necessary in acquiring technology, is a company dedicated to developing and maintaining a growing stable of industry-leading design products. The company takes its leadership position seriously enough to know the dangers of out-pacing the market. Revit is Autodesk's proposition to those architectural firms that also take leadership seriously.

## REVIT AS THE NEW KID ON THE BLOCK

Autodesk has come out with more than 15 releases of AutoCAD in its 20-year history and 5+ Architectural Desktop releases. Revit is currently in its sixth major version after its first public release in April of 2000. With no underlying familiar CAD framework in prior release (as AutoCAD underlies Architectural Desktop) providing an existing user base, Revit is still new to the broad market and has modest adoption to date—currently about one-twentieth as many licensed seats as Architectural Desktop. To reach the majority of the market will clearly take years. Whether your firm or institution is an aggressive front runner or a trend follower, you should take some issues into account in your implementation planning.

### FILE COMPATIBILITY

Given these relative numbers, the issue of file compatibility is important, both to take advantage of existing DWG files that a firm may have in inventory and to coordinate successfully with consultants and other team members operating in CAD. This need not be a reason to avoid or delay exploring Revit's capabilities. Revit is built to interface easily with AutoCAD products. "Interoperability" is the word Autodesk uses to emphasize the two-way connectivity. Revit can link to AutoCAD files as easily as AutoCAD does. Revit imports (reads in) 2D DWG files with full layer, linetype, and color information. Revit also outputs 2D or 3D DWG files that are tidy and complete, with layers logically assigned.

### TRAINING

Executives facing the decision to begin using Revit will need to budget for training as part of the implementation process. Revit ships with a series of tutorials and the company provides free live training sessions online. So, for those individuals who can successfully self-teach using the provided lessons, there will be no cost for materials or instruction. Autodesk will provide a training team for larger firms making an investment in Revit. As of now, only a few resellers and Autodesk Training Centers provide scheduled classes or on-site instruction sessions by certified instructors.

### PROJECT SIZE

Large building projects are difficult to manage no matter what software you use to design and document them, and with Revit's single-file building model, larger projects have proportional challenges. Your first projects in Revit should be relatively modest, perhaps in the 100,000 square foot range. With careful configuration of template files, worksets, and model linking, firms are presently doing projects in Revit approaching a million square feet. Each release of Revit has improved performance in project size, and with Revit 6 an experienced Revit team can confidently tackle just about any building design.

Revit's file is based on a single building model that carries all the relevant information about the building structure and contents, rather than a system of individual files created per floor, for instance, and brought back by reference into a shell or assembly file. Revit's system of Worksets allows for a growing team to work on a model as a design develops by "checking out" parts of the model with a central file holding and coordinating data. This system has the advantage of simplifying what can grow into a bewildering nest of references in a large model.

## DATA OUTPUT LIMITATIONS

Revit is very efficient at presenting building or component information in schedules. The data tabulated in schedules can be exported to a spreadsheet or database such as MS Access via ODBC, the Microsoft-developed process for sharing field information between applications. While this exported data is powerful, complete and useful for many applications outside Revit, the ODBC process is limited and cumbersome process compared to the Data Extract function in the latest release of AutoCAD. While ODBC export allows for updating an external file based on changes in the Revit model, the process at present cannot be reduced to a one- or two-click operation, nor automated. Revit's Export Schedule capability does allow for quick conversion of any scheduled information into *.txt* format for import into other applications, and holds promise of developing into a live link between the Revit file and external data-parsing applications.

## NO PROGRAMMING INTERFACE

Revit does not contain a programming interface, as does AutoCAD. One of Auto-CAD's strengths from the very beginning was the decision early on to embed a LISP interpreter in the application and make this interpreter available. This allowed developers of all stripes to augment basic "vanilla" AutoCAD with routines of sometimes startling ingenuity to aid in creating CAD content, to automate repetitive tasks, and to read values from the drafted geometry. In recent years with the rise of Microsoft Visual Basic as a programming language common to many Windows applications, AutoCAD has developed Visual Basic for AutoCAD as a way for the application to interface with and react to other applications.

As a single-purpose building modeler rather than a general-purpose drafting tool, Revit does not have the same need for external hooks as AutoCAD, and it has been carefully designed to eliminate many of the repetitive tasks that AutoLISP was often used to speed up. There is currently debate in the Revit user community as to the need for an API (advanced programming interface), with some feeling that the base program will become less highly developed if Autodesk's programmers know that there are third-party developers working on tweaks, routines, or application capabilities to sit on top of Revit. Based on Autodesk's experience with Inventor, which debuted with no API but which now has a growing number of outside developers creating an industry around it, we can expect to see Revit showing a programming interface in future releases—probably based on a version of Microsoft Visual Basic.

## PREPARATION FOR SUCCESSFUL IMPLEMENTATION

### SHOCK TROOPS

Only the foolishly optimistic would expect to proclaim a new company standard without working to overcome inertia and objections. Introducing new concepts and tools to a work arena is a process, not an event, and takes preparation and education even more than vision and enthusiasm.

Most firms have members who are eager to learn, willing to push the limits of their knowledge, ready to try new things. All firms have members who are most productive when comfortable with routine, not very interested in expanding their skills if risk is potentially involved, and slow to adapt to new circumstances. Quick and ready learners are not necessarily better designers or more useful team members than steady-as-she-goes types (and may not be the company's AutoCAD or ADT experts), but can be valuable as "early adopters" who spearhead an effort to learn, utilize, and broadcast an important shift in work standards.

- Assemble a team whose members are willing and capable of working with new software and techniques
- Develop a plan for training
- Set a schedule for individual and group training
- Establish a forum for communication: team members to each other, to CAD manager, and to project manager
- Use an initial project as a focus for the training and development of team work practices

### PROJECT BY PROJECT

Decide beforehand on the first project for the spearhead team to work on in Revit. Considerations for this project may involve building size, client requirements, and schedule.

- Understand your company's design process, including your current standards
- Understand the scope of the project
- Clearly define and communicate the project goals
- Establish project and team evaluation criteria for later review
- Expect and allow for differences in workflow from current company process

### REMEMBER YOUR ROOTS

Revit is built to take maximum advantage of AutoCAD—"interoperability" is the word Autodesk likes to use. Revit can read in AutoCAD files either via link (equal to external reference) or import, and will retain layer, color, and linetype information. It will also export linework from its models to AutoCAD on a by-view basis. Its export DWG files use layer, color, and linetype information per default built-in standards, which are fully customizable to your current company standards.

- Team members can import existing site data in DWG, DGN (MicroStation), or DXF format. Use Revit's topographic tools to build the 3D site surface.
- Link or import existing plans for reference in creating new project content
- Scan in and trace over hand concept sketches
- Link or import DWG/DGN/DXF files received from consultants
- Export DWG/DGN/DXF files to consultants
- Link or import standard details and symbols from company archives during CD phase

## EXPAND THE TEAM

Once a first project has been completed or nearly completed in Revit, the company can build on that experience to develop the use of Revit as useful enterprise software.

During post-project analysis from the first Revit project, let other members of the company know about the successes of the spearhead team, and be sure to evaluate it honestly and communicate problem areas. Was training adequate? Were expectations clearly expressed? Was support available in a timely way?

Target the next projects, divide the Revit team between the projects, and plan and schedule training for new team members. In this way, knowledge and experience with Revit can spread the most effective way—from user to user. People learn most efficiently when actually working with their new knowledge, and when there is a free flow of information—tips, tricks, questions, and answers pertaining to work in hand.

## QUICK TOUR—DESIGN, VIEW, AND DOCUMENT A BUILDING

Revit installs with a substantial set of tutorial exercises that cover all aspects of the program, plus the company provides live seminar-type exercises on specific topics online. This book contains exercises and lessons to introduce Revit's commands and methods that are substantially different from the ones provided with the software. First up is a short exercise to give you a very quick look at Revit's most basic tools and integrated display capabilities.

 **Tip:** Autodesk makes Revit available as a free download from the www.autodesk.com/ revit website, and will send a CD on request. Revit will work in demo mode (no saving or printing, viewing and experimenting only) without a license. A fully functional time limited trial is available at no cost.

## REVIT COMMANDS AND SKILLS

| | |
|---|---|
| *Walls* | *Roof* |
| *View Properties* | *Section* |
| *Doors* | *Sheet—Drag and Drop Views* |
| *Windows* | *View Scale* |
| *Editing—Copy, Mirror, Element Properties* | *Activate View* |
| | *3D View* |
| *Dimensions—EQ toggle* | *Shade View* |
| *Add Level* | |
| *Floor* | |

## Exercise 1. Lay out a simple building

1.1   Launch Revit from the desktop icon. (You can also start Revit by selecting Start>Programs>Autodesk Revit.) Revit will open an empty project. The View Window (drawing area) will open to a Level 1 plan view (see Figure 1–7). Four elevation symbols will be visible in the View Window. Menus on the Menu Bar at the top of the screen and in the tabs of the Design Bar control modeling and other tools. The Project Browser located between the View Window and Design Bar shows available views in customizable tree form.

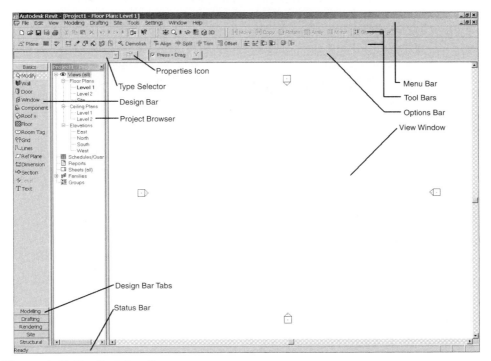

**Figure 1–7** *The Revit interface*

1.2 From the Basics tab on the Design Bar, select Wall. The cursor will change appearance to a pencil symbol and the Type Selector drop-down box on the Options Bar will display a generic wall type, shown in Figure 1–8.

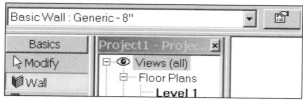

**Figure 1–8** *The Wall tool on the Basics tab*

1.3 Select the down arrow at the right of the Type Selector, as shown in Figure 1–9, to expand the list.

**Figure 1–9** *Expand the list of default wall types*

1.4 Scroll up the list and select Basic Wall: Exterior – Brick and CMU on MTL Stud as shown in Figure 1–10. Accept the default Height and Location Line values on the Bar in Figure 1–10.

**Figure 1–10** *Brick and CMU wall type*

1.5 Check the Chain Option so that your selecting will draw walls continually until you stop the command. Leave the Straight Line tool depressed. Leave the Offset value at 0' 0". (See Figure 1–11.)

**Figure 1–11** *Settings for the walls*

1.6 Place the cursor in the View Window left of center and above the middle, but inside the elevation markers. If you pause the cursor, a Tooltip will appear that reads Click to enter wall start point, as shown in Figure 1–12. This message will also appear in the Status Bar at the bottom of the screen.

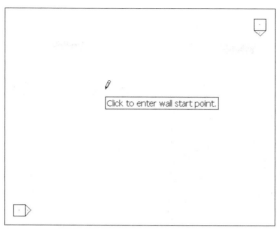

**Figure 1–12** *About to draw walls*

1.7 Left-click to start the wall. Pull the cursor to the right. This will create the wall with its exterior face to the top of the screen. A temporary dimension will appear showing the length of your wall, and an angle value, as shown in Figure 1–13. If you pull directly right, a dashed green reference line will appear and a Tooltip will verify that the wall is horizontal. Pull the cursor until the dimension reads 40'-0" and left-click to set the wall endpoint.

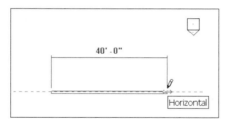

**Figure 1–13** *The first wall*

1.8 Pull the cursor down to start the second wall. If you need to zoom in closer to see your work, right-click to get a context menu with zoom options, and choose Zoom In Region. Then select two points as appropriate (see Figure 1–14). You can also use the scroll wheel if your mouse is so equipped.

**Figure 1–14** *Right-click to access zoom controls*

1.9     As you pull the cursor down, the temporary dimension, angle readout, reference line, and Tooltip all appear, as shown in Figure 1–15. Make this wall 40' long as well, and then left-click.

**Note:** The actual dimensions of the walls we are creating in this exercise do not matter. Nor do we really care if the walls are perpendicular to one another. Revit is designed for quick sketching and contains many tools to establish/revise dimensions and relationships later in the design process. These tools are covered in the tutorials and will be illustrated in later lessons.

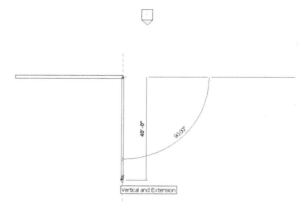

**Figure 1–15** *The second wall is perpendicular to the first*

1.10    Pull the cursor to the left 40' and left-click. Revit will pick up alignment points, snap to them, and display them in the Tooltip, as shown in Figure 1–16.

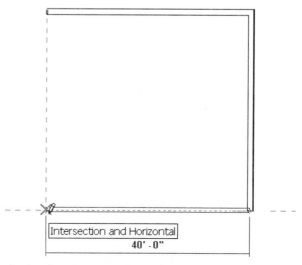

**Figure 1–16** *The third wall ends even with the first one*

1.11 Pull the cursor up and Revit will snap to your start point. Left-click to finish this simple exterior (see Figure 1–17). To exit the Wall tool, hit ESC twice, or choose Modify from the Design Bar, as shown in Figure 1–18. This puts the cursor back into select mode.

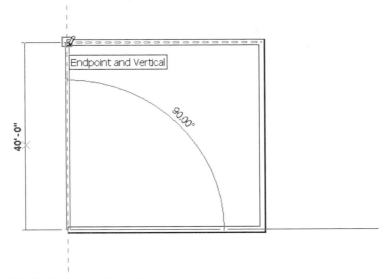

**Figure 1–17** *Finish the outside walls*

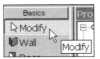

**Figure 1–18** *The Modify command is at the top of all the Design Bar tabs*

1.12 Put the cursor in the View Window and right-click. The bottom option is View Properties (see Figure 1–19). Select (left-click) View Properties to bring up Revit's main visibility control, as shown in Figure 1–20.

**Figure 1–19** *Select View Properties from the right-click menu*

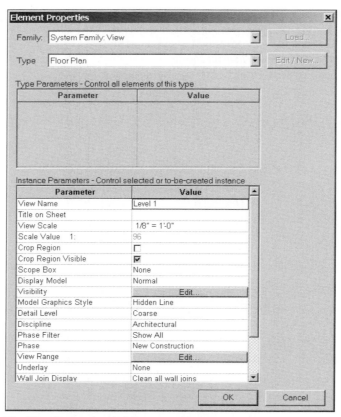

**Figure 1–20** *The View Properties dialog*

1.13 Click in the Value field for Detail Level and change it from Coarse to Fine, as displayed in Figure 1–21.

| Visibility | Edit... |
|---|---|
| Model Graphics Style | Hidden Line |
| Detail Level | Fine |
| Discipline | Coarse |
| Phase Filter | Medium |
| Phase | Fine |

**Figure 1–21** *Three levels of view detail*

1.14 Choose the Edit button in the Visibility value field to bring up the Visibility/Graphic Overrides dialog, shown in Figure 1–22. It has three tabs. Choose the Annotation Categories tab and uncheck Elevations in this view. Select OK twice to exit the Visibility and View Properties dialogs to reveal the model.

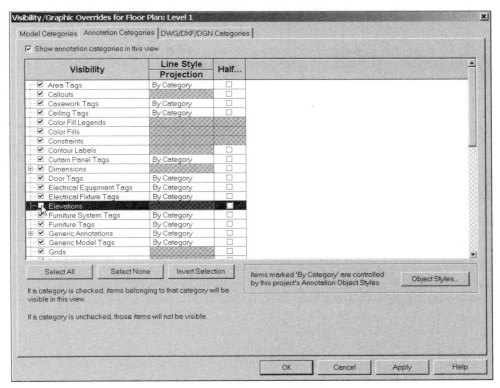

**Figure 1–22** *Turning off the elevation symbols in this plan view*

1.15  Right-click the context menu and choose Zoom to Fit, as shown in Figure 1–23. Figure 1–24 shows the adjusted view, a complex wall structure with hatching.

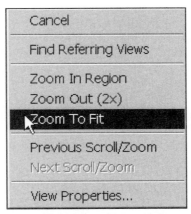

**Figure 1–23** *The quickest way to zoom to the edges of the model*

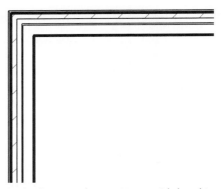

**Figure 1–24** *The walls now display complex structure with hatching*

## Exercise 2. Add doors and windows

2.1 Select Door on the Basics tab of the Design Bar. Select Single Flush: 36" x 84" from the available choices. Locate one in the left wall 4' from the upper wall. Revit will snap to the walls and read the wall location as you move the cursor. Holding the cursor to the interior or exterior side will set the swing and hand (see Figure 1–25).

**Figure 1–25** *Locate an exit door near the upper wall*

2.2 Set a second door 4' from the bottom wall. If you need to adjust the swing of a door, control arrow sets appear when it is selected. Select the appropriate control arrow to toggle the swing or hand (see Figure 1–26).

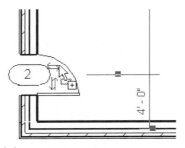

**Figure 1–26** *Locate a second door*

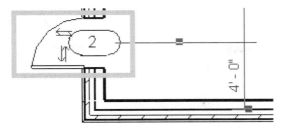

**Figure 1–27** *Adjust the door swing if necessary*

2.3    The default list of doors does not include a double door. While the door tool is active, select the Load button (shown in Figure 1–28) on the Options Bar to bring up a content file browser dialog as shown in Figure 1–29. Select the *Doors* folder in the *Imperial Library* folder, and then choose Open.

**Figure 1–28** *Look for content from the library*

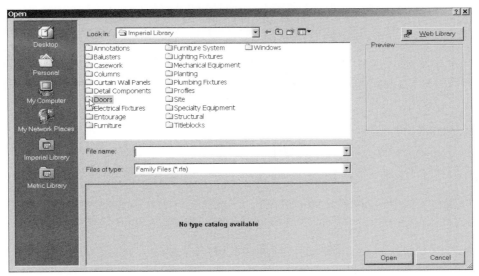

**Figure 1–29** *The content library browser dialog. You can create your own folders, or browse the Web from here.*

2.4    In the *Imperial Library/Doors* folder, select *Double-Glass 1.rfa* as shown in Figure 1–30, and then choose Open.

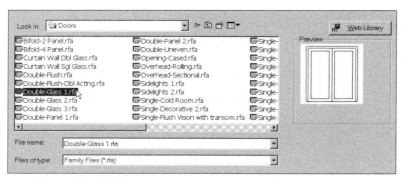

**Figure 1–30** *The double glass door type*

2.5 Locate an instance of the double glass door midway between the other two doors (Revit will snap to their center points) as shown in Figure 1–31, then hit ESC or choose Modify to terminate the Door tool.

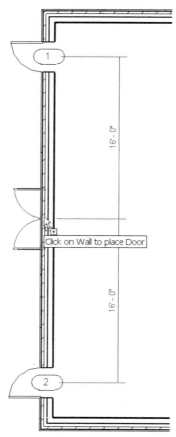

**Figure 1–31** *Three doors in the west wall*

2.6   Choose the Window tool button in the Basics tab on the Design Bar.

2.7   Select Fixed: 24" x 48" from the list of available windows.

2.8   Locate an instance of this window in the west wall, 8' from the centers of the lower and middle doors (see Figure 1–32). If you hold the cursor to the exterior side of the wall, the dimensions will appear to the right and the glazing will be on the left (exterior) side.

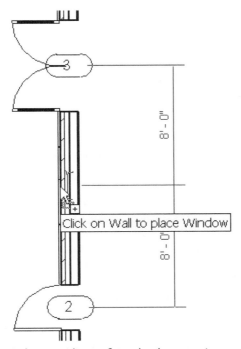

**Figure 1–32** *The first window, equidistant from the door opening centers*

2.9   Add a second window midway between the center and upper doors.

2.10  Change the window selection to Fixed: 36" x 72" and place 5 instances in the lower wall. Do not be concerned about their location during placement. Use the control arrows if necessary to keep the glazing on the exterior side of the wall. Hit ESC or choose Modify to terminate the window placement.

## Exercise 3. Edit the model

3.1   Choose Dimension on the Basics tab of the Design Bar, shown in Figure 1–33.

**Figure 1–33** *The Dimension command, used for locating components precisely*

3.2   The dimensioning symbol will appear next to the cursor (see Figure 1–34). Hold the cursor over the left-hand wall. It will snap to the wall centerline, which will highlight, and the Tooltip will read out the wall properties. Select the wall, and then hold the cursor over each window in turn from left to right—a center mark for each window will highlight to show when the window is selectable—and select the right wall last to complete the dimension string.

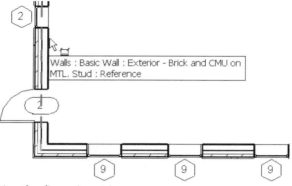

**Figure 1–34** *Starting the dimension string*

3.3   Pull the cursor above the wall, inside the building, to place the dimension string. After you select, the grips for the dimensions will appear, and the letters EQ with a slash through them (see Figure 1–35). Select the EQ symbol, it is a toggle to lock the objects dimensioned in an equidistant relationship (see Figure 1–36).

Architectural drafting practice locates dimension strings on the exterior of building walls for purposes of clarity in busy plan pages. In our simple model we want to keep the selections easy for right now. It would be a simple matter to relocate the string later, in the construction document phase of a real project.

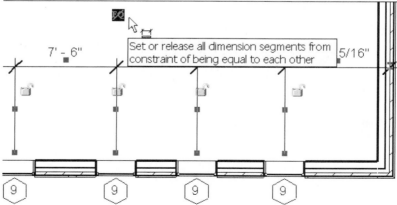

**Figure 1–35** *Setting the EQ toggle*

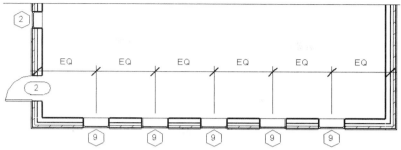

**Figure 1–36** *The results of the EQ toggle*

3.4 Choose Modify from the Basics Tab of the Design Bar to terminate the Dimension tool.

In the Select mode, where the cursor appears as an arrow, Revit allows you to select more than one entity by holding down the CTRL key while selecting. This builds a selection set upon which you can perform an action. Selected items will appear in red. To remove an item from a selection set, hold down the SHIFT key and deselect it.

Revit 6 follows the Autodesk standard for selection windows—selecting a point in the view window, *not* over an entity, and holding down the cursor left button starts a selection window with different properties, depending on its orientation. Dragging *left to right* starts an enclosing window that selects only entities completely enclosed within its border, which is a solid line on the screen. Dragging from *right to left* starts a crossing window that selects any entity within or touching its border, which is a dashed line.

3.5 Place the cursor below and to the left of the left-hand window. Drag the cursor up and to the right to make a window around the windows in the lower wall, but do not include the entire wall. Release the cursor button to make the selection, as shown in Figure 1–37.

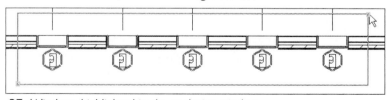

**Figure 1–37** *Windows highlighted in the enclosing window*

3.6 When the windows are highlighted, choose the Mirror command on the Toolbar, shown in Figure 1–38.

**Figure 1–38** *The mirror command*

3.7 Select the center of the center door in the left wall to define the mirror line (see Figure 1–39). A reference line will appear to assist your selection. Windows will appear in the upper wall.

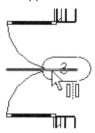

**Figure 1–39** *Use the door as the mirror midpoint*

## Exercise 4. Add another level and a floor

4.1 Double-click on the West Elevation view in the Project Browser to open up that view.

4.2 Right-click and Zoom in Region around the left door. Note the appearance of the vertically compound wall structure at close range.

4.3 Right-click and choose Previous Scroll/Zoom (or Zoom to Fit) to restore the entire elevation on the screen.

4.4 Choose the Level tool on the Basics tab of the Design Bar, as shown in Figure 1–40.

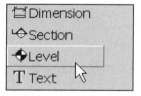

**Figure 1–40** *The Level tool*

4.5 Place the cursor over the left end of the Level 2 marker line. Revit will snap to it and read out an offset dimension as you move the cursor up, as shown in Figure 1–41.

4.6 Select to start a new level 8' above Level 2. You can type **8** to set the vertical offset at 8' if you have any problem getting the temporary dimension to read the desired distance. Pull the cursor to the right and it will snap to an alignment point above the elevation symbol for Level 2. Left-click and the new Level 3 will appear. Note that the Project Browser now shows a floor plan and ceiling plan for Level 3.

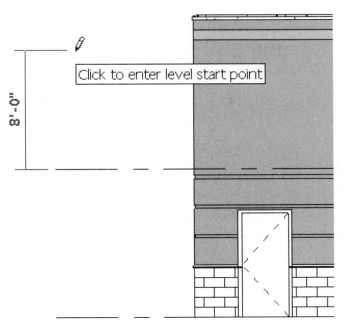

8'-0"

Click to enter level start point

**Figure 1–41** *The new level started*

4.7 Hit ESC or choose the Modify button to terminate the command.

4.8 Double-click on Floor Plan Level 2 in the Project Browser to open that view.

4.9 Choose the Floor tool on the Basics tab of the Design Bar, shown in Figure 1–42.

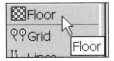

**Figure 1–42** *The Floor tool*

4.10 The walls in the View Window will gray out. The Design Bar will go into Sketch mode, with the Pick Walls tool activated. The Status Bar reads Pick Walls to Create Lines. Leave the Offset value at 0' 0" and the Extend into wall (to core) option checked on the Options Bar (see Figure 1–43).

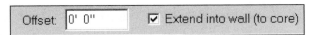

**Figure 1–43** *Floor Options*

4.11 Hold the cursor over the interior surface of the left wall—the wall will highlight and the wall properties will appear in a Tooltip and in the status bar, as shown in Figure 1–44.

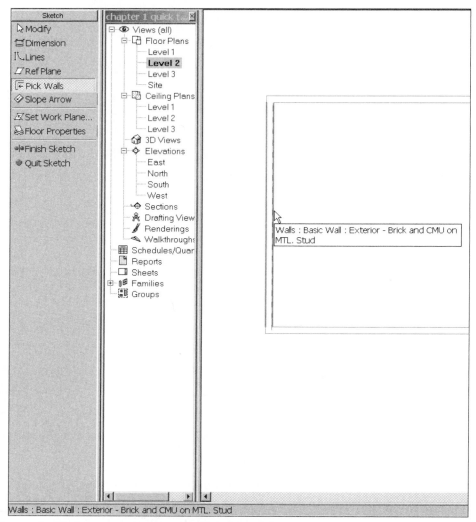

**Figure 1–44** *Ready to select a wall in Floor Sketch mode*

4.12 Select the wall and a red sketch line will appear on its interior surface. Use the control arrows that appear to adjust its position if necessary.

4.13 Select the remaining three walls to complete a sketch of the floor edges, as shown in Figure 1–45.

**Figure 1–45** *The floor sketch is complete*

4.14   Choose Finish Sketch (see Figure 1–46). Choose Yes in the question box that appears. When sketch mode completes the Basics tab will reappear in the Design Bar. The View Screen will not change appearance.

**Figure 1–46** *Finish Sketch*

## Exercise 5. Add a roof and view it in section

5.1   Double-click on Floor Plan Level 3 in the Project Browser. Choose the Roof>>Roof by Footprint tool in the Design Bar, as shown in Figure 1–47.

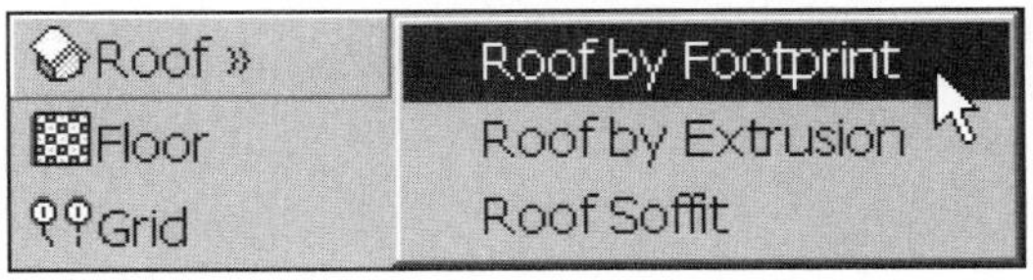

**Figure 1–47** *Roof by Footprint*

5.2    Revit shifts into Sketch mode, as with the Floor tool. This will be a flat roof below the parapet cap on the exterior walls. On the Options Bar, uncheck Defines Slope and check Extend into wall (to core), as shown in Figure 1–48.

**Figure 1–48** *No slope for this roof, which will extend into the wall past the finish layer*

5.3    Select the interior surface of the walls, as with the Floor command. The walls will highlight in turn as the cursor passes near them, and reference lines will appear. Adjust the position of the sketch lines with the control arrows, as before.

5.4    Choose Finish Roof. Choose No in the Attach Walls to Roof question box. Choose Yes to the overlap/join geometry/cut volume question.

5.5    Double-click on view Floor Plan Level 1 in the Project Browser to open that view in the View Window.

5.6    Choose the Section tool on the Basics tab of the Design Bar, shown in Figure 1–49.

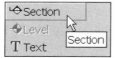

**Figure 1–49** *The Section tool*

5.7    Select a start point for the Section line to the left of the upper window in the left wall (see Figure 1–50).

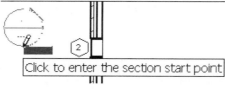

**Figure 1–50** *Starting the Section*

5.8    Pull the cursor to the right horizontally, and select a spot to the right of the wall to locate the Section (see Figure 1–51).

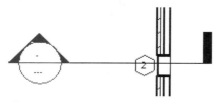

**Figure 1–51** *Finishing the Section*

5.9    The Project Browser will show a + sign to the left of Sections. Select the + sign and Section 1 will appear in the Project tree, as shown in Figure 1–52.

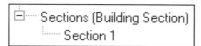

**Figure 1–52** *The new Section in the Browser*

5.10    Double-click on Section 1 to open that view.

5.11    Right-click and choose View Properties (always the bottom option in the right-click context menu). Change the Detail Level to Fine.

5.12    Select the Level 2 floor. Use the Type Selector drop-down list to change its type from Generic to Steel Bar Joist 14" – VCT on Concrete.

5.13    Select the Roof. Use the Type Selector drop-down list to change its type from Generic to Steel Truss – Insulation on Metal Deck – EPDM. Note the new appearance of these components.

## Exercise 6. Documenting the model

6.1    Right-click on Sheets (all) in the Project Browser, and choose New Sheet, as shown in Figure 1–53.

**Figure 1–53** *Creating a new sheet for plotting*

6.2    Choose OK to accept the default E1 30 x 42 Horizontal title block for the new sheet.

6.3    The new sheet appears in the View Window. Select Floor Plan Level 1 in the Project Browser (do not double-click), hold down the left mouse button, and drag it into the sheet (see Figure 1–54). Release the button and the outline of the view at scale with a title below it will appear (see Figure 1–55).

**Figure 1–54** *Starting to drag a view onto the sheet*

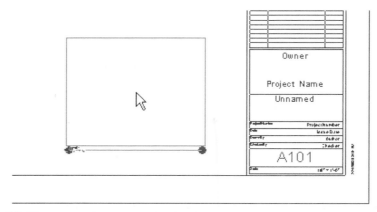

**Figure 1–55** *The view outline appears under the cursor*

6.4 The viewport is small on the page. Press ESC to terminate the placement. Select Floor Plan Level 1 in the Project Browser again, right-click and select Properties.

6.5 In the Element Properties dialog shown in Figure 1–56, change the View Scale to 1/4" = 1'-0" and choose OK.

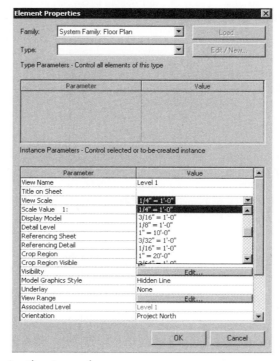

**Figure 1–56** *Changing the view scale*

Pcl#: 27180659   Typ:R4   Tm: WPK515 Order # 64639159SM   Ctn: 1 of 1

Pl#:   1769   Date: 12/18/04   Wve:001   Tote: 27196156   SLA: 7

| Location | ISBN | Pieces | Description |
|---|---|---|---|
| 515-106-022 | 1401850499 | 1 | INTRO/IMPLEMENT ADESK REV |

The enclosed materials are compliments of
MIMI EGLIN 800 998-7498X2351

SHC: 0   THOMSON LEARNING

Page: 1 of 1

#00005

6.6 Repeat this scale change for the West Elevation and Section 1 in the Project Browser.

6.7 Select Floor Plan Level 1 in the Project Browser and drag it onto the sheet, as before. Note the new size. Locate it in the lower right corner of the sheet.

6.8 Select and drag views West Elevation and Section 1 from the Project Browser to the sheet and arrange them as shown in Figure 1-57. Revit will provide alignment snap lines for placement. Align the Elevation view title with the title of the Plan view, and align Level 2 of the Section with Level 2 of the Elevation view.

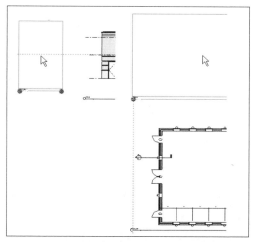

**Figure 1–57** *Revit will snap to view titles and level lines for viewport alignment*

6.9 Zoom in over the plan viewport so you can see the left wall components clearly. Place the cursor over the viewport so that its border becomes highlighted. Right-click and choose Activate View, as shown in Figure 1–58. This will allow you to change the model from the sheet view.

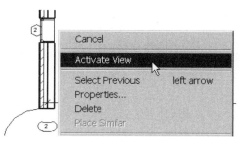

**Figure 1–58** *Making the viewport active to work in it*

6.10 Select one of the windows in the left wall, hold down the CTRL key, and select the other so both are selected.

 **Tip:** If you have difficulty selecting the windows you can zoom in closer. You can also use the TAB key on your keyboard to cycle through the available choices until the window you want to modify becomes highlighted and you can select it.

6.11 When both windows are selected, use the Type Selector drop-down list to change their type to Fixed 16" x 72" as shown in Figure 1–59.

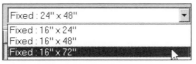

**Figure 1–59** *Changing the window type*

6.12 Choose the Properties button to bring up the Element Properties dialog shown in Figure 1–60. Change the Sill Height for these two windows to 1' (you can simply type **1**—Revit's default Imperial unit is the foot). Choose OK.

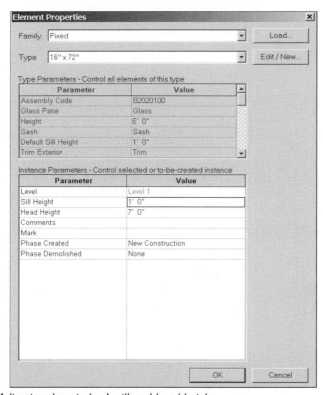

**Figure 1–60** *Adjusting the window's sill and head height*

6.13 Right-click and select Deactivate View.

6.14 Right-click and select Zoom to Fit.

6.15 Examine the Elevation and Section views to verify that the windows have changed.

6.16 Choose the Default 3D View button on the Menu Bar, as shown in Figure 1–61. This will open a Southeast isometric view of the model.

**Figure 1–61** *The button to create a 3D view*

6.17 Choose View>Shading from the Menu Bar, as shown in Figure 1–62.

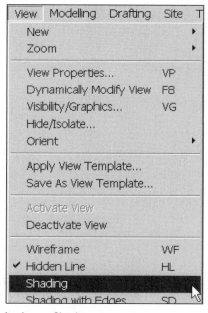

**Figure 1–62** *Change the display to Shading*

6.18 Select View>Orient>Southwest from the Menu Bar.

6.19 Double-click Sheets>A101 – Unnamed from the Project Browser to return to the plotting sheet. Note that 3D Views now has a + sign to the left of it in the Browser. Select the + sign to revel the new 3D view, named simply {3D}.

6.20 Select the new {3D} view name in the Project Browser. Select the Properties icon from the Options Bar and change its scale to 1/4" = 1'-0", as you did before. Choose OK.

6.21 Select and drag the {3D} view onto Sheet A101. Locate it in the lower left corner of the sheet. Its title will snap into alignment with the title of the Floor Plan view, as shown in Figure 1-63.

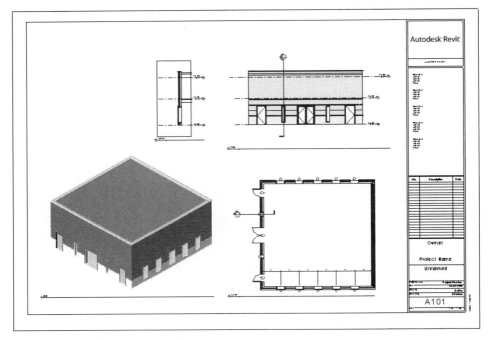

**Figure 1–63** *The plotting sheet with plan, elevation, section, and shaded isometric views*

6.22 Save the file as *quicktour.rvt*.

## SUMMARY

Congratulations! You have created, edited, and documented a building shell in Revit, using its intelligent components and their innate properties entirely. This sheet is ready to print and you have not drawn any 2D lines or manipulated any layers, but worked entirely with objects, views, and their associated properties.

Admittedly, this model is quite basic, but you have gotten a taste of Revit's simplicity, ease of use, and power. The remaining chapters will show how Revit's strengths in modeling, representation, documentation, data extraction and change propagation function, and prepare you to implement this new architectural design tool in a work environment.

## REVIEW QUESTIONS

### Multiple Choice

1. To make objects visible or not, use controls found in

     a) the Design Bar

     b) the properties of each view

     c) the Project Browser

     d) the Options Bar

2. Hatch lines inside walls in a plan

     a) are always visible if you zoom in close

     b) are only visible in the Level 1 Floor Plan view

     c) are visible when the View Detail level is set to fine

     d) have to be drawn in by hand

3. Zoom controls are found

     a) on the upper Toolbar

     b) in the View menu

     c) in the right-click context menu

     d) all of the above

4. Revit will allow you to edit the swing and hinge controls of a door

     a) during placement

     b) if it is selected by clicking after placement

     c) by using the View properties

     d) a and b, but not c

5. For construction documentation purposes Revit provides

     a) Sheets with Views placed on them at scale

     b) Annotations that keep track of their Sheet placement

     c) both a and b

     d) Revit can't produce documents

## True/False

6. Sections and Elevations require update procedures to reflect model changes.

7. When creating floors and roofs, Revit uses Sketch mode to allow the user to select walls or draw lines.

8. Views placed on Sheets do not require update procedures to reflect model changes.

9. A floor object can be offset from its base Level.

10. Entities cannot be selected in 3D views.

 Answers found on the accompanying CD.

# CHAPTER 2

# Leadership Strategies

## INTRODUCTION

This section is addressed to managers, supervisors, and team leaders in design firms. College professors or those leading courses of instruction will also find the concepts and exercises useful.

Students will benefit from examining content and structure in template files, importing a CAD file, creating a custom titleblock, and exporting a file to CAD.

## OBJECTIVES

- Understand the mechanics and value of template files in Revit
- Experience Revit's input/output interface with AutoCAD
- Discuss Revit's file management systems: import/export, linking, worksets

## CAD MANAGER/INSTRUCTOR

If your responsibilities include deploying the design software your organization uses, keeping designers equipped with software tools, organizing files, and maintaining standards, you have headaches enough without having to adopt (or worse, half-adopt) a new standard. As your company or institution considers using Revit, what does Revit have in store for you?

The bad news is that Revit is different enough from AutoCAD and Architectural Desktop that even the people who listen to you and make an effort to follow your directives, instructions, specifications, and standards will be confused at times by this new software. The good news is that Revit is much simpler than AutoCAD and ADT in significant ways—*not* having to use layers to govern the appearance of objects on screen and paper, for example, does away with an entire, often complicated, set of standards that nobody ever follows completely anyway.

Revit does contain many of the same mechanisms that you currently use—text and dimensions have styles; reference files, content files, and central files all use coherent path setups; drawing sheets work the same way as Paper Space layouts, only more efficiently. Once you understand the strengths and operational underpinnings of this new modeler you will be better able to serve the needs of your designers, project leaders, clients, and students.

## TEMPLATE FILES HOLD COMPANY STANDARDS

Revit uses template files to hold an array of settings, exactly as does AutoCAD. It works with two types of template files: *.rft* files for family templates and *.rte* files for project (building model) files. Families are a major organizing system in Revit. They do the work both of Styles in AutoCAD or Architectural Desktop and Multi-View blocks in ADT. Families will be treated in later exercises.

When first opening, Revit uses a *default.rte* template file that corresponds to the *acad.dwt* template file for AutoCAD—the very basics. Just as AutoCAD and Architectural Desktop allow the user to specify and modify other template files with enhancements for specific purposes, Revit has template files with significantly developed views, content, and sheet setups for specific types of building models. These templates are ready to create company-standard, project-specific, or client-specific templates. As with AutoCAD products, completed Revit project files can be converted to templates.

## TEMPLATE OVERVIEW

As a CAD Manager or instructor you may or may not be familiar with Revit at this point. Whether you have created and documented more building models than the designers or students you are responsible for or not, in this next exercise we'll look under the hood and examine some of the settings and setups that will make your job easier.

## REVIT COMMANDS AND SKILLS

| | |
|---|---|
| *Templates* | *Import File* |
| *Predefined Views* | *Import Settings—Lineweights* |
| *Predefined Sheets* | *Revit Lineweights* |
| *View Phase Filters* | *Parameters* |
| *View Graphic Overrides* | *Labels* |
| *Predefined Schedules* | *Export File* |
| *Titleblock* | *Export Settings—Layers* |

### Exercise 1. Examine templates

## The default template—the very basics

    1.1    Launch Revit. Revit will open to a Floor Plan View of Level 1 (the words Level 1 will be highlighted under the Floor Plans section of the Project Browser tree under Views (all), as shown in Figure 2–1). Study the contents of the Project Browser to the left of the View Window. You can open predefined Floor Plans for Levels 1 and 2, Ceiling Plans for Levels 1 and 2, and Elevations (graphically indicated on the Level 1 Floor Plan by elevation boxes). There are no schedules or sheet views defined.

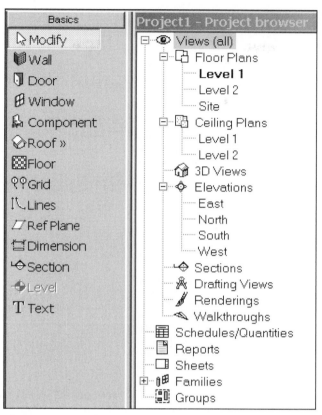

**Figure 2–1** *The Project Browser for an empty file using the default template*

1.2  Select the + symbol next to Families in the Project Browser. Scroll down the list of Families and expand (click the + symbols next to) Structural Columns, Structural Framing, and Walls. Expand W-Wide Flange-Column under Structural Columns, W-Wide Flange under Structural Framing, and Basic Wall under Walls (see Figure 2–2).

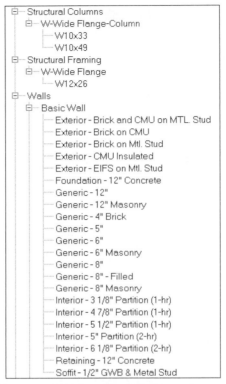

**Figure 2–2** *Structural Content and Walls in the default template*

1.3 Place the cursor over the point of the West Elevation symbol in the View Window.

1.4 Right-click and choose Go to Elevation View to open the West Elevation view, as shown in Figure 2–3). This has the same effect as double-clicking the West View listing in the Project Browser.

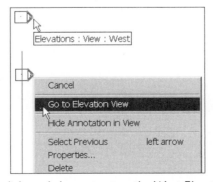

**Figure 2–3** *Locate, right-click, and choose to open the West Elevation*

1.5  The West Elevation shows the two Levels in this file, at 0'-0" and 10'-0", shown in Figure 2–4.

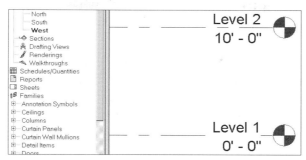

**Figure 2–4** *The levels in this empty file*

## Examine the residential (imperial) template—predefined views and sheets

1.6  From the File menu, select File>New>Project, as shown in Figure 2–5, to open the New Project dialog.

**Figure 2–5** *Opening a new project*

1.7  In the New Project dialog, choose the Browse button to choose a template file for the new project, as seen in Figure 2–6. Note that the path to your templates folder will be different than what is shown here.

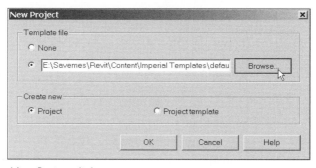

**Figure 2–6** *The New Project dialog*

1.8  In the Choose Template dialog, select *Residential-Default.rte,* as shown in Figure 2–7. Choose Open to return to the New Project dialog. Choose OK to create a new project using this residential template.

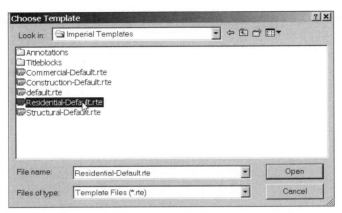

**Figure 2–7** *Choose the Residential-Default template*

1.9 When the new residential project file opens, study the Project Browser (shown in Figure 2–8). There are many more floor plans than in the default file, including footing, framing, and electrical plans. The levels have been renamed to First Floor and Second Floor, and there is a Roof Plan. There are schedules predefined as well: Door, Lighting Fixture, Room Finish, and Window.

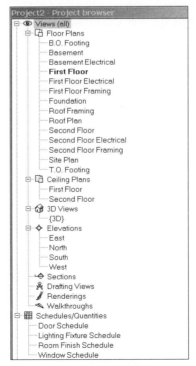

**Figure 2–8** *Floor plans and schedules in the Residential template*

1.10 Choose the + next to Sheets to expand the list of sheets. This template includes plans, elevations, detail sheets, sections, and a schedule page.

1.11 Choose the + next to A0 – Basement Plan and A1 – First Floor Plan, as shown in Figure 2–9. These expand to show that the Basement Plan and First Floor Plan have already been placed on these two sheets. Similar views have been placed on other floor plans, reflected ceiling plans, roof plans, and electrical plans.

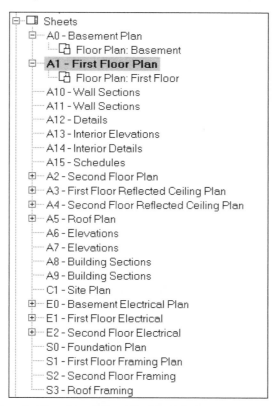

**Figure 2–9** *The expanded list of predefined Residential sheets*

1.12 Double-click on A1 – First Floor Plan in the Project Browser to open that view. It has been defined to hold a default Revit D size 22 x 34 titleblock, with the sheet name and number entered.

1.13 Choose the + symbol next to Families in the Project Browser. Scroll down the list of Families and expand Structural Columns, Structural Framing, and Walls (see Figure 2–10). Expand Pipe-Column under Structural Columns, and Basic Wall under Walls. Study the different content available in this template compared to the default file.

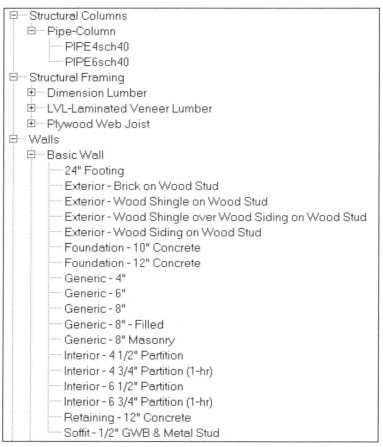

**Figure 2–10** *Structural Columns, Structural Framing, and Walls in the residential template*

1.14 From the Window menu, select Window>Project 1 - Floor Plan: Level 1 to open up this view. Note that the Project Browser now shows the views, sheets, and families appropriate to this file.

1.15 From the Window menu, select Window>Project 2 – Floor Plan: First Floor to return to the residential file.

1.16 Double-click on East Elevation to open that view. With the cursor in the View Window, right-click and select Zoom in Region. Choose two points around the Elevation symbols at the right of the screen to see the pre-defined Levels in this empty file (see Figure 2–11)

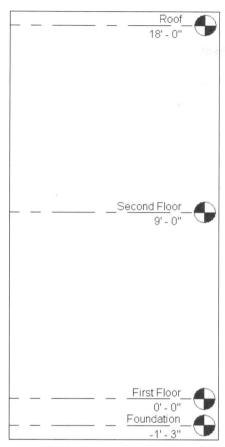

**Figure 2–11** *Default Residential levels: Foundation, First Floor, Second Floor, and Roof*

## Examine the construction (imperial) template—views with phases, and schedules

1.17    From the File menu, select File>New>Project.

1.18    In the New Project dialog, select Browse.

1.19    In the Choose Template dialog, select *Construction-Default.rte*. Select Open.

1.20    In the New Project dialog, click OK.

1.21    When the new construction project opens, note that the elevation symbols have been turned off in the Level 1 View window. Study the Project Browser: note the Floor Plans located below Level 1 and the list of 3D Views.

1.22    Select 3D View 02 – Demo in the Project Browser, right-click, and select Properties (always located at the bottom of the right-click menu), as shown in Figure 2–12.

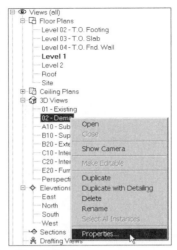

**Figure 2–12** *Opening the View Properties box*

1.23    In the Element Properties dialog that appears, as in Figure 2–13, note the
        Phase Filter and Phase that have been applied to this view. This file has been
        set up to use Phases—building elements will be differentiated according to
        the time they come into existence or disappear. Phases can be named and
        filtered, in this case by function: structure, enclosure, exterior/interior, and
        construction/finish. Each Phase can appear differently (or not at all) in each
        view set up for the project file. Choose Cancel to leave the Element Prop-
        erties dialog for the current view

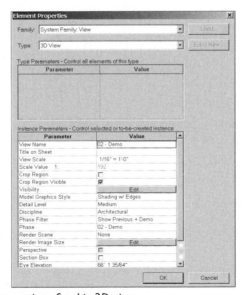

**Figure 2–13** *The Phase settings for this 3D view*

1.24    From the Settings menu, select Settings>Phases, as shown in Figure 2–14.

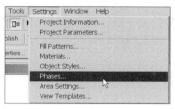

**Figure 2–14** *Opening Phase Settings*

1.25    In the Phasing dialog, study the first tab, Project Phases, as shown in Figure 2–15. The other two tabs are Phase Filters and Graphic Overrides. This file has nine phases defined in it: Existing, Demo, five New Construction phases, Furnishings, and Project Completion.

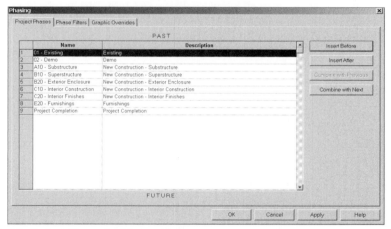

**Figure 2–15** *Project Phases in this template*

1.26    Choose the Phase Filters tab as shown in Figure 2–16. Phase Filters sets up view conditions in which phases are displayed according to their categories, displayed with overrides, or not displayed. In this way each element can be viewed in different ways in different views. Revit supplies the temporary phase for elements that are created and demolished during the project.

| | Filter Name | New | Existing | Demolished | Temporary |
|---|---|---|---|---|---|
| 1 | Show All | By Category | Overridden | Overridden | Overridden |
| 2 | Show All | By Category | Overridden | Overridden | Overridden |
| 3 | Show Complete | By Category | By Category | Not Displayed | Not Displayed |
| 4 | Show Demo + New | By Category | Not Displayed | Overridden | Overridden |
| 5 | Show Demo Only | Not Displayed | Not Displayed | Overridden | Not Displayed |
| 6 | Show New Only | By Category | Not Displayed | Not Displayed | Not Displayed |
| 7 | Show Previous + Demo | Not Displayed | Overridden | Overridden | Not Displayed |
| 8 | Show Previous + New | By Category | Overridden | Not Displayed | Not Displayed |
| 9 | Show Previous Phase | Not Displayed | Overridden | Not Displayed | Not Displayed |

**Figure 2–16** *Phase Filters*

1.27 Choose the Graphic Overrides tab as shown in Figure 2–17. Here the visual properties of Phases are applied—line weight in plan and section, color, linetype, and material for 3D views.

| Phase Status | Line Weight | | Line Color | Line Pattern | Material |
|---|---|---|---|---|---|
| | Projection | Cut | | | |
| Demolished | 1 | 3 | Black | Demolished 3/16" | Phase-Demo |
| Existing | 2 | 3 | RGB 127-127-127 | Solid | Phase-Exist |
| New | 4 | 5 | Black | Solid | Phase-New |
| Temporary | 1 | 3 | RGB 000-000-127 | Dash 1/16" | Phase-Temp |

**Figure 2–17** *Graphic Overrides for Phasing*

1.28 Choose Cancel to exit the Phasing dialog.

1.29 Choose the + symbol next to Schedules/Quantities to expand the list in the Project Browser, as seen in Figure 2–18. This template includes examples of nearly all the available standard schedules.

**Figure 2–18** *Predefined schedules in this template*

1.30 Double-click Elevations>East in the Project Browser to open that view. Right-click and select Zoom In Region. Select two points around the Elevation symbols in the right of the View Window. Study the predefined Levels and their heights in this template as shown in Figure 2–19 .

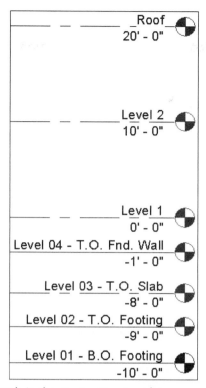

**Figure 2–19** *Predefined levels in the construction template*

1.31 Choose the + symbol next to Families in the Project Browser to expand the list. Choose the + symbol next to Structural Columns and Walls to expand those two Family Categories.

1.32 Choose the + symbol next to Concrete—Rectangular—Column, W— Wide Flange—Column, and Basic Wall to expand the list of preloaded content in those families. Study the list of available components in this template.

## Examine the commercial (imperial) template

1.33 From the File menu, select File>New>Project to open the New Project dialog.

1.34 Choose the Browse button to choose another template.

1.35 In the Choose Template dialog, select *Commercial—Default.rte* in the list of templates in the *Imperial Templates* folder and choose the Open button. Choose OK in the New Project dialog.

1.36 In the Project Browser, study the list of default floor plans. Double-click Elevations>East. Zoom in Region around the elevation symbols in that view (see Figure 2–20). Note the location of the predefined levels in this template.

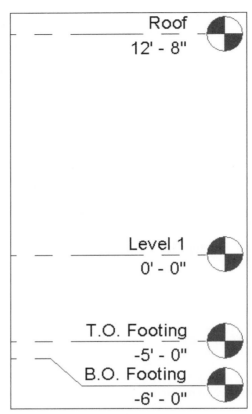

**Figure 2–20** *Levels in the commercial template*

1.37 In the Project Browser, note the list of predefined sheets. Select the + symbol next to Families to expand the list of categories.

1.38 Choose the + symbol next to Structural Columns, Structural Foundations, Structural Framing, and Walls to expand those categories.

1.39 Select the + symbol next to W—Wide Flange—Column, Pile Cap—Rectangular, W—Wide Flange, and Walls>Basic Walls to expand those families (shown in Figure 2–21). Study the list of available components in this template.

**Figure 2–21** *Structural components and walls in the commercial file*

1.40 Double-click Level 1 to return to that view. You are still in the commercial project file. Do not close any open Revit files.

## Exercise 2. Compare specific content in different templates

2.1 From the Window menu, select Window>Close Hidden Windows, as shown in Figure 2–22. This will close all the views that you have open except the current one in the current file and the last open view in other open files. Keeping open windows to a minimum will free up display system resources.

**Figure 2–22** *How to close hidden view windows*

2.2 In the Project Browser, choose the + symbol next to Families>Profiles to view the list of predefined and named profile shapes available for sweeps or edges in this template. If necessary, widen the Browser pane by holding the cursor over its right border until the cursor changes to a two-headed arrow, then click and pull the pane border to the right so you can see the expanded tree view (see Figure 2–23).

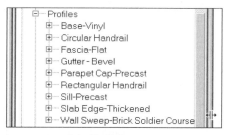

**Figure 2–23** *Profiles in the commercial template*

2.3 From the Window, select the available view from *Project 1* (the default template). When that file is loaded, choose the + symbol next to Profiles to view the list of profile shapes in the default template, as shown in Figure 2–24.

**Figure 2–24** *Profiles in the default template*

2.4   From the Window menu, select the available view from *Project 2* (the residential template). When that file loads, choose the + symbol next to Families>Profiles to view the list of profile shapes in the residential template.

2.5   From the Window menu, select the available view from *Project 3* (the construction template). When that file loads, choose the + symbol next to Families>Profiles to view the list of profile shapes in the construction template.

## Examine view properties in a template

2.6   From the Window menu, select the available view from *Project 2* (the residential template). When that file loads, double-click Floor Plan>First Floor to make that view active if it is not already the open view.

2.7   On the Basics Tab of the Design Bar, choose Wall to start the Wall tool.

2.8   In the Type Selector on the left of the Options Bar, choose the down arrow in the drop-down box and select Basic Wall: Exterior – Wood Shingle on Wood Stud, as shown in Figure 2–25. Do not change the Height or Location Line properties.

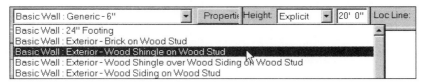

**Figure 2–25** *Exterior wall properties*

2.9   In the sketch lines area of the Options Bar, choose the Rectangle button so that by selecting two points in the View Window you can draw four connected walls with their exterior sides correctly situated.

2.10   Select a point to the upper left of the View Window and pull the cursor down and to the right to start defining the new walls. Use the temporary dimensions that appear to make the new walls approximately 30' wide by 25' long (see Figure 2–26). The actual wall lengths do not matter.

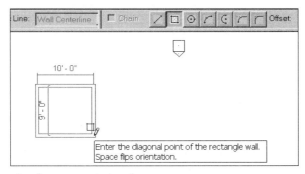

**Figure 2–26** *Drawing four connected walls*

2.11    With the Wall tool still active, change the wall type to Basic Wall: Interior – 4 1/2" Partition. Change the sketch line type to Lines and choose the Chain checkbox to sketch new walls continuously until you exit the command.

2.12    Draw a wall from the midpoint of the left wall (approximately) horizontal to the middle of the interior space, then up vertically to the middle (approximately) of the upper wall (see Figures 2–27 and 2–28). The actual wall dimensions or locations do not matter.

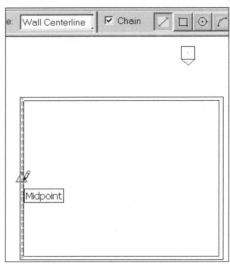

**Figure 2–27** *Starting interior walls*

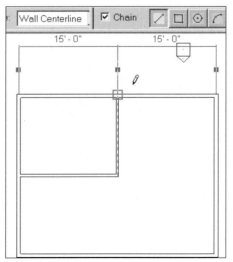

**Figure 2–28** *Interior walls in place*

2.13    Select Component from the Design Bar to terminate the Wall tool and start the Component placement tool. Use the drop-down list to select Duplex Outlet: Single, as shown in Figure 2–29.

**Figure 2–29**  *Select Duplex Outlet in the Component command*

2.14    Place an outlet at the approximate midpoint of the right side interior wall, facing left into the smaller room you have created with the new walls (see Figure 2–30). The actual placement does not matter.

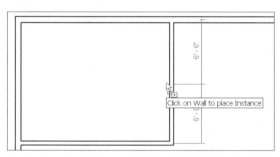

**Figure 2–30**  *Locating an electrical outlet*

2.15    A Revit warning, shown in Figure 2–31, will appear to let you know that the settings for this Floor Plan view will not let you see the new electrical fixture. Choose OK to close this warning.

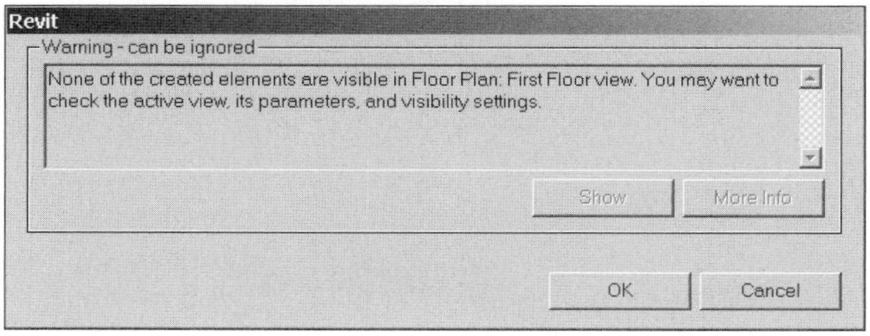

**Revit**
Warning - can be ignored
None of the created elements are visible in Floor Plan: First Floor view. You may want to check the active view, its parameters, and visibility settings.

Show    More Info

OK    Cancel

**Figure 2–31**  *Revit warns you about this view*

2.16 Double-click Floor Plans>First Floor Electrical to open that view. You will see the new electrical outlet correctly placed. Note the light appearance of the walls.

2.17 With the cursor in the View Window but not directly over any part of the model, right-click and select View Properties (always the bottom choice in this menu).

2.18 In the View Properties dialog that opens, choose the Edit button in the Value field to the right of the Visibility Parameter field.

2.19 In the Visibility/Graphics Overrides dialog that opens, study the Model Categories tab (see Figure 2–32). Note that for all component categories except Electrical Components and Electrical Fixtures, the Half (Halftone) graphic override has been checked. This accounts for the dim appearance of the walls in this view.

| Visibility | Line Style | | Half... | Detail ... |
| | Projection | Cut | | |
| --- | --- | --- | --- | --- |
| ⊞ ☑ Casework | By Category | By Category | ☑ | By View |
| ⊞ ☑ Ceilings | By Category | By Category | ☑ | By View |
| ☑ Columns | By Category | By Category | ☑ | By View |
| ⊞ ☑ Curtain Panels | By Category | By Category | ☑ | By View |
| ☑ Curtain Wall Mullions | By Category | By Category | ☑ | By View |
| ⊞ ☑ Detail Items | By Category | | ☑ | By View |
| ⊞ ☑ Doors | By Category | By Category | ☑ | By View |
| ☑ Electrical Equipment | By Category | | ☐ | By View |
| ☑ Electrical Fixtures | By Category | | ☐ | By View |
| ☑ Entourage | By Category | | ☑ | By View |
| ⊞ ☑ Floors | By Category | By Category | ☑ | By View |
| ⊞ ☑ Furniture | By Category | | ☑ | By View |
| ☑ Furniture Systems | By Category | | ☑ | By View |
| ⊞ ☑ Generic Models | By Category | By Category | ☑ | By View |
| ☑ Lighting Fixtures | By Category | | ☑ | By View |
| ⊞ ☑ Lines | By Category | | ☑ | By View |
| ☑ Massing | By Category | By Category | ☑ | By View |
| ☑ Mechanical Equipment | By Category | | ☑ | By View |
| ⊞ ☑ Parking | By Category | | ☑ | By View |

**Figure 2–32** *Halftone settings in the Electrical Plan*

2.20 Choose Cancel twice to exit the View Graphics and View Properties dialogs.

2.21 Choose Floor Plans>First Floor Plan in the Project Browser to highlight it. Right-click and select Properties, as shown in Figure 2–33, to open the Properties for the First Floor Plan view without having to open the view itself.

**Figure 2–33** *Properties of the First Floor Plan*

2.22   In the View Properties dialog that opens, choose the Edit button in the Value field to the right of the Visibility Parameter field.

2.23   In the Visibility/Graphics Overrides dialog that opens, study the Model Categories tab, shown in Figure 2–34. Note that no component categories have the Halftone graphic override checked. Also note that the Visibility of the Electrical Fixtures category has been unchecked. In this template the Floor Plan view will not show outlets and electrical switches; a separate Electrical Plan has been created for that purpose. This is similar to the standard Ceiling Plans in all templates.

| Visibility | Line Style | | Half... | Detail ... |
|---|---|---|---|---|
| | Projection | Cut | | |
| ☑ Casework | By Category | By Category | ☐ | By View |
| ☑ Ceilings | By Category | By Category | ☐ | By View |
| ☑ Columns | By Category | By Category | ☐ | By View |
| ☑ Curtain Panels | By Category | By Category | ☐ | By View |
| ☑ Curtain Wall Mullions | By Category | By Category | ☐ | By View |
| ☑ Detail Items | By Category | | ☐ | By View |
| ☑ Doors | By Category | By Category | ☐ | By View |
| ☑ Electrical Equipment | By Category | | ☐ | By View |
| ☐ Electrical Fixtures | By Category | | ☐ | By View |
| ☑ Entourage | By Category | | ☐ | By View |
| ☑ Floors | By Category | By Category | ☐ | By View |
| ☑ Furniture | By Category | | ☐ | By View |
| ☑ Furniture Systems | By Category | | ☐ | By View |
| ☑ Generic Models | By Category | By Category | ☐ | By View |
| ☑ Lighting Fixtures | By Category | | ☐ | By View |
| ☑ Lines | By Category | | ☐ | By View |
| ☑ Massing | By Category | By Category | ☐ | By View |
| ☑ Mechanical Equipment | By Category | | ☐ | By View |
| ☑ Parking | By Category | | ☐ | By View |

**Figure 2–34** *Halftone and visibility controls for the First Floor Plan*

2.24   Choose Cancel twice to exit the View Graphics and Properties dialogs.

2.25   From the File menu, select File>Exit (always the bottom selection on this menu) to exit Revit. Do not save any of the open files.

## Explore templates on your own—use what's already made for you

Revit contains other templates—Imperial and metric—besides the ones we have just studied. All the templates are worth examining for useful view setups, sheets, families, and schedules that have been pre-loaded for users and managers to adopt and adapt. At present there is no mechanism in Revit—such as the Design Center file/content browser in AutoCAD 2000 and later releases—to allow drag-and-drop transfer of settings or content from one file to another. Revit does support standard copy/paste from one file to another, and has a Transfer Project Standards command to move settings, family types, line weights, and rendering materials maps from one open file to another. We'll look at this tool in future exercises.

## INTERFACING WITH CAD–REUSE WHAT YOU HAVE ALREADY CREATED

One of the cardinal rules for productivity with CAD is to draw as little as possible, and use editing commands to populate your designs. That is, draw an object once and copy or array it many times, or draw half of a symmetrical design and mirror it to complete the pattern. This approach is not only quicker than repetitive drawing, but less error-prone.

Nearly every design firm on the planet has a library or archive of drawn work that designers mine repeatedly for reuse in new projects. Revit's ability to import from and export to AutoCAD and MicroStation means that firms setting out to use the power of new technology can make effective, leveraged use of their existing files and applications. If you have already drawn designs, components, symbols, or details in CAD that pertain to your new Revit project, there is at least one effective way to use them in your Revit project model or documentation.

### Imports and Exports: Imports First

Revit does not use layers or levels, the fundamental organizing mechanism of AutoCAD and MicroStation. Revit will readily read in layers/levels in linked or imported files, and exports DWG, DXF, or DGN files with properly constructed layers/levels.

There are limits to link/import; Revit will not support AutoCAD ACIS solids or regions but will import 3D surfaces. The import process creates a single object in the Revit file, with control over the appearance of its contained elements—exactly like the Insert Block command in AutoCAD. You can Explode imported objects, also with limits—into lines, curves, text, and filled regions. Revit makes particular use of 3D line/polyline information from imported *.dwg*, *.dxf*, or *.dgn* files to create editable Toposurface site objects. Revit does not recognize attribute definitions, but does recognize attribute values of inserted blocks in an imported *.dwg* file and will explode those values to text. Revit uses Windows fonts, and contains an editable mapping file for routing AutoCAD *.shx* font information (an older format for lettering contained in AutoCAD from the days before Windows) to Windows fonts. Revit will read proxy graphics for AEC Objects from Architectural Desktop, such as walls and floors. These proxy graphics will come in 3D and explode to 2D Model lines in Plan views.

Revit will export 2D information from 2D views or 3D information from 3D views using mapping files to assign layers or levels to Revit objects by category. Imported 3D information comes into AutoCAD as 3D polyline meshes that explode to 3D faces.

Revit's tutorials contain exercises for creating titleblocks from scratch. The next exercises will examine Revit's standard titleblocks and then use an imported AutoCAD file to create a highly formatted titleblock without extensive drawing or tracing.

## Exercise 3. Import and reuse an AutoCAD titleblock

### Examine a standard titleblock—edit parameters

3.1 Open Revit to a new default project. It will have no sheets defined. Select Sheets in the Project Browser. Right-click and choose New Sheet, as shown in Figure 2–35.

**Figure 2–35** *Adding a sheet to a new file*

3.2 In the Select a Titleblock dialog that opens, choose the Load button.

3.3 In the Open file browser dialog, make sure the Look In: value is the *Imperial Library* folder and select *Titleblocks*. Choose Open.

3.4 Select each of the default titleblocks in turn to examine its appearance in the preview window. Select Cancel to return to the previous dialog.

### Examine the default selections

3.5 Choose OK to create a sheet with the E1 30 x 42 Horizontal default title-block. Revit will open the new sheet view.

3.6 Click on the new sheet titleblock to select it. On the Options Bar, choose the Properties icon. Fill in the Parameter information as shown in Figure 2–36. Choose OK.

| Parameter | Value |
|---|---|
| Sheet Name | Site Plan |
| Sheet Number | A101 |
| Date/Time Stamp | 05/20/04 |
| Sheet Issue Date | 05/20/04 |
| Drawn By | Chris Fox |
| Checked By | Michael Peterson |
| Designed By | Danielle Browning |
| Approved By | Lucy Van Pelt |
| Scale | |

*Instance Parameters - Control selected or to-be-created instance*

**Figure 2–36** *Titleblock parameter fields*

3.7 With the cursor in the view window, right-click and choose Zoom in Region. Select two points around the lower-right corner of the title block. Note that some, but not all of the information fields (labels) have updated. Select the titleblock, and the linework will turn red to show it has been selected. Labels will turn blue.

3.8    From the Settings menu, select Project Information. This will open the
       Type Properties Project Information dialog. Fill in the project information
       as shown in Figure 2–37. You will need to choose the Edit button in the
       Project Address field to open the entry box for that information. When
       you have entered the address information, choose OK. Choose OK again to
       close the Project Information dialog.

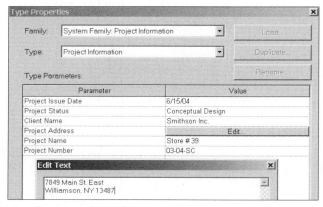

**Figure 2–37** *Project Settings information*

3.9    Study the labels in the titleblock again. You can edit the information in
       any label by selecting it. Select the A101 Sheet Number label and hold the
       cursor over it. The Tooltip will show Edit Parameter. Select the label again
       and the data entry field will open. Change the Sheet Number from A101 to
       A1, as shown in Figure 2–38; left-click outside the entry field (or hit ENTER)
       and the label will be updated.

**Figure 2–38** *Editing a titleblock label*

3.10   Click the drop-down arrow next to the Undo arrow on the Standard Tool-
       bar, under the Drafting menu button. Scroll down the list of actions to Load
       Family. Your undo list may differ slightly from the one shown in Figure 2–39.
       Left-click inside the Undo drop-down list to accept the Undo command.
       Revit will delete the titleblock and sheet.

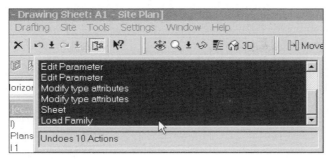

**Figure 2–39** *Undoing a series of actions*

## Create a new titleblock that will use an old file

3.11   From the File menu, select File>New>Titleblock, as shown in Figure 2–40.

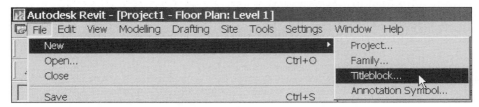

**Figure 2–40** *Starting a new titleblock*

3.12   In the New Titleblock dialog, select the *E1 – 42 x 30.rft* template, as shown in Figure 2–41. Choose OK.

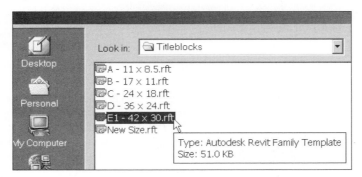

**Figure 2–41** *Select the E size template*

3.13   Revit will enter the Family Editor. Lines defining a 42" x 30" outline will appear. These lines cannot be deleted. If you select any of the lines, a dimension will appear. If you select the value of a dimension an edit field opens, allowing you to change the size of the outline. Do not make any changes at this time.

3.14    Save the file in a convenient place as *sample titleblock.rfa*. Save titleblocks your company will use in projects in the *Content\Imperial* (or *Metric*) *Library\ Titleblocks* folder on your network.

So far in this exercise we have examined a default titleblock supplied by Revit, and started to create a new one. Rather than draw this new titleblock from scratch, we will import an AutoCAD file. This file contains a titleblock consisting of lines, arcs (from filleted corners), text, and block attributes. The titleblock, which was originally created in Model Space, has been inserted into a Paper Space layout—the normal use for titleblocks in AutoCAD. Block attribute information is filled in by the user according to prompts built into the attribute definitions.

The imported file is empty of content except for the titleblock, to keep things simple. The titleblock linework is created in three colors, so it will plot heavy, medium, and light line weights. Revit has a mechanism to capture this intent and display AutoCAD lines in their desired line weights.

## Import settings—line weights

3.15    From the Settings menu, select Settings>Line Weights. Since we are in the Family Editor for a titleblock there is only one tab, for Annotation Line Weights.

3.16    Change the line width for Line Weight 3 to **0.009"**. Change the line width for Line Weight 5 to **0.020"**. See the left side of Figure 2–43. Choose OK.

3.17    From the File menu, select File>Import/Export Settings>Import Line Weights DWG/DXF, as shown in Figure 2–42.

**Figure 2–42** *Open the Import Line Weights dialog*

3.18    Revit will open the Import Line Weights dialog, which reads the contents of a line weights mapping file. The default file is *importlineweights-dwg-default.txt,* inside *Program Files\Autodesk Revit 6.1\Data*. This file gives all AutoCAD colors—listed by number—a default line weight value of 0, which when a file is saved and loaded, means Line Weight 1. Change the line weight for color 1 (red—medium lines in the file we will import) to **3**; change the line weight of color 4 (cyan) to **2**; change the line weight of color 5 (blue in AutoCAD) to **5**. (See the right side of Figure 2–43.)

This step applies Revit's line weights, which we adjusted to our own values in step 3.16, to incoming AutoCAD lines.

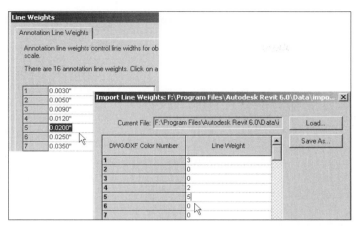

**Figure 2–43** *Editing the line weights in a titleblock file, and applying them to imported colors*

3.19 Select the Save As button to open the *Data* folder, shown in Figure 2–44. Save the line weight changes in a file named *importlineweights-dwg-sampletb.txt*. Choose Save. Choose OK.

 **Note:** This file then becomes the basis for imported line weights. You can use the Load button in the Import Line Weights dialog to load other files.

**Figure 2–44** *Save the line weight settings file*

3.20 From the File menu, select File>Import/Link>DWG, DXF, DGN, as shown in Figure 2–45.

**Figure 2–45** *Import a file*

## Import the file

3.21 Navigate to the folder where you placed AutoCAD content files from the CD included with this book. Select *sample titleblock inserted.dwg* as the file to import (shown in Figure 2–46). Select the radio button to select Black and white under Layer/Level Colors for the display. Under Import or Link, make the Layers selection read All, and leave the Link box unchecked. Under Positioning, select Automatically place and Center-to-center. Choose Open.

**Figure 2–46** *Import settings*

3.22 If a warning appears, choose OK. When the imported file loads, use the Zoom In Region command to examine the lower-right corner of the title-block closely. The line weights we specified are visible, as you can see in Figure 2–47.

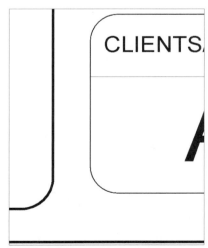

**Figure 2–47** *Imported lines with weight*

The imported file is a single object, and therefore not editable. If we explode it to make the lines and text editable the linework will lose width information. Rather than do that, we will use the imported lines to make new linework that we can edit, then remove the imported linework as we no longer need it.

### Adjust the visibility of imported entities

3.23    Zoom to Fit to see the whole titleblock. Click on the titleblock to select it. On the Options Bar, operation buttons appear. Select Delete Layers.

3.24    In the Select Layers/Levels to Delete dialog, check layer 0 and layer Defpoints to eliminate them (shown in Figure 2–48), then choose OK.

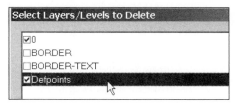

**Figure 2–48**  *Removing unneeded layers*

3.25    Right-click and select Properties from the right-click menu.

3.26    In the Properties dialog, select the Edit value for the Visibility parameter.

3.27    In the Visibility/Graphics Overrides dialog for the Sheet view, select the DWG/DXF/DGN Categories tab. Choose the + symbol next to Imports in Families to expand that category.

3.28    Uncheck layer BORDER-TEXT to turn it off in this view. Choose OK twice to close the dialog.

### Using imported lines to make new lines

3.29    The text will no longer be visible. Select the Lines command on the Family tab of the Design Bar to put Revit in sketch mode. Choose the Pick (arrow) icon on the Options Bar.

3.30    In the Type Selector drop-down list, make Wide Lines the current line type.

3.31    Select, in turn, the four sides of the large inner border of the title block that defines where graphics will go. Each time you select a line in the imported file, dimensions and a lock symbol will appear, to enable you to edit the line you have just created in Revit. Carefully select the fillet arcs at the corners to finish the outline. Do not select the outermost border lines.

3.32    In the Type Selector drop-down list, make Medium Lines the current line type.

3.33    Select, in turn, the sides of the five title block text sections. Carefully select the fillet arcs at the corners to finish the outlines. Zoom and Pan as necessary. See Figures 2–49 and 2–50.

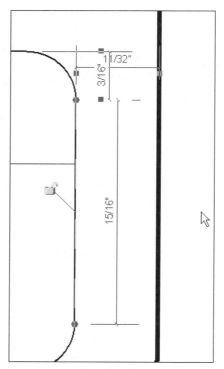

**Figure 2–49** *A new line is created by selecting a line in the imported file*

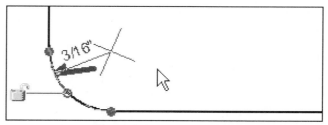

**Figure 2–50** *Creating the filleted corners*

3.34    In the Type Selector drop-down list, make Thin Lines the current line type.

3.35    Select, in turn, the dividing lines inside the title block text sections.

3.36    When you have completed selecting imported linework to create new lines, check yourself. Zoom to Fit in the View Window. Type **VG** to open the Visibility/Graphics Overrides dialog. Select the DWG/DXF/DGN Categories tab. Select the + symbol next to Imports in Families to expand that category. Uncheck layer BORDER to turn it off in this view. Choose OK to close the dialog.

3.37   Examine the linework for completeness. Note that the four outside lines making up the titleblock border (supplied by Revit) do not have fillet corners. Choose the Modify tool at the top of the Family tab on the Design Bar.

3.38   Select each of the four outside lines in turn and use the Type Selector drop-down list to change them from Title Block lines to Wide Lines.

3.39   Select the Lines tool on the Design Bar. Select the Fillet arc tool on the Options Bar, as shown in Figure 2–51. Use the Type Selector to make the current linetype (which will be applied to the fillets) Wide Lines.

**Figure 2–51**   *Starting the Fillet arc line tool*

3.40   Zoom in around the upper-left corner of the title block. Select the two outside lines in turn, close to their intersection point. A radius arc will appear and adjust its radius as you draw the cursor in and out from the intersection. When the radius value reads 1/2", as shown in Figure 2–52, left-click to create the fillet.

   **Tip:** If you have trouble making the radius value read correctly by dragging the cursor, place the fillet where convenient, click the radius dimension value that displays to make it editable, and type in **.5** or **1/2**.

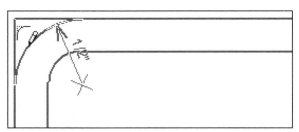

**Figure 2–52**   *Creating the first fillet*

3.41   Repeat this process for the other three corners of the titleblock. Choose Modify from the Design Bar to terminate the fillet tool.

3.42   Zoom to Fit to check your linework. When the lines look complete, type **VG** to open the Visibility/Graphics Overrides dialog. Select the DWG/DXF/DGN Categories tab. Select the + symbol next to Imports in Families to expand that category. Check layer BORDER-TEXT to turn it on in this view. Select OK to close the dialog.

3.43   Select the titleblock text to select the import object. From the Options Bar, select Delete Layers. In the Select Layers/Levels to Delete dialog that opens up, check layer BORDER to eliminate, then choose OK.

3.44    Select Explode to make the text editable.

3.45    Zoom in Region around the lower-right corner of the titleblock. Select the A101 text to select it, then choose the Properties icon on the Options Bar. Revit has given this piece of text a name and assigned a font and size based on (but not necessarily the same as) the AutoCAD font. Choose the Edit/New button to make the value field cells editable. Choose Cancel twice to exit the dialog without making any edits.

 **Tip**: If your firm uses older AutoCAD *.shx* fonts and you want to have them map automatically to Windows fonts during imports, open the file *shxfontmap.txt* in *Program Files\Autodesk Revit 6.1\Data* and enter your own mapping preferences.

## Working with parameters for labels

3.46    We want to replace text in this titleblock with Revit labels to give editable fields when this titleblock is loaded into files. Select the Label tool from the Family tab on the Design Bar. Click in an open area of the drawing to start to define a label.

3.47    The Select Parameter dialog will open up, as shown in Figure 2–53. Study the list of predefined parameters for titleblock labels. This particular titleblock design uses two approval fields and a file name field, so we will have to add parameters to the default list. Choose Add.

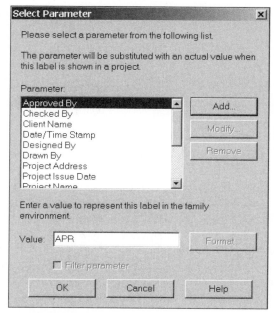

**Figure 2–53** *The Select Parameter dialog*

3.48 In the Parameter Properties dialog, click Select. If you get a warning that a shared parameter file has not been specified, choose Yes to choose a file.

3.49 In the Edit Shared Parameters dialog, choose Create to make a shared parameter file to hold the new parameters we are about to define. This file will be a *.txt* file that you may very well edit later—save it in a convenient network location. Give it the name *Shared Parameters* and choose the Save button.

3.50 In the Shared Parameter dialog, choose the New button under Groups to define a parameter group inside the new file. In the New Parameter Group dialog shown in Figure 2–54, type in the name **Titleblock labels**. Choose OK.

**Figure 2–54** *Naming the new parameter group*

3.51 In the Edit Shared Parameters dialog, the name Titleblock Labels appears in the Parameter Group drop-down list. In the Parameters section select New.

3.52 In the Parameter Properties dialog shown in Figure 2–55, give the first new parameter the name **Approved By 2**. Leave it as a Text field. Choose OK.

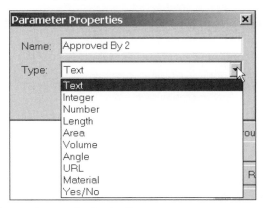

**Figure 2–55** *Possible new parameter properties*

3.53 Add the following new text parameters: Approved By 2 Date; Approved By Date; Filename. Choose OK.

3.54 In the Shared Parameters dialog, Approved By 2 will be highlighted. Choose OK.

3.55 In the Parameter Properties dialog, Approved By 2 will appear in the Name field. It will not be available to indicate that it is not editable in this dialog. Choose OK. The Select Parameter dialog will open, with Approved By 2 now in the list of parameters available for our new label. We are now ready to add labels to the family. Choose Cancel to exit the dialog. Hit ESC to terminate the label tool for now.

3.56 Delete the text entities for the sheet number and file name in the lower-right corner of the titleblock.

## Defining and working with labels

3.57 Click on the Label tool in the Design Bar to place a new label. Select a point inside the lower section of the box you just emptied of text, as shown in Figure 2–56. Select Sheet Number from the list, as you see in Figure 2–57. Leave the Value field at A101. Select OK.

**Figure 2–56** *Preliminary location for the first label*

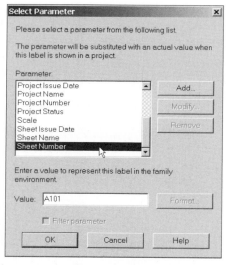

**Figure 2–57** *Selecting the parameter for the first label*

3.58 In the Type Selector drop-down, the new label will be identified as Label: Tag 1. Choose the Properties button.

3.59 In the Element Properties dialog, choose Edit/New. In the Type Properties dialog, select Rename.

3.60 In the Rename dialog shown in Figure 2–58, give the label type the name **1/2" Tag**. Choose OK.

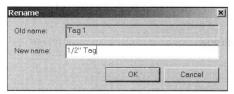

**Figure 2–58** *Renaming the default label type*

3.61    In the Type Properties dialog, change the Text Size to **1/2"**. Make the Background value Transparent. Choose OK.

3.62    In the Element Properties dialog, choose Edit/New. In the Type Properties dialog, select Duplicate. Give the new label type the name 1/4" Tag. Choose OK.

3.63    In the Type Properties dialog, change the size to 1/4". Choose Apply to set the field value.

3.64    Choose Duplicate. Give the new label type the name **3/16" Tag**. Click OK.

3.65    Change the size for the new label type to 3/16". Click on Apply. You now have three label sizes defined.

3.66    Make the 1/4" Tag the current type and choose OK twice to return to the label placement task.

3.67    When you move the cursor into the upper section of the line box, Revit will find alignment planes from the previous label. Align the new label with the first one and left-click.

3.68    In the Select Parameter box, choose Add, then Select, then Filename. Choose OK three times to place the label.

3.69    Select Modify on the Family Design Bar. Select the Sheet Number label. Choose Properties from the Options Bar.

3.70    Change the Horz. Align value to Center. Change the Vert. Align value to Middle, as shown in Figure 2–59. Choose OK. Move the label to a suitable place in the linework. Repeat for the Filename label (see Figure 2–60). Note how Revit still finds alignment planes.

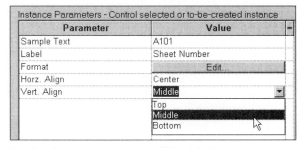

**Figure 2–59** *Setting the alignment properties of the label*

**Figure 2–60** *Aligning the Filename label with the Sheet Number label*

3.71    Erase the imported text objects reading ENGINEERING, XX/XX/XX (there are two of these), and ARCH in the box immediately above the area in which you have been working. We will replace this text with labels.

3.72    Select the Label tool. In the Type Selector, make 3/16" Tag the current type. Select in a convenient place away from linework and text for a preliminary location for this label.

3.73    In the Select Parameter dialog, select Approved By and choose OK. The label will appear with control points and a rotation control. Put the cursor over the rotation control and rotate the label until it is vertical. Revit will snap to a vertical alignment.

3.74    Select another spot to start placing a second label. In the Select Parameter dialog, choose the Add button, then Select in the Parameter Properties dialog. In the Shared Parameters dialog select Approved By Date, then choose OK twice to add this parameter to the list. The Select Parameter dialog will still be active. In the Value field type **APR Date** for the place-holder text for this new label as shown in Figure 2-61, then choose OK. Rotate the new label to a vertical position as with step 3.73.

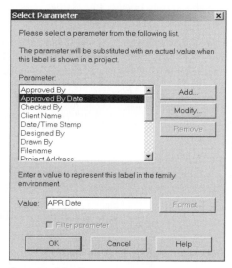

**Figure 2–61** *Adding the Approved By Date parameter*

3.75    Select Modify and move the new labels to their correct positions, as shown in Figure 2-62.

3.76    Select Label and add two new labels with these Parameters: Approved By 2 (Value **APR2**) and Approved By 2 Date (Value **APR2 Date**). Rotate and align these labels as with the first two.

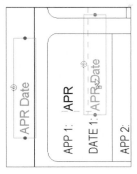

**Figure 2–62**  *Locating the approval labels*

3.77    Add a 1/2" Tag label with the Parameter: Sheet Name to the right of the company name and address. Rotate it to a vertical position. Edit its horizontal alignment to Center and vertical alignment to Middle, and locate it appropriately, as shown in Figure 2–63.

**Figure 2–63**  *The lower half of the titleblock information panel*

3.78 Add 3/16" Tag labels for Designed By, Drafted By, Scale, and Project Issue Date in the appropriate areas on the title block.

3.79 Add a 1/2" Tag label for the Client Name. To the right of it add a 1/4" Tag label for the Project Name and a 3/16" Tag label for the Project Address. Change the horizontal alignment of all those labels to Center and the vertical alignment to Middle. Place the labels carefully.

3.80 Adjust the remaining pieces of text as necessary. See Figure 2–64.

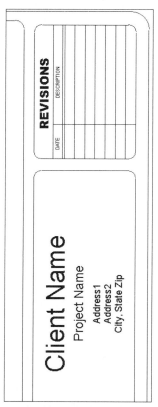

**Figure 2–64** *The upper half of the title block information panel*

3.81 From the File menu, select Purge Unused. In the Purge Unused dialog, you can expand the Other Styles items to see what you will be eliminating from the file and uncheck items you may want to keep. Choose OK.

3.82 Save the Family. Close the family file.

3.83 Revit returns to the open project file. In the project browser, select Sheets, right-click, and select New Sheet. In the Select a Titleblock dialog, choose Load and browse to the location where you just saved the sample title block. Select your new title block and choose Open, then OK.

3.84   The labels with shared parameters that we created earlier in the exercise will be represented by question marks, but they are valid labels, oriented properly. If you choose either Settings>Project Parameters or File>Shared Parameters you will see that the new parameters have come into the project.

3.85   Close this file without saving it.

## Exports from Revit

Revit writes out DWG, DXF, and DGN files to complete the interchange cycle with today's major CAD applications. You can export an individual view or multiple views. Exported 2D views produce 2D linework and 3D views produce a 3D model. Revit's 3D view settings (visibility, hidden line mode) are ignored—a hidden line view opens in 3D Wireframe mode in AutoCAD. The exported model will hide and shade in AutoCAD.

Revit maps object categories to preconfigured layer names. As with layer import map files, you can create your own export mapping files for projects or clients. We will examine this in the next exercise.

Revit exports schedule field information to delimited files for import into spreadsheet applications (such as Microsoft Excel and Lotus 123). Revit can export model components to ODBC databases (such as Microsoft Access) using specific table relationships (between type and instance, for example). ODBC export can then be updated throughout the project to the same database file to give progress reports. Revit ships with a separate PDF writer. Once this is installed, you can export views to PDF format as a print process. There is an available DWF (Drawing Web Format) creation module available free from Autodesk as a plug-in. DWF is Autodesk's proprietary format for Web-based viewing and sharing of CAD file information across many platforms. Revit will export various image types directly from any view or selection of views. These images will import into other applications, such as Microsoft Word. Revit has a walk-though creation routine that will export an AVI animation file, which will play on any media player. Revit will export Cost Reports in HTML format for inclusion in other applications. We will treat these output options in future exercises.

These export capabilities make Revit powerful and useful throughout a building design project. If you need to send DWG or DGN information to outside consultants using their layer standards, Revit makes this operation straightforward. If you need number-crunching information, Revit will supply this from the very beginning of conceptual design work, and gives you a consistent way to update the figures. If you need images, you can generate them at any time in more than one format. Once you have set up your various export routines the process is simple enough so that non-experts can readily generate supplementary files from a Revit model. (At present there is no one-button scripting capability to make admin-type output generation bulletproof.) This will save designer time for designing.

## Exercise 4. Export layer settings

4.1 Launch Revit. From the File menu, choose File>Close to take the open empty project file out of memory. Select File>Open. Open the *c_rvt6_ WBC_Condominium.rvt* file from the *Program Files\Autodesk Revit 6.1\Training\ Common* folder. The file will open to the default 3D view.

4.2 From the File menu, choose File>Import/Export Settings>Export Layers DWG/DXF.

4.3 The Export Layers dialog will open, displaying the current export layer map file (see Figure 2–65). *Exportlayers-dwg-AIA.txt* is the default. All of Revit's object categories are shown with layer and color information. This dialog will allow you to create exact export specifications for the most demanding client, consultant, or government office. Once you have examined the categories and mappings, choose Cancel to exit the dialog.

| Category | Projection | | | Cut |
|---|---|---|---|---|
| | Layer name | Color ID | Layer name | |
| Area Tags | A-AREA-IDEN | 2 | | |
| Callouts | A-ANNO-SYMB | 6 | | |
| Casework | A-FLOR-CASE | 3 | A-FLOR-CASE | 3 |
| Casework Tags | I-ELEV-IDEN | 5 | | |
| Ceiling Tags | A-CLNG-IDEN | 6 | | |
| Ceilings | A-CLNG-SUSP | 5 | A-CLNG-SUSP | 5 |
| Surface Pattern | A-CLNG-GRID | 1 | A-CLNG-GRID | 1 |
| Color Fill Legends | G-ANNO-NPLT | 5 | | |
| Color Fills | A-AREA-IDEN | 2 | | |
| Columns | A-COLS-ENCL | 4 | A-COLS-ENCL | 4 |
| Constraints | A-ANNO-NPLT | 5 | | |
| Contour Labels | C-TOPO-TEXT | 3 | | |
| Curtain Panel Tags | A-WALL-IDEN | 4 | | |
| Curtain Panels | A-WALL-CWMG | 1 | A-WALL-CWMG | 1 |
| Door Elements | {A-WALL-CWMG} | 1 | {A-WALL-CWMG} | 1 |
| Door Elements Cut | {A-WALL-CWMG} | 1 | {A-WALL-CWMG} | 1 |
| Door Panel | {A-WALL-CWMG} | 1 | {A-WALL-CWMG} | 1 |
| Door Swing Elevation | {A-WALL-CWMG} | 1 | {A-WALL-CWMG} | 1 |
| Door Swing Plan | {A-WALL-CWMG} | 1 | {A-WALL-CWMG} | 1 |
| Glass | {A-WALL-CWMG} | 1 | {A-WALL-CWMG} | 1 |
| Glass Cut | {A-WALL-CWMG} | 1 | {A-WALL-CWMG} | 1 |
| Glass Projection | {A-WALL-CWMG} | 1 | {A-WALL-CWMG} | 1 |
| Mullion | {A-WALL-CWMG} | 1 | {A-WALL-CWMG} | 1 |
| Panel Cut | {A-WALL-CWMG} | 1 | {A-WALL-CWMG} | 1 |

**Figure 2–65** *The Export Layers dialog*

## Examine the Revit file views and export

4.4 In the Project Browser, double-click on Floor Plans>First Floor to open that view. Look it over and make yourself familiar with it quickly.

4.5 Open the West elevation and look it over. Note the dashed green reference planes that are visible in this view. They will not plot, but will export.

4.6   In the Project Browser, select the + symbol next to Sections to expand the list in that view category. Double-click Section 1 to open that view, and examine it briefly.

4.7   From the File menu, choose File>Export>DWG, DWF, DGN. In the Export dialog that opens up, use the Save in drop-down browser panel to locate a folder for the export files you are about to create.

4.8   In the File Name drop-down list, edit the file name prefix to read **Condo**. Revit will create files with this prefix and the name of the view as the rest of the file name in the folder you specified in the upper part of the dialog. If you do not have AutoCAD 2000 or later available on your system, use the Save as type list to select the application you will use to check the export files. Leave the Export Layers File to the default *exportlayers-dwg-AIA.txt*.

4.9   In the Export range section of the Export dialog, select Selected views/ sheets and choose the Select button, as shown in Figure 2–66.

**Figure 2–66** *Preparing the export file names and location*

4.10   In the Views list panel, hold down the CTRL key and select on 3D View: {3D}, Elevation: West, Floor Plan: First Floor, and Section: Section 1, as shown in Figure 2–67. Choose OK to return to the Export dialog.

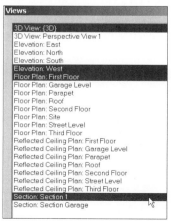

**Figure 2–67** *Select the views to export*

4.11 Choose the Save button to generate the exports. Revit will work for a minute or two, depending on your system resources.

4.12 When Revit has finished the exports, choose File>Exit to leave Revit.

4.13 In your File Manager or local network Explorer, find the folder you specified for the dwg exports from Revit. Note that Revit has also created a *.pcp* (plot configuration parameters file) for each export file (see Figure 2–68). These are editable text files, so that you can specify plotted lineweights based on the colors used in the Export layers file. Note also that the 3D file is considerably larger then the others.

| Name | Size | Type |
|------|------|------|
| Condo - 3D View- {3D}.dwg | 9,292 KB | AutoCAD Drawing |
| Condo - Elevation- West.dwg | 464 KB | AutoCAD Drawing |
| Condo - Floor Plan- First Floor.dwg | 92 KB | AutoCAD Drawing |
| Condo - Section- Section 1.dwg | 313 KB | AutoCAD Drawing |
| Condo - 3D View- {3D}.pcp | 22 KB | AutoCAD Plot Configuration Parameters |
| Condo - Elevation- West.pcp | 22 KB | AutoCAD Plot Configuration Parameters |
| Condo - Floor Plan- First Floor.pcp | 22 KB | AutoCAD Plot Configuration Parameters |
| Condo - Section- Section 1.pcp | 22 KB | AutoCAD Plot Configuration Parameters |

**Figure 2–68** *Revit creates .dwg and .pcp files*

## Exercise 5: Examine the exported files in AutoCAD

The following exercise assumes that you have AutoCAD installed on your system

5.1 Launch AutoCAD. Open the four new files Revit created in turn to examine their contents and layer structure.

5.2 Zoom Extents in Model Space in the *Condo-3D View- {3D}* drawing. Change the view from Shaded to Hidden Line to 2D Wireframe and view the results each time. In the 2D Wireframe view, select a railing section on the interior stairwell and note its layer in the Object Properties toolbar. Right-click, select Properties from the context menu, and note that the railing section is a block. Select a few more entities and note what they consist of: blocks and polyline meshes.

5.3 Open the Layer Properties Manager dialog box for this file and note the layers Revit has created (see Figure 2–69).

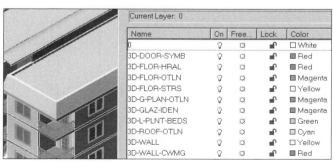

**Figure 2–69** *Layers in the 3D file*

5.4    Close the Layer Properties Manager. Close this file without saving and examine the *Condo-Floor Plan-First Floor* file.

5.5    Use a crossing window to select a section of wall including a window and window tag. Type **LI**, and then hit ENTER to list the contents of your selection set: it consists of lines, blocks, and Mtext.

5.6    Open the Layer Properties Manager for this file. Note the layers that Revit has created (see Figure 2–70). One layer has a dash linetype. All layers except layer 0 have lineweights applied. The drawing has been set up to use color-based plot styles.

| Name | On | Fre... | Lock | Color | Linetype | Lineweight | Plot Style |
|------|----|--------|------|-------|----------|------------|-----------|
| 0 | ♀ | ☼ | 🔓 | ☐ White | Continuous | —— Default | Color_7 |
| A-ANNO-NPLT | ♀ | ☼ | 🔓 | ■ Blue | Continuous | —— 0.05 mm | Color_5 |
| A-ANNO-SYMB | ♀ | ☼ | 🔓 | ■ Magenta | Continuous | —— 0.05 mm | Color_6 |
| A-AREA-IDEN | ♀ | ☼ | 🔓 | ☐ Yellow | Continuous | —— 0.05 mm | Color_2 |
| A-DOOR-IDEN | ♀ | ☼ | 🔓 | ■ Magenta | Continuous | —— 0.05 mm | Color_6 |
| A-DOOR-SYMB | ♀ | ☼ | 🔓 | ■ Red | Continuous | —— 0.13 mm | Color_1 |
| A-FLOR-HRAL | ♀ | ☼ | 🔓 | ■ Red | Continuous | —— 0.05 mm | Color_1 |
| A-FLOR-OTLN | ♀ | ☼ | 🔓 | ■ Magenta | Continuous | —— 0.13 mm | Color_6 |
| A-FLOR-STRS | ♀ | ☼ | 🔓 | ☐ Yellow | Continuous | —— 0.05 mm | Color_2 |
| A-GLAZ-IDEN | ♀ | ☼ | 🔓 | ■ Magenta | Continuous | —— 0.13 mm | Color_6 |
| A-SECT-MBND | ♀ | ☼ | 🔓 | ■ Blue | DASH | —— 0.05 mm | Color_5 |
| A-WALL | ♀ | ☼ | 🔓 | ☐ Yellow | Continuous | —— 0.13 mm | Color_2 |
| A-WALL-CWMG | ♀ | ☼ | 🔓 | ■ Red | Continuous | —— 0.05 mm | Color_1 |
| G-PLAN-OTLN | ♀ | ☼ | 🔓 | ■ Magenta | Continuous | —— 0.05 mm | Color_6 |
| L-PLNT-BEDS | ♀ | ☼ | 🔓 | ■ Green | Continuous | —— 0.05 mm | Color_3 |

**Figure 2–70** *Layers in the Floor Plan*

5.7    Close the Layer Properties Manager. Close the *Condo-Floor Plan-First Floor* file without saving and open the *Condo-Elevation-West* file.

5.8    Select one of the dashed blue lines created from the reference planes in the Revit file. Note its layer in the Object Properties toolbar drop-down list—it has been created on an AIA-standard nplt layer, to indicate that it should not be plotted.

5.9    Select an area of brick facing and list it—a hatch on layer A-WALL-PATT, an AIA standard layer name.

5.10    Open the Layer Properties Manager and note the names and properties of layers in this file. One layer has an aligning_line linetype. Select the name of the linetype to open the Select Linetype dialog box shown in Figure 2–71, and note the linetypes Revit has created in this file. Choose Cancel to close this dialog. Choose Cancel to close the Layer Properties Manager.

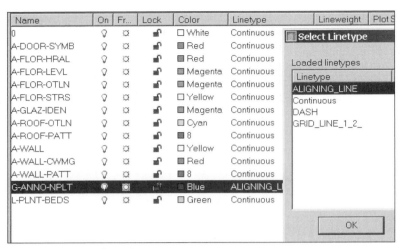

**Figure 2–71** *Layers in the Elevation*

5.11 Close the *Condo-Elevation-West* file without saving and examine the *Condo-Section-Section 1* file.

5.12 In the *Section1* file, select the text label for the Parapet elevation line. Right-click and choose Mtext Edit. Change the font, justification, or content of the text inside the editor to verify that it is editable Mtext. Close the Mtext editor.

5.13 Select a section of stair or stair railing. Note that it is a block. Explode it and check the linework—everything has been drawn with color ByLayer. Each section of stair or railing and each window symbol, whether in elevation or section, has been created as a block. The blocks are named according to type, but each instance has a slightly different name as an identifier. The blocks will explode to linework as we just saw, and they will respond properly to the REFEDIT command.

5.14 Open the Layer Properties Manager for this file. Note the layer names and properties.

5.15 Close the file without saving and exit AutoCAD.

## CONCLUSIONS TO DRAW: REVIT WORKS WITH CAD

CAD managers or others responsible for maintaining the flow of CAD information through an office should take heart from this demonstration. Revit readily accepts AutoCAD content, and just as readily creates output to AutoCAD that captures in a legitimate and completely adjustable form the content of the Revit model. Firms with experienced AutoCAD drafters or detailers, or a valuable inventory of standard detail files, can proceed with implementing Revit, knowing that Revit's modeling capabilities will not bypass the strengths of their workforce or archives.

## Teamwork Tools: Links and Worksets

One of the early criticisms of Revit came from its single building model structure. Most users of AutoCAD are familiar with the xref system, whereby an unlimited number of files can be linked to compose a complex building model or a campus, with very little strain on system resources. References also allow a team to work on parts of an assembled whole.

We have seen that Revit will import AutoCAD and other files. An import is not a live connection between files—in that way it corresponds to a block insertion in AutoCAD. Revit uses a system of links that correspond exactly to external references in AutoCAD. Revit will create and maintain links with DWG, DXF, DGN, and RVT (Revit) files. Links with other Revit files are intended to be links between entire building models, as in a campus or site plan. Revit warns specifically that linking a vertical assembly of floors together to make a single building is *not* supported. This is the opposite of the way firms use AutoCAD and Architectural Desktop to create assembled building models, so be aware that designers moving from AutoCAD-based design to Revit will have to learn some new methods.

Revit has developed the Workset system for design teams working on a single building model. Worksets—divisions of the model using all sorts of criteria—provide a mechanism for individuals to check out parts of the model from a central storage location and return these parts to update the central file. This is also an opposing concept to the usual xref setup, and one that CAD managers will have to understand to make effective use of Revit and prevent loss of time or work on a project.

AutoCAD xrefs usually consist of many separate files, often on completely separated company systems and servers. The individual part files are collated in one or more files to create an assembly model or plotting setup, and often the designers or detailers that work on the parts and pieces never see the entire assembly. Revit's Worksets work the other way around. There is always a central file; worksets have to separate from and return to the central file without gaps or lapses in order for information to update correctly. This system is not a return to the days of mainframes; Revit has mechanisms to allow work to take place remotely.

We shall look specifically at links and worksets, their intended and actual uses, in future exercises.

CAD Managers are often responsible for scheduling training or enhancement instruction to keep designers aware of progress with CAD software. People who become relatively efficient with a particular set of tools or methods often become hide-bound and not interested in learning new practices and techniques, which results in loss of efficiency over time. Any manager who supervises drafters who may never have mastered Paper Space in AutoCAD release 14 or earlier, and tried to convince them of the superiority of multiple Layouts, will understand the difficulty of keeping workers up-to-date in this era of constant change.

Revit comes with simple, comprehensive tutorials; its interface and tools make Revit considerably easier to pick up and start working with than nearly any CAD drafting program. It is quite possible for individuals to teach themselves enough in a short time to create the basics of designs and visualizations that are far more advanced than what they could have created in CAD in the same amount of time, even after training. However, Revit is a powerful, sophisticated apparatus that requires more than a rudimentary amount of knowledge to work effectively and efficiently. More and more firms are realizing that training is essential. The amount of training for experienced designers to become efficient in Revit is measured in hours or days, not weeks, but a training budget should be considered part of any implementation strategy.

Along with the necessary training, CAD managers can help quite a bit by offering designers who are tackling the job of learning an entire new software system in the stressful context of a live project the encouragement that brave souls deserve. This will also be a helpful characteristic for the Project Managers who will oversee Revit teams.

## PROJECT MANAGER

So far in this chapter we have talked about the concerns of CAD managers, who have to maintain a filing system and set of company standards that exists quite apart from what may or may not happen with Revit in the organization. CAD managers, expert though they are in CAD, may not get the chance to become familiar with Revit during the period when their company is implementing its use. Managers must make sure that those who do use Revit can communicate effectively with the network, other departments or offices, and those in the company who will continue to use CAD. There is an individual in most firms implementing Revit who needs to be aware of its effect on work habits and methods—the Project Manager.

As a Project Manager your primary responsibility is to assemble and shepherd the design team through its paces. No matter what your previous training and experience with Revit, there are bound to be a few surprises when people new to the program go live on a project. As with any foray into the unexpected, preparation and planning are key. We can't emphasize enough the value of organized training. Reasonable expectations, given the size, complexity, and deadlines of the trial project, are also critical for team success. Team makeup will play an important role in reaching project goals and developing the efficiency benefits that Revit makes possible.

## Training

Firms implementing Revit are reporting with regularity their conclusions that on the job training is not as effective as 16 to 32 hours of dedicated training time with Revit's online lessons and tutoring from users with previous experience.

## Manage Expectations

Successful implementers also report that keeping the first project or two to a proper size for new users to handle goes a significant way to predicting success. Revit is a complex, comprehensive design package. Nobody ever became proficient with Auto-CAD, MicroStation, or Architectural Desktop overnight. Despite Revit's relative ease in creating complex models quickly, project and company management must allow the inevitable learning curve to run its course. Smart building models take a little more time for new users to create than line drawings, but provide downstream benefits.

## Resistance and Skepticism

The willingness of designers and drafters to take on new challenges will also play a part in their ability to deal with the tensions of producing a project with new tools. Early adopters who relish the chance to learn and expand their skills will provide a different atmosphere than skeptics or those who work best entirely within their comfort zones. Change always meets some resistance, and new methods have to prove effective under fire, so to speak. It's the project manager's challenge to keep designers focused on solving problems, not on the perceived difficulties of working with unfamiliar tools.

## Versatility is Valuable

Revit does not remove the need for designers to understand drafting. It does eliminate the primacy of drafting skills in today's building model design process, and will hasten the disappearance of the CAD jockey whose sole function is as a scribe for designers. A Revit model and documentation package requires far fewer repetitive tasks—designers spend more time designing, and the software takes care of providing views, updating annotations, extracting numbers, and the like. Designers with versatility rather than specialists in CAD skills will play larger roles in Revit projects.

## Changing roles

What is a manager to expect from Revit? While each project will be different, and results will vary from firm to firm, certain trends are already clear.

Revit's ability to produce renderings within the application, and the relative ease of exporting models to other rendering programs, mean that design illustrations and visualizations will come into play earlier than before. This will mean earlier client response to conceptual design options, hardly ever a bad thing. Revit includes a suite of site development tools that allow building designers to specify site design more accurately and completely than before, without having to export plans to other applications.

Revit relies heavily on families, which function as combinations of AutoCAD Styles and 3D Blocks. Families manage functional aspects such as titleblocks and building components such as walls or furniture. Revit provides a number of family templates for the creation of custom content. Skill at creating families for company use will prove as valuable as was skill at composing AutoCAD library material. Creating and managing family content is a subject large enough to merit its own book.

Revit provides tools for data handling, including schedules for nearly every type of component, with customizable parameters for holding information that can be exported from schedules or shared between projects. Once schedules are set up in a project, updating and handling the information they contain becomes more of an administration task than a job requiring CAD or design skills to initiate. Revit's space-planning tools also lend themselves to use by administrative personnel.

The designed effect of Revit's modeling, documentation, and data handling structure is to move nearly all project tasks down the food chain, to eliminate repetitive, low value, but specialist work, and to return control of design projects to designers. This will eventually mean changes in company workforce structure. For managers implementing Revit in a first design project or two, you can expect to be pleasantly surprised at the ease with which designers will produce complex building models and documentation. The longer lasting value of those complex models will only become apparent over time.

## SUMMARY

This chapter has taken a look at some of the more technical aspects of Revit—templates and titleblocks are important parts of any firm's design standards. Proper understanding and setup of templates will result in significant benefits. We examined Revit's titleblock structure both to illustrate labels, data fields, and parameters, and to show how Revit will accept and work with vector information from other CAD formats. Lastly, we saw how easily and effectively Revit will export model information to CAD.

# REVIEW QUESTIONS

## Multiple Choice

1. Template files can contain

   a) customized Levels

   b) customized views and sheets

   c) customized family content

   d) all of the above

2. Template files cannot contain

   a) custom menus

   b) custom Design Bar tabs

   c) custom lineweights

   d) a and b, but not c

3. Revit will export

   a) to specified layers

   b) the current view or selected views/sheets

   c) 2D and 3D information

   d) all of the above

4. Revit's name for an editable text field to place on a titleblock is

   a) dimension

   b) parameter

   c) label

   d) note

5. An imported CAD file comes in as

   a) individual line, dimension, and text entities

   b) a single object that can be exploded to individual entities

   c) a single object that can't be modified

   d) an image

## True/False

6. Revit allows you to delete Layers from imported AutoCAD files

7. Once you save a template file, you can't edit it later

8. Revit does not recognize text in imported files that have been exploded

9. Text labels can be rotated

10. Project information automatically appears in certain titleblock label fields

 Answers found on the accompanying CD.

# CHAPTER 3

# Sharing with CAD Applications

## INTRODUCTION

This section begins with a series of related exercises using basic Revit techniques to create a project file. The next exercises show you how to bring existing AutoCAD files into Revit, and export specific parts of the Revit model out to AutoCAD in ways that maximize the strengths of both programs.

## OBJECTIVES

- Import CAD files as a reference for design project setup
- Create a site and building shell using the imported information
- Start detailing the shell using the imported information
- Export a view of the model into AutoCAD for development by others

## REVIT COMMANDS AND SKILLS

*Settings*

*File import*

*Toposurface*

*Split Toposurface*

*Levels*

*Reference Planes*

*Walls*

*Roofs*

*Phases*

*File Linking*

*Drafting View*

*Windows*

*Web content search*

*Type Editing*

*Profile Editing*

*Sketching: Line, Arc, Tape Measure*

*Editing commands: Move, Copy, Rotate, Align, Trim, Split*

*File export layer settings*

*File export*

## ORGANIZING FOR TEAMWORK

The exercises in the rest of this book will follow a hypothetical design project. The program of this project is as follows:

- Project setup tasks: Work with an existing lakeside site containing one multistory building, an access road, and a parking lot. The building holds collegiate classrooms and offices. Model the existing building shell and site in Revit. The reference material consists of AutoCAD files—a site plan in 3D, and 2D exterior elevations of the existing building.

- Project Phase 1 tasks: Design a second multistory building near the first one. The new building is to hold offices, classrooms, labs, and a kitchen/cafeteria facility. Transform one wing of the existing building into an elevated passage to the new building. Move the parking lot underground; turn the space between the two buildings into a pedestrian courtyard; create a pedestrian plaza over the new parking garage. The reference materials consist of scanned sketches.

- Project Phase 2 tasks: Design a second building to contain offices, a dining facility, and a residence tower. Move the occupants of the offices in buildings 1 and 2 to building 3. Convert the office space in building 2 to classrooms. Convert building 1 into a residence hall. There is no reference material for this phase.

- Project Phase 3 tasks: Design a performing arts building to sit between buildings 2 and 3. This building is to have offices; rehearsal halls; an art gallery; and theater space with a stage, wing, overhead space, a projection booth, and raked seating. The building will feature a central atrium with skylights, cafe with seating, and a ceremonial staircase. There is no reference material for this phase.

The exercises will be designed to cover all the basic Revit skills necessary to complete a team design and documentation project using multiple files. We encourage readers to "warm up" for the book exercises by working through the excellent tutorials included with Revit. Our exercises may touch on material that has been covered in a similar way in Revit's tutorials, but we have made efforts to avoid duplication. The exercises will develop skills and techniques as they would be used in the course of a typical design project; for this reason the exercises are best done in order. Since design projects of commercial or institutional size usually involve teams, the exercises are designed to be modular to a certain extent. In a class or training situation one group of students could be assigned the Phase 2 exercises and one group to Phase 3, for instance, and then the groups should exchange assignments.

Teamwork is an important part of every design project not produced by a single individual. Autodesk long ago developed wblock insertions and the external reference system for AutoCAD so that more than one person could work on a project and share information using a couple of different modes. Revit contains a similar import/link system for CAD files, and the single building model includes links to Revit files, so that designers can now coordinate models in a multi-building or campus development. Revit contains Worksets and Phases, specifically team-oriented mechanisms for dividing and coordinating design tasks.

A project of the size and scope we have outlined takes many hours of work to complete, and a large portion of that work is inevitably repetitive, even with a design application as carefully designed and complete as Revit. Our exercises will touch on each phase of work, but we will not carry these phases through to completion, once the appropriate concepts and techniques have been illustrated.

 **Instructor's Note:** The scope of this hypothetical project will provide opportunities for auxiliary exercises for projects or extra credit. Students can explore alternate design ideas or flesh out areas of the model (classroom, office, or residence layouts).

For those individuals working these exercises on their own and not as part of a group, we admire your gumption and encourage you to work through the file linking and Workset exercises so that you will understand how files and projects are subdivided and linked together. Even if you will work entirely by yourself for the foreseeable future, you will certainly find a use for a multi-file project, or Worksets.

To begin, in this chapter we will open a blank imperial file and import a series of AutoCAD files—a site plan and elevations for the existing building on the site. We'll use those files to make a Revit model of the existing building, and use that model to export an elevation of one face of the building that incorporates new design information out into AutoCAD.

## Exercise 1. Create a project—establish settings

1.1   Launch Revit. Revit will open to a Floor Plan view of Level 1 in an empty project file using the default Imperial template. Four Elevations appear in the View window. Place the cursor anywhere in the View window (but not over an Elevation or Elevation View control arrow), and right-click. This will open the right-click context menu (see Figure 3–1). View Properties is always the bottom command. Select View Properties to open the Element Properties Dialog for the System Family: View/Floor Plan for Level 1 (see Figure 3–2). Study the properties of this view, noting the default View Scale (1:96, or 1/8" = 1'-0") and Phase information Choose OK to exit the dialog.

**Figure 3–1** *You can reach View Properties with a right-click*

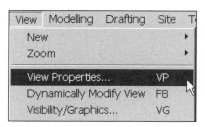

| Instance Parameters - Control selected or to-be-created instance | |
|---|---|
| **Parameter** | **Value** |
| View Name | Level 1 |
| Title on Sheet | |
| View Scale | 1/8" = 1'-0" |
| Scale Value    1: | 96 |
| Crop Region | □ |
| Crop Region Visible | ☑ |
| Scope Box | None |
| Display Model | Normal |
| Visibility | Edit... |
| Model Graphics Style | Hidden Line |
| Detail Level | Coarse |
| Discipline | Architectural |
| Phase Filter | Show All |
| Phase | New Construction |
| View Range | Edit... |
| Underlay | None |
| Wall Join Display | Clean all wall joins |

**Figure 3–2** *The View Properties for the Level 1 Floor Plan in an empty Imperial file*

1.2 Double-click on Site in the Project Browser to open that view. This view also shows the four Elevations. Type **VG** to open the Visibility/Graphics dialog for the current view.

**Tip:** Revit contains a number of two-letter keystroke shortcuts for common commands, and the user can create more. Shortcuts appear on drop-down menus (see Figure 3–3). We will introduce a number of them in the exercises in this and future chapters.

| View | Modelling | Drafting | Site | T |
|---|---|---|---|---|
| New | | | | ▶ |
| Zoom | | | | ▶ |
| View Properties... | | | VP | |
| Dynamically Modify View | | | F8 | |
| Visibility/Graphics... | | | VG | |

**Figure 3–3** *Menu items and their associated keystroke shortcuts*

**Note:** Views are crucial in Revit, and managing them is an important skill to develop early. The View Properties dialog contains overall controls for the current view plus the Visibility/Graphics dialog, which contains visibility settings for every object type in the building model within the current view.

1.3 On the Annotation Categories tab of the Visibility dialog for the Site view, uncheck Elevations (see Figure 3–4). Select OK to exit this dialog. The Elevations will disappear from the screen.

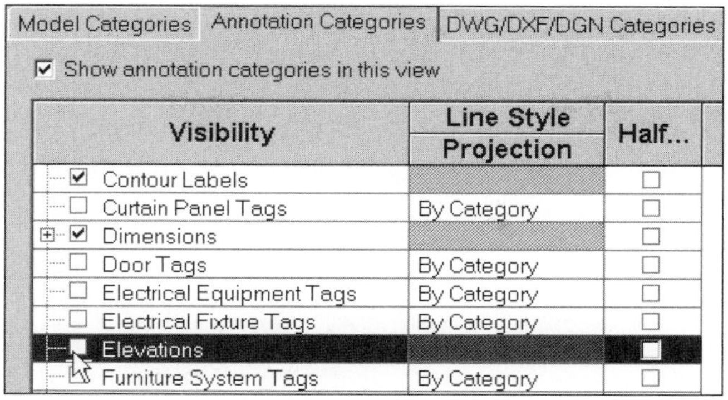

| Model Categories | Annotation Categories | DWG/DXF/DGN Categories |

☑ Show annotation categories in this view

| Visibility | Line Style Projection | Half... |
|---|---|---|
| ☑ Contour Labels | | ☐ |
| ☐ Curtain Panel Tags | By Category | ☐ |
| ⊞ ☑ Dimensions | | ☐ |
| ☐ Door Tags | By Category | ☐ |
| ☐ Electrical Equipment Tags | By Category | ☐ |
| ☐ Electrical Fixture Tags | By Category | ☐ |
| ■ Elevations | | ■ |
| ☐ Furniture System Tags | By Category | ☐ |

**Figure 3–4** *Making the Elevations not visible in the Site View*

1.4   Type **VP** to open the View Properties dialog for this view. Note the default View Scale (1:240, or 1" = 20' – 0") and Phase settings. Select Cancel to exit this dialog without making changes.

From the Menu Bar, select Settings>Project Information (the top selection on the Settings menu) as shown in Figure 3–5. This will open a Type Properties dialog for the Project Information System Family. Edit the value fields for Project Information Parameters as follows (or as specified by your instructor if in a class):

1. Project Issue Date  12/31/04 (or the ending date of your training)
2. Project Status      Conceptual Design
3. Client Name         Lakeshore College (or the name of your business institution)
4. Project Address     4800 Madison Road, Diamond Lake TN 37052
5. Project Name        Dudley Drive Renovations
6. Project Number      004R1

Choose OK to apply the values and close the dialog. This information will appear on Titleblocks for Sheets.

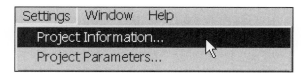

**Figure 3–5** *Reaching Project Information*

1.5   From the Settings menu, select Settings>Options (the bottom selection). In the Options dialog that opens, on the General tab, set the Save Reminder interval value in the Notifications section to 30 minutes. Revit does *not* auto-save (see Figure 3–6).

**Figure 3–6** *Setting the Save Interval*

1.6 On the File Locations tab, set the Default path for user files as directed by your instructor if you're in a class (see Figure 3–7). Your path will be different from the one pictured. Choose OK to exit the dialog.

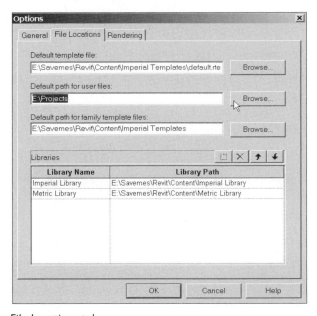

**Figure 3–7** *The File Locations tab*

1.7 From the File menu, select Save As. In the Save As dialog, choose the Options button. In the File Save Options dialog that opens, set the number of backup files to save to 1, or as specified by your instructor (see Figure 3–8). The default setting is for 3 backups, identified by a 00x addition to the file name. Revit will save up to 50 copies of Workset enabled projects. This option is not a global Revit setting, and *must be adjusted for each new file.*

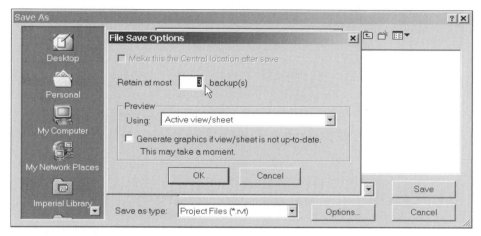

**Figure 3–8** *Setting the number of backups*

1.8 Choose OK to close the File Save Options dialog. In the Save As dialog, give this new project the name **Chapter 3 Phase 1**, and save it as a Project Files (*.rvt) in the location specified by the class instructor (see Figure 3–9).

**Figure 3–9** *Give the file a name and location*

## Exercise 2. Import external content

2.1 Open the file *Chapter 3 Phase 1.rvt* from the previous exercise if you closed it after step 1.9, or continue with the open file. From the File menu, choose File>Import/Link>DWG, DXF, DGN. In the Import/Link dialog that opens, use the Look in: drop-down list to navigate to the folder that contains the file *site plan chapter 3.dwg*. Select that file so that the name highlights in the folder contents window and the name appears in the File name: field. A preview image will appear, as shown in Figure 3–10.

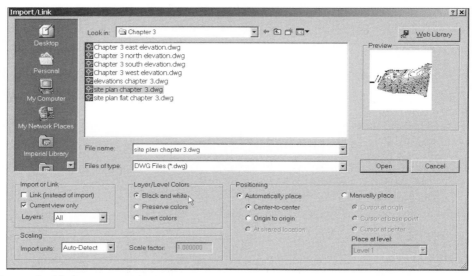

**Figure 3–10** *Import settings*

2.2 Adjust the settings in the lower half of the dialog: Check Current view only under Import or Link. This will allow Revit to import text. Select Black and white under Layer/Level Colors. Make the Import units under Scaling read Auto-Detect, and select Automatically Place>Center-to-Center under Positioning. Choose Open to start the import.

2.3 When the screen shows linework, type **ZF**, the keyboard shortcut for Zoom to Fit.

2.4 When the cursor is placed anywhere over the imported object, a tool tip will appear that identifies it as an Import Symbol, and a border will appear around the import.

Left-click to select the imported symbol. The cursor will change appearance to a double-headed arrow, to signify that you can now drag it to another position. The contour lines will change from black to red, Revit's color for selected items. The Type Properties drop-down list on the Options Bar will show the type category of the import (Import Symbol: site plan chapter 3.dwg), the Properties icon will become active, and other action buttons become visible on the Options Bar.

With the cursor over the selected symbol, right-click and study the right-click context menu. With an object selected, commands available on the Options Bar are also available by using right-click. In this case, Delete Selected Layers and Explode Imported Instance appear in the context menu (see Figure 3–11). Move the cursor off the linework of the import symbol and left-click to deselect it.

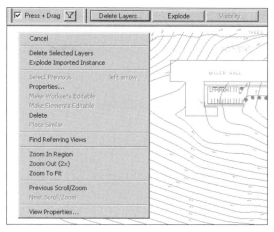

**Figure 3–11** *Available commands for an Imported Instance*

2.5　Place the cursor in the Design Bar pane to the left of the screen. If Site is not in the stack of command tabs at the bottom of the Design Bar, right-click to bring up the list of available tabs and choose Site. This will add the tab to the project and make it the active one (see Figure 3–12).

**Figure 3–12** *Making the Site tab available*

## Create a Toposurface using the imported file

2.6　From the Site tab on the Design Bar, select Toposurface. The Design Bar shifts into Sketch mode. The top of the Design Bar now reads Toposurface, and the command list changes. Select Use Imported, then left-click over the Import Symbol to select it.

2.7　The Add Points from Selected Layers dialog opens. All the layers in the AutoCAD drawing will be checked. Uncheck layer topo lines, then select Invert. This will isolate the topo lines layer as the only layer to use for creating 3D points in a Toposurface mesh, as shown in Figure 3–13. Choose OK to create the Toposurface.

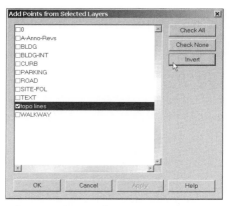

**Figure 3–13** *Using points from one layer only*

2.8 Choose Properties in the Toposurface Sketch commands on the Design Bar. In the Element Properties dialog, change the Material to Site – Grass and change Phase Created to Existing (see Figure 3–14). Choose OK to exit the dialog.

**Figure 3–14** *Properties of the new Toposurface*

2.9 Select Simplify Surface. In the Simplify Surface dialog, set the Accuracy value to 12", or simply 1 with no units of measurement, and Revit will change it to 1' 0" as shown in Figure 3–15. The default unit in project files from an Imperial template is the foot. Choose OK. Select Finish Surface.

**Figure 3–15** *Surface Accuracy*

2.10 Place the cursor in an open part of the View window and right-click. Choose Zoom in Region from the right-click context menu, and select two points around the outline of the building.

2.11 Note the labels of the contour lines at the parking lots (see Figure 3–16). These are in feet. The contour lines in the imported file were created using AutoCAD polylines with their Z-elevations set to the corresponding value.

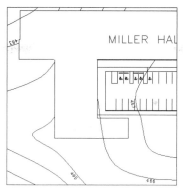

**Figure 3–16** *Imported contour lines and their labels*

2.12    Type **VG** to bring up the Visibility dialog. On the DWG/DXF/DGN Categories tab, choose the + symbol next to *site plan chapter 3.dwg* to expand the
list under it. Uncheck topo lines to turn off visibility of that layer, as shown
in Figure 3–17. Choose OK twice.

| ⊟ ☑ site plan chapter 3.dwg | By Category | By Category | ☐ |
|---|---|---|---|
| ☑ 0 | By Category | By Category | |
| ☑ A-Anno-Revs | By Category | By Category | |
| ☑ BLDG | By Category | By Category | |
| ☐ BLDG-INT | By Category | By Category | |
| ☑ CURB | By Category | By Category | |
| ☑ PARKING | By Category | By Category | |
| ☑ ROAD | By Category | By Category | |
| ☑ SITE-FOL | By Category | By Category | |
| ☑ TEXT | By Category | By Category | |
| ☐ topo lines | Override... | Override... | |
| ☑ WALKWAY | By Category | By Category | |

**Figure 3–17** *Turning off the AutoCAD layer topo lines*

## Adjust Levels to fit the Toposurface Elevations

2.13    The contour lines in the import symbol will disappear, as that layer is no
longer visible in this view. The new Toposurface is not visible. In order to
make use of the newly created Toposurface, which is sitting at a height of
450' to 470', we need to adjust the elevations of the two default Levels in
the project file.

2.14    Double-click on Elevations>East in the Project Browser to open that view.
The Toposurface will occupy most of the View window, with the default
Levels 1 and 2 visible at the bottom. Right-click and Zoom In Region at the
lower-right corner of the elevation view, so you can read and adjust the
labels for the levels.

2.15    Select the label text 0' 0" for Level 1. The level line will highlight red and a
box will appear around the label. Select the text 0' 0" again and the box will
become an editing field. Change the Level's Elevation by typing in **466** (Revit
will read this as 466') and hitting ENTER.

2.16 The Level will disappear from the screen. Adjust the Elevation for Level 2 to **476**, as shown in Figure 3–18. It too will disappear from the screen when you hit ENTER.

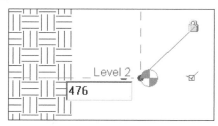

**Figure 3–18** *Adjust the elevation of both levels*

2.17 Type **ZF**. Note that the Level lines are now above the contours of the Toposurface in this view.

2.18 Double-click on Site in the Project Browser. Contour lines are now visible in the Site view. Hold the cursor over a contour line and the tooltip will identify the object as Topography: Surface.

2.19 Type **VG** to open the Visibility dialog. On the DWG/DXF/DGN tab, expand the list under *site plan chapter 3.dwg* and uncheck all layer names except BLDG. Choose OK. Save the file if you intend to stop at this point.

## Exercise 3. Split the Surface at the building outline

3.1 Open the file from the previous exercise or continue with the open project. On the Site tab in the Design Bar, choose Split Surface, as shown in Figure 3–19. Click a contour line to select the new surface.

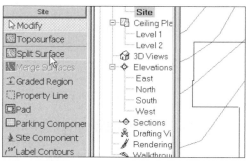

**Figure 3–19** *Split the new surface*

The Design Bar will switch to Sketch mode. The Lines tool will be active by default. The Options Bar will show drawing commands. The cursor will appear as a pencil symbol; the Status bar and tooltip will display information pertinent to drawing. Make sure the Draw icon is active (depressed), and place a check in the Chain option on the Design Bar.

3.2 Right-click in the View window, choose Zoom In Region, and select two points at the corners of the building outline to make that fill the screen. Trace the building outline starting in the upper left corner, going counterclockwise.

As you draw, Revit will display temporary dimensions to show you the length of the line you are creating, plus a direction vector that will snap to horizontal and vertical planes (see Figure 3–20). Revit will also snap to geometric points such as endpoints and midpoints as the cursor passes over them. You can type in distance values directly and Revit will apply them to lines or walls while you are sketching.

 **Tip:** You can also right-click to reach transparent Zoom commands: Zoom In Region, Zoom Out (2x), Zoom to Fit, and Previous Scroll/Zoom. This enables you to keep drawing while adjusting the View window. The scroll wheel (if your mouse has one) will also provide zoom and pan controls.

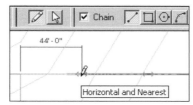

**Figure 3–20** *Dimensions assist your sketch*

3.3 Sketch around the building outline but do not draw the last segment. Choose Finish Sketch in the Design Bar. Since your sketched lines do not meet completely, Revit will display an Error – cannot be ignored message, as seen in Figure 3–21. The Expand>> button provides a tree view of the error or errors, which can be helpful for advanced users. Pressing the Show button directs you to the standard View commands and does not usually provide an informative picture of what you might have missed. Select the Cancel button instead.

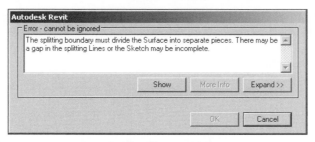

**Figure 3–21** *Error message—your sketch will not work*

3.4 Select the Modify command from the Design Bar (always the top selection). As you move the cursor over the building outline, the lines you drew previously will highlight in magenta. This way you can test for missing lines (our deliberate fault in this sketch) or lines that overlap.

3.5 Choose the Lines command on the Sketch tab in the Design Bar. On the Options Bar, select the Pick icon. Hold your cursor over the line at the right side of the building outline that you did not trace over before. The line will highlight green to indicate that it is selectable.

3.6 Left-click to select the line. It will highlight with grips and a lock symbol (open) to indicate that you can lock the sketch line to the underlying symbol.

3.7 Choose Finish Sketch. If you now place your cursor inside the building outline, the tooltip will identify that surface and it will highlight separately from the surface outside the building outline.

Click inside the building outline. The inner surface will highlight. On the Options Bar, choose the Properties icon. In the Element Properties dialog, change the Material Value to Concrete – Cast-In-Place Concrete. Choose OK twice, and again on the warning message that appears (see Figure 3–22).

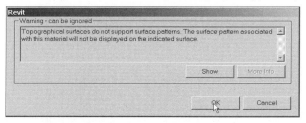

**Figure 3–22** *This warning refers to pattern display in shaded views*

3.8 Type **VG** to reach the Visibility dialog. Choose the DWG/DXF/DGN Categories Tab. Expand the list under the imported dwg file. Uncheck layer BLDG and check layer ROAD to isolate that portion of the imported file.

## Split the Surface at the road outline

3.9 Choose Split Surface on the Site tab of the Design Bar. Click the main Toposurface (not inside the building outline) to select it. The Design Bar will switch to Sketch mode.

3.10 On the Options Bar, select the Pick icon. The Toposurface will become gray, to make it easier to select the lines on the ROAD layer in the imported object. Select the lines that make up the outline of the roadway and parking lot. You may have to use Zoom In Region and Pan to make your selections accurate.

Be careful not to select the same line twice when selecting existing linework or objects to create sketch lines. Revit will place <Sketch>: Model Lines on top of each other, which can make editing difficult.

When you have selected the existing linework of the roadway outline, select Draw and draw a line to close the sketch at the upper right end of the road. Choose Properties on the Design Bar to access the properties of the Toposurface you are sketching. In the Element Properties dialog, change the Material Parameter value to Site – Asphalt (see Figure 3–23). Click OK.

| Instance Parameters - Control selected or to-be-created instance | |
|---|---|
| **Parameter** | **Value** |
| Material | Site - Asphalt |
| Surface Area | 0 SF |
| Comments | |
| Mark | |
| Phase Created | Existing |
| Phase Demolished | None |

**Figure 3–23** *Assigning a material to the new surface*

3.11    Choose Finish Sketch. If Revit displays an Error – cannot be ignored message as in Figure 3–24, choose Continue to return to your sketch. Select Modify from the Sketch tab on the Design Bar. Move your cursor over the roadway outline and Revit will highlight the new sketch lines in turn, so you can find any overlapping lines or gaps.

 **Note:** If the error message concerns lines overlapping or intersecting, choose the Trim tool from the Toolbar and use it to trim back lines that cross rather than meet (see Figure 3–25). If you have trouble selecting lines while using this command, hit ESC to return to the Modify mode and try deleting one or more of the lines. You may find you have inadvertently placed lines on top of one another.

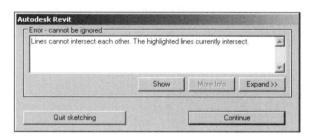

**Figure 3–24** *Lines intersect or line on top of one another*

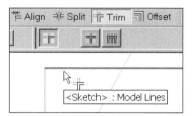

**Figure 3–25** *The trim tool in action—note the appearance of the cursor*

 **Note:** If the error message says that there may be a gap in the lines or the sketch may be complete, you can use the Trim tool to have lines meet, draw a new line between the endpoints of lines that gap, or select a line on the imported object.

3.12    When you have corrected any drawing errors, select Finish Sketch.

## Crop the Site view

3.13 Select Site in the Project Browser. Right-click and select Properties. In the Element Properties dialog, check the box in the Crop Region value field, as shown in Figure 3–26. Choose OK.

3.14 Type **ZF**. You will see the Crop Region around the site plan Toposurface. The cursor tooltip will identify it when you place the cursor over the Crop outline. Select the Crop Region to activate its control points (see Figure 3–27).

| Instance Parameters - Control selected or to-be-created instance | |
|---|---|
| **Parameter** | **Value** |
| View Name | Site |
| Title on Sheet | |
| View Scale | 1" = 20'-0" |
| Scale Value    1: | 240 |
| Crop Region | ☑ |
| Crop Region Visible | ☑ |

**Figure 3–26** *Activating the Site view Crop Region*

3.15 Use the Stretch controls at the midpoints of each side of the Crop Region to make it smaller, as shown in Figure 3–27. The area below and to the right of the existing building will be the location of the three phases of new work on this hypothetical project.

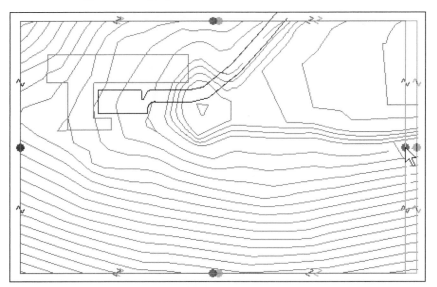

**Figure 3–27** *Adjusting the Crop Region size*

3.16 Click outside the Crop Region once it has been condensed to your satisfaction to deselect it. Right-click and use Zoom In Region with two points inside the corners of the Crop Region to make the site plan fill the View Window.

3.17   Choose Site in the Project Browser, right-click, and choose Properties. In the Element Properties dialog, uncheck Crop Region Visible to simplify the Site view so you will not inadvertently select the Crop Region during further work in this view. Choose OK. Save the file.

## Exercise 4. Import another file for Floor Level Elevation information

4.1   Open the file from the previous exercise or continue in the open project. Double-click on Elevations>South in the Project Browser to make that view active. Right-click and use Zoom in Region to make the Level tags readable.

4.2   From the File menu, choose File>Import/Link>DWG, DXF, DGN. In the Import/Link dialog, navigate to the folder that contains the file *Chapter 3 south elevation.dwg*. Select that file so the name appears in the File Name field of the dialog. Adjust the Import or Link setting as before—check Current view only to bring in text. Select Black and white in the Layer/Level Colors section. In the Positioning setting portion of the dialog, select Manually place and Cursor at origin (see Figure 3–28). Choose Open.

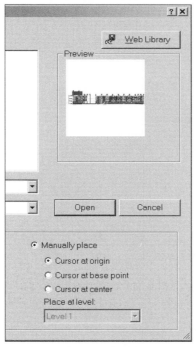

**Figure 3–28** *Place this import manually*

4.3   The imported file object will appear with the cursor at its lower-left corner. Place the import object to the right of the Level tags.

**Tip:** If you select the import object it will highlight red and the cursor will become the double-headed Move arrow. You will get better control for short moves by choosing the Move command from the Toolbar. Revit will ask for a start and end point of the move, and provide temporary dimensions and the angle readout. Revit will snap to points and extensions of lines on the import and model, so it's easy to align them both.

4.4 Select the elevation field for Level 2. Select it again to make the edit window active. Change the elevation value to 479'-4 (Revit will supply the inches) to match the specification on the import, and hit ENTER.

**Note:** If the level bubble disappears, Revit has relocated it to the right of the import. Zoom to Fit and Zoom Region so you can see the level end. Select it so that blue grips appear. Put the cursor over the right grip, and a tooltip noting that dragging this grip will relocate the end of the level will appear. Left-click and you can drag the level back to its original end point.

4.5 Use the same technique to edit the name values of Levels 1 and 2 to 1ST FLOOR MILLER HALL and 2ND FLOOR MILLER HALL. Choose Yes in the question box to rename views that reference the level names throughout the Project Browser (see Figure 3–29).

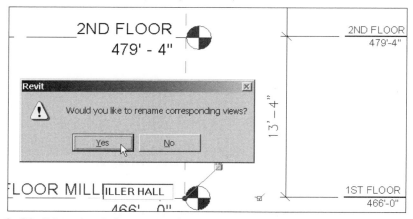

**Figure 3–29** *Editing Level elevations and names*

4.6 Choose the Basics tab in the Design Bar to open its commands. Select Level to add more Levels to the Project. Move the cursor to the left of the View window above the 2ND FLOOR MILLER HALL level line. Revit will read from that line and show a temporary dimension up from the level line. Type **13'4** on the keyboard to match the dimension in the import object. Revit will apply that dimension and show the level's elevation bubble (see Figure 3–30). Move the cursor to the right and Revit will snap to an alignment point above the existing bubbles. Left-click to accept that alignment point and Revit will start another level.

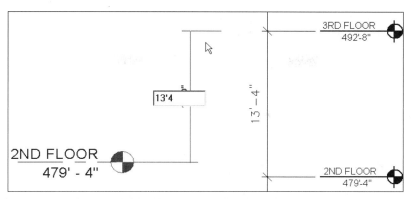

**Figure 3–30** *Specify the Level offset from the one below*

4.7 Type the dimension offset between the 3RD FLOOR and ROOF levels as shown in the import file (**14'8**) and either hit ENTER or left-click outside the new level's elevation value field. Align the tag as before (see Figure 3–31.

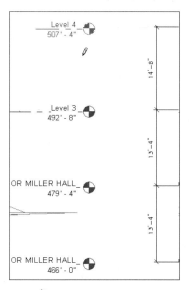

**Figure 3–31** *Bubbles will snap to align*

4.8 Choose Modify from the top of the Basics tab in the Design Bar to exit the Level tool.

 **Tip:** You can also hit ESC twice to exit the Level command completely. Many commands in Revit are multi-level routines, which allow the user to cancel a certain part of the procedure while still remaining in the command set. AutoCAD and Architectural Desktop users should be well familiar with the ESC-ESC reflex.

4.9    Edit the names of Levels 3 and 4 TO 3RD FLOOR MILLER HALL and ROOF MILLER HALL. Accept the rename question each time.

4.10    Edit the length of the new elevation lines. Select the 3RD FLOOR level and use the blue control dot at its left end to drag the line to the left. Repeat for the ROOF level. Note that the lines will align and you can move more than one.

 **Tip:** While dragging the level ends you can type **ZF** to zoom transparently and keep control over the grips.

4.11    If Revit has not yet reminded you to save your file, save it now.

## Exercise 5. Create Revit walls using the imported plan

5.1    Open the file from the previous exercise or continue with the open file *Chapter 3 Phase 1.rvt*. Double-click on Site in the Project Browser to return to that view. Type **VG**. In the Visibility dialog, select the DWG/DXF/DGN Categories tab. Expand the list under the site plan imported dwg. Uncheck layer ROAD. Check layer BLDG.

5.2    Select the Model Categories tab. Uncheck Topography, as shown in Figure 3–32. Choose OK.

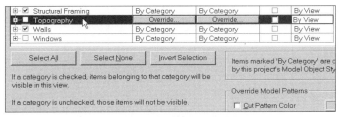

**Figure 3–32** *Topography is on the Model Categories tab*

5.3    The imported building outline is now isolated in the Site view. Choose Wall on the Basics tab of the Design Bar. The cursor will change to a pencil symbol, indicating sketch mode. The status bar at the bottom of the screen will read "Click to enter wall start point," which will also appear as the tooltip. Change the Height value for the walls you are about to create to ROOF MILLER HALL using the drop-down list. Check the Chain option. Leave other options to their defaults (see Figure 3–33).

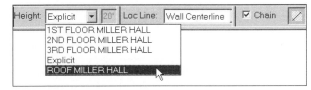

**Figure 3–33** *Set the height of the walls to a level*

5.4    Trace walls around the building outline clockwise. This will place the defined exterior face of the walls to the outside.

 **Note:** There are sections of the building outline where lines do not terminate end to end. This is deliberate—all designers and drafters have at one time inherited incompletely translated files, unfinished models, or badly drawn sketches.

Revit's snap and alignment tools will allow you to draw a horizontal wall that terminates perpendicular to a desired line or wall even if the underlying elements you may be tracing over are not horizontal or do not define complete corners.

5.5    When the new walls make a complete outline, choose Modify on the Basics tab of the Design Bar. Type **VG** to reach the Visibility dialog. On the DWG/DWF/DGN Categories tab, expand the site plan list and uncheck BLDG. Choose OK. This will leave the new walls isolated in this view.

## Adjust the Elevations to fit the walls

5.6    Double-click Floor Plans>1ST FLOOR MILLER HALL in the Project Browser to open that view. Type **ZF** to expand the view. Note the location of the Elevations. Since we created walls according to a location on an imported object, and not in the center of the project file, the Elevations are now inaccurate and should be relocated.

 **Note:** If the Elevations are not visible in this view, type **VG** and use the Visibility dialog to turn them on.

5.7    Use the cursor to select two points around the North and South Elevations. The two Elevations will turn red to indicate they have been selected. Select the Move command from the toolbar. A dashed boundary will appear around the selected items, along with a temporary reference line, and the cursor will change to the Move symbol (see Figure 3–34).

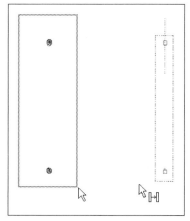

**Figure 3–34** *Preparing to move the North and South Elevations*

5.8 Select a start point above the North Elevation, then a second point above the middle of the north wall of the building outline to move the two Elevations. As you move the cursor to select a starting point, Revit will display a temporary dimension. As you move the items, Revit will show an angle vector and temporary dimension, which will allow you to enter a distance directly if you desire. Place the Elevations by eye.

5.9 Repeat the Move on the East and West Elevations, placing them outside the building outline roughly in the middle of the figure. Note that the South Elevation does not include the North–South wing and its extension (see Figure 3–35).

Revit will readily duplicate Plan and Section Views by selecting them in the Project Browser and using the right-click context menu, but not Elevations. You can create a copy of an Elevation in a Plan view and adjust the properties of the new Elevation separately.

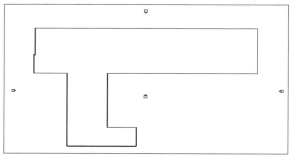

**Figure 3–35** *The Elevations moved around the building outline*

5.10 Select two points around the South Elevation to select it. Choose the Copy command from the Toolbar. Create a copy of the South Elevation 120' below the original, so that the copy includes the wing and extension, as shown in Figure 3–36. Revit will create a new Elevation named South1 in the Project Browser.

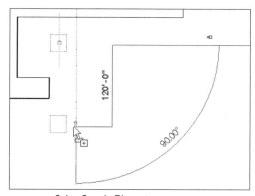

**Figure 3–36** *Creating a copy of the South Elevation*

5.11    Double-click Elevations>South in the Project Browser to open that view. Type **VG** to open the Visibility Dialog. On the Model Categories tab, uncheck Topography. Choose OK. Type **ZF** to see the new walls and import symbol centered in this view.

5.12    Repeat this process in each of the other Elevation views. Note the difference between the views South and South1. Return to the South Elevation view. From the Window menu, select Window>Close Hidden Windows to save system resources (see Figure 3–37).

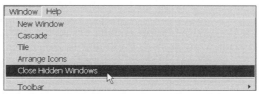

**Figure 3–37** *Closing Windows on views you won't use for a while*

## Alignment using Reference Planes

5.13    Select the import symbol. It will highlight red. Choose the Move tool and then select two points to place the import symbol above the new walls, as shown in Figure 3–38. You will adjust the actual placement later.

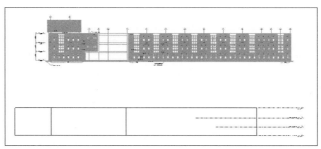

**Figure 3–38** *The import moved above the model*

5.14    Type **ZF** to enlarge the model and imported file in the View window. Select one of the Level lines so that it highlights red and a blue grip (control dot) appears at the left end. Use the control dot to pull the end of the Level line to the left, past the end of the building model. Two or more Level lines will travel together. Select another Level line and adjust its length the same way. Revit will snap-align the ends of the Level lines, and then all Level lines will move as one.

5.15    One of the Level lines will still be highlighted, and an empty blue checkbox will appear to the left of its control dot. Check that box to turn on an Elevation bubble and tag for that end of the Level line.

5.16    If the text for the bubble runs over the building model, uncheck the box and adjust the length the lines as necessary. Turn on the left end bubbles and tags for all four levels in the building in this view (see Figure 3–39).

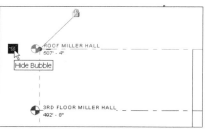

**Figure 3–39** *Hide or Display Level Bubbles with the checkbox*

5.17    Type **ZF** to enlarge the model and import object in the View Window. Select Ref Plane from the Basics tab of the Design Bar. The cursor will change to a pencil symbol; the Status Bar and Tooltip will display information relative to the active command.

5.18    Select the upper-left corner of the model to start the Reference Plane. Pull the cursor straight up, so that the angle vector reads 90° and the tooltip shows Vertical. Select a second point to create the reference plane.

5.19    Choose Modify from the Basics tab on the Design Bar. Select the new Reference Plane, right-click, and choose Properties from the context menu. In the Element Properties dialog, give the Reference Plane the Name WEST END MILLER HALL. Choose OK.

5.20    Zoom in Region around the left end of the import symbol and include a section of the Reference Plane. Click the import to select it. The linework will turn red. Type **M**. The Status Bar will read Hotkeys:M menu:edit|modify. Use arrow keys for more choices. Press ENTER or SPACE to select. Hit the DOWN ARROW key twice and the Status Bar will cycle the menu:edit choice to menu:edit|move. Press the SPACE BAR to enter the Move command. Select a point on the left edge (Revit will probably select the endpoint of one of the horizontal lines that represent brick.) Pull the cursor to the left or right so it snaps to the Reference Plane. Select again to locate the import at the Reference Plane (see Figure 3–40). Hit the ESC key twice.

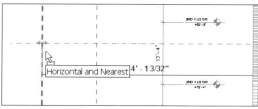

**Figure 3–40** *Move the import object to the Reference Plane*

5.21    Pan to the right so that the section part of the imported building elevation fills the View window. Choose Ref Plane from the Basics tab of the Design Bar. Create two vertical Reference Planes at the edges of the section (representing the exterior face of upper story walls), next to column lines N and L (see Figure 3–41).

**Figure 3–41**  *Reference Planes at the Section*

5.22   Name the Reference Plane at column line **N WEST SIDE OF WALK-WAY**. Name the Reference Plane at Column **L EAST SIDE OF WALK-WAY**, as shown in Figure 3–41. Use Zoom to Fit to see the model. Study the building profile of the section—the first story is wider than the upper stories.

5.23   Double-click on Floor Plans>2ND FLOOR PLAN MILLER HALL in the Project Browser to open that View. Zoom In Region around the T-junction between the horizontal and vertical wings of the building. Note that the right-hand wall is aligned Interior Face with the Reference Plane, not Exterior Face. Choose the Align tool from the Toolbar. The cursor will show alignment arrows at the point of the cursor; the Status Bar and tooltip will show directions appropriate for the command sequence.

5.24   Select the right EAST SIDE OF WALKWAY (right-hand) Reference Plane, then select the Exterior (right-hand) face of the right wall. A blue lock symbol will appear at the midpoint of the wall. Select the symbol to change it from unlocked to locked. This will fix the wall's exterior face at the Reference Plane (see Figure 3–42). If you move the plane, the wall will move with it.

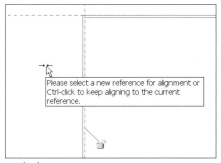

**Figure 3–42**  *The alignment lock*

**Note:** Revit's parametric capabilities allow the designer to establish a nearly unlimited set of relationships between building elements. We will explore ways to take maximum advantage of some of these possible relationships in future lessons. One of the first, most basic parametric utilities is the alignment lock. When combined with Reference Planes–Revit's 3D construction lines–in a carefully planned design, locking and other orientation specifications can make even late design changes straightforward to manage—closer to wholesale than retail.

5.25    Exit the Align command. Select the horizontal wall to the left of the screen. Use the blue control dot to stretch the wall to the WEST SIDE OF WALK-WAY Reference Plane. Lock the lock that appears. Hit ESC to terminate the stretch. Save the file.

## Exercise 6. Adjust walls to match the imported elevation

6.1    Open the file from the previous exercise or continue in the open project file. Right-click and use Zoom Out (2X) as many times as necessary to see the North–South wing of the building. Select the left-hand wall of the North–South wing (to the left of the WEST SIDE OF WALKWAY Reference Plane). Select Properties. Set the Top Constraint value to Up to Level: 2ND FLOOR MILLER HALL. Select OK. Hit ESC twice and the wall will change color from red (highlighted) to light gray, indicating that it does not reach the current level but is visible as an underlay.

6.2    Select the Split command from the Toolbar. The cursor changes to a scalpel icon. Select the bottom (East–West) wall of the I-wing at the intersection with the WEST SIDE OF WALKWAY Reference Plane. Revit will snap to the Reference Plane to make the selection easier. Select Modify from the Basics tab of the Design Bar to terminate the command.

6.3    Select the same East–West wall to the left of the Reference Plane. Only the part of the wall to the left of the Reference Plane will highlight. Right-click and choose Properties. Set the Top Constraint value to Up to Level: 2ND FLOOR MILLER HALL. Choose OK. Hit ESC twice and the wall will change color from red (highlighted) to light gray.

6.4    Choose the Wall command from the Basics tab of the Design Bar. The Type Selector should read Generic – 8". Set the Height to ROOF MILLER HALL. Change the Location Line value on the Options Bar to Finish Face Exterior (see Figure 3–43). We are going to create the west wall for the upper stories of the North–South wing of the building, which will become a bridge-way between the existing classroom hall and the first new building.

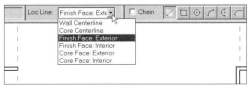

**Figure 3–43** *Prepare to draw the wall at its face*

6.5    Put the cursor over the intersection of the WEST SIDE OF WALKWAY (left-hand) Reference Plane and the 2nd Floor East–West wall at the bottom of the building that we just split and adjusted for height. Pull the cursor vertically to draw a wall along the Reference Plane to the East–West wall we lengthened previously. Select the upper (interior) face of the existing wall as the second selection so the two walls join correctly. The wall joins will not be identical to the other walls, since the new wall starts at the 2ND FLOOR level. Click the

Lock symbol that appears to lock the two walls together. Choose Modify on the Basics tab of the Design Bar to terminate the command.

## Define walls as existing-to-remain and existing-to-be-demolished

6.6   Use Zoom to Fit. Select two points outside the building outline to create a selection box that surrounds all the walls. The second story walls will all highlight red, along with the South Elevation and possibly the WALKWAY Reference Planes. Choose the Filter Selection icon from the Options Bar. In the Filter dialog that opens, uncheck Elevations, Reference Planes and Views to remove them from the selection set. See Figure 4-34.

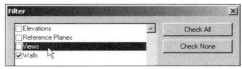

**Figure 3–44** *Filter the selected objects*

Choose Properties. Change the value of the Phase Created Parameter to Existing. Choose OK. Hit ESC twice to clear the command and selection set.

6.7   Select one of the two gray (1st floor) walls. Hold down the CTRL key and select the other gray wall. Select the bottom wall. Select the short North–South and short East–West walls that make up the ell to the right of the walkway (see Figure 3–45). Right-click and choose Properties. Set the Phase Created Parameter to Existing and set the Phase Demolished Parameter to New Construction. Hit ESC twice to clear the command. The walls will display with a dashed linetype.

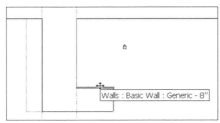

**Figure 3–45** *The walls that will be demolished*

## Place roofs on the building—single-story section first

6.8   Type **ZP** as many times as necessary to return to the previous close-in view of the vertical wing. Select Roof>>Roof by Footprint from the Basics tab of the Design Bar. The Design Bar will switch to Sketch mode with drawing commands. The Pick Walls tool will be active (depressed) by default.

6.9   On the Options Bar, uncheck Defines Slope. This will be a flat roof. Leave the Extend into wall (to core) option unchecked. We are creating this roof for representational purposes only, and so do not need Revit to show it accurately in section.

6.10　Select the interior face of the two one-story walls. Select the exterior face of the wall at the LEFT SIDE OF WALKWAY Reference Plane and the exterior face of the East–West wall to its left (at the top of the screen). We are placing a roof on the one-story construction only.

**Note:** Revit will snap to either face of each wall. As you select you create a line in the roof sketch (shown in magenta). Revit will display information about the sketch line—length, offset from wall face, and a double-headed orientation toggle arrow. If you mistakenly select the wrong side of a wall, you can move the sketch line to the other side right away. You can also select the line later to make adjustments.

6.11　Note that because the East–West wall at the top of the sketch extends past the other wall, so does the sketch line. Type **T** to activate the T hotkey. The only command beginning with T is trim/extend, so hit the SPACE BAR to start the command.

**Tip:** Hotkeys and their associated keystrokes, such as arrow key cycling or SPACE BAR/ENTER command acceptance, should be typed within about six seconds of each other, or Revit clears the hotkey. For those who like to use the keyboard, hotkey stroke combinations will be worth developing for speed. Forefinger-T>Thumb-SPACEBAR, with just a bit of practice, can become an automatic and extremely quick way to enter the important Trim command, for example.

6.12　Use Trim to make the sketch lines meet at the corners with no gaps or overlaps, as shown in Figure 3–46. Choose Roof Properties from the Sketch tab of the Design Bar. Change the Phase Created value to Existing and the Phase Demolished value to New Construction. Choose OK.

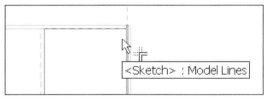

**Figure 3–46** *Using Trim on sketch lines for the Roof Footprint*

6.13　Choose Finish Roof from the Sketch tab of the Design Bar. The roof will appear with a dashed linetype. (Where the roof lines overlap other lines, the dashes may not be visible.)

## Get more information—import another file

6.14　Double-click Elevations>South in the Project Browser to return to that view. Use Zoom in Region around what appears to be a standing seam gable roof at the left side of the imported elevation drawing. Select the Tape Measure tool on the Toolbar. This will read the distance and angle between any two points. Select the intersection of the WEST SIDE OF WALKWAY Reference Plane and the line identified as METAL ROOF EDGE for the first point. Pull the cursor to the left, horizontally to the wall line, under the right side of the metal roof, and select the intersection of that line and the METAL ROOF EDGE line. The Tape Measure will read 27'-0". Select to terminate the measurement and get ready for another.

6.15 Select two points to measure the overhang of the upper roof. The Tape Measure will read 2'-0". Select two points to measure the height of the wall under the upper roof. The Tape Measure will read 2'1 3/4". Write down these measurements for future use. Pick Modify on the Basics tab of the Design Bar to exit the command.

6.16 Select the WEST SIDE OF WALKWAY Reference Plane. Type **C** and hit the SPACE BAR to enter the Copy command. Select any convenient point, then pull the cursor to the left until the temporary dimension reads 27'-0" (see Figure 3–47). Left-click to accept the distance.

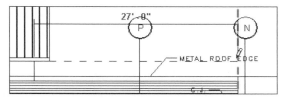

**Figure 3–47** *Making a copy of the Reference Plane*

**Tip:** Revit's temporary dimensions read out in distance increments that are keyed to the zoom scale of the current view. At a close zoom, distance will display in inches, and the cursor will snap at each inch. When zoomed out at larger scale, the temporary dimension will read and snap in 4", 6", or 1' increments. You can enter a specific distance from the keyboard at any time.

6.17 Select the new Reference Plane. Open its Properties dialog. Give it the name **EAST SIDE GABLE WALL**, shown in Figure 3–48. Choose OK.

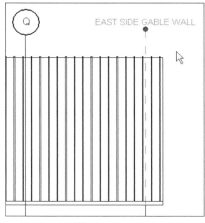

**Figure 3–48** *Name the new Reference Plane*

6.18 Double-click Elevations>West in the Project Browser to open that view. From the File menu, choose File>Import/Link>DWG, DXF, DGN. Navigate to the folder with the file *Chapter 3 west elevation.dwg* and select that file. Adjust the settings in the Import/Link dialog as before: Check Current view only; select Black and white Layer/Level Colors; and choose Manually Place/Cursor at origin.

6.19 Locate the imported symbol above the building model. You can align it exactly later. Zoom in Region around the gable roof area of the import. Use the Tape Measure tool to check the angle of the roof—select the peak and then an eave edge. The Tape Measure will display a distance of 24'-5" and an angle of 45° or 135°—a 12 in 12 pitch. We will use this information in an upcoming exercise. Save the file.

## Exercise 7. Create two flat roofs—one to be demolished

7.1 Open or continue working in the file from the previous exercise. Open the ROOF MILLER HALL Floor Plan view. Zoom In Region to the North–South wing and the short East–West leg. Choose Roof>>Roof by Footprint from the Basics tab of the Design Bar. The Design Bar will change to Sketch mode with the Pick Walls tool active. On the Options Bar, check Extend into wall (into core). This will be a flat roof, so leave Defines slope unchecked.

7.2 Select the interior face of the left-hand wall. Select, in turn, the interior faces of the other walls that define the two building wings. Select the Lines tool to draw a line between the endpoint of the sketch line at the top of the sketch where there is no third story wall drawn (see Figure 3–49).

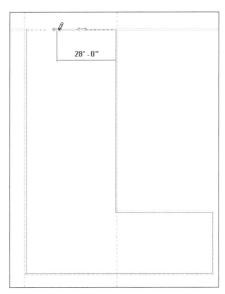

28' - 0"

**Figure 3–49** *Draw the last line for the roof outline*

7.3    Choose Roof Properties from the Sketch tab of the Design Bar. In the Element Properties dialog, set the Phase Created value to Existing. Set the Phase Demolished value to New Construction. (This roof will be replaced by a gable to echo the roof on the main wing of the building.) Set the Base Offset From Level value to −1'; since we are placing a generic roof of 12" thickness, we will set it down 12" from its Base Level so its top surface and the tops of the walls align (see Figure 3–50). Choose OK.

| Phase Created | Existing |
|---|---|
| Phase Demolished | New Construction |
| Base Offset From Level | -1 |
| Cutoff Level | None |

**Figure 3–50** *Set the roof properties*

7.4    Choose Finish Roof on the Sketch tab of the Design Bar. Choose Yes to the question about cutting the roof volume out of the walls. The roof will appear in dashed lines.

7.5    Zoom or Pan so the East–West wing of the building fills the View window. Select Roof>>Roof by Footprint. Check Extend into wall on the Options Bar. Accept the Pick Walls tool on the Sketch tab. Choose the interior faces of the southeast, east, north, and northwest walls of the building wing.

7.6    Select the Lines tool on the Sketch tab. Zoom In Region if necessary to draw the next lines. Check Chain on the Options Bar. Start the lines by selecting the lower endpoint of the last sketch line you created. Pull the cursor horizontally right to the EAST SIDE GABLE WALL Reference Plane so that the plane highlights green, and left-click.

7.7    Pull the cursor down vertically to start a line at a 90° angle from the previous one. Revit will snap to either face (or the location line) of the exterior wall. Place the line endpoint at the interior face of the East–West wall. Select the endpoint of the first sketch line you created to close the loop (see Figure 3–51). This loop does not enclose the southwest corner of the building, which we will work on in the next exercise.

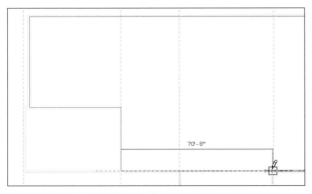

70' - 8"

**Figure 3–51** *The sketch loop for the flat roof section*

7.8 Choose Roof Properties on the Sketch tab. Set the Phase Created value to Existing. Leave the Phase Demolished value at None. Leave the Base Offset From Level value at –1' 0". Choose OK. Choose Finish Roof. Choose Yes on the question.

## Create walls for the gable roof section

7.9 Zoom in Region to the lower-left corner of the East–West wing. Choose the Wall tool from the Basics tab. Choose Properties. Note that Revit sets the Base Constraint for new walls to ROOF MILLER HALL, since the current view is of that Level. Set the Unconnected Height value for the new walls to 8'—we will adjust them to fit under the gable roof later. Choose OK. Check Chain on the Options Bar.

7.10 Select the lower-left corner of the exterior walls to start the new walls. Pull the cursor vertically to the exterior side of the short jog in the walls below, and select that point. Pull the cursor horizontally to the right to the Reference Plane and select the plane when it highlights.

7.11 Pull the cursor down vertically until the exterior face of the lower East–West wall highlights, and left-click. Pull the cursor left horizontally to the starting point and click to finish the walls (see Figure 3–52).

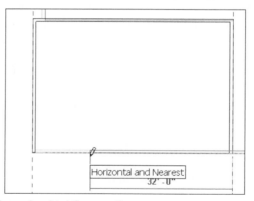

**Figure 3–52** *Walls above the third floor roof*

## Create the final existing roof section

7.12 Choose Roof>>Roof by Footprint on the Basics tab. Choose Roof Properties on the Sketch tab that appears. Set the Base Offset From Level value to 2'1.75" (the distance we measured earlier). Revit will convert decimal inch entry to fractions. Set the Phase Created value to Existing. Choose OK.

7.13 On the Options Bar, set the Overhang value to 2' (the other distance we measured from the imported elevation). Check Defines Slope and place the cursor over the bottom new wall. As you move the cursor from the interior to the exterior face, Revit will locate a green highlighted line to show you where the sketch line will be. Select the exterior face. A magenta sketch line will appear, with a length dimension, offset dimension, double-headed control arrow to flip orientation, and slope arrow (two lines making an angle symbol). You can select either of those dimensions at this point to edit the properties of the sketch line. Choose the Properties icon from the Options Bar. In the Element Properties dialog for the Model Line that you just created, change the Rise/12" value to 12". Choose OK to return to the sketch.

7.14 Select the exterior face of the upper wall to define the other slope of the roof. Choose Properties to check the Rise/12" value and change it if necessary. Choose OK. Uncheck Defines slope and select the exterior faces of the left and right walls to complete the sketching part of the roof creation (see Figure 3–53).

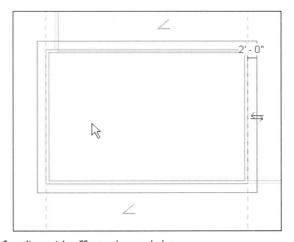

**Figure 3–53** *Roof outline with offset value and slope arrows*

7.15 Choose Finish Roof. Choose Yes in the question about attaching the walls to the Roof. Revit will trim the walls to the underside of the new Roof.

## Adjust views of the model

7.16 The roof will display with cutlines. You will not see the ridge of the roof. In order to see the ridge line in the roof plan, we need to adjust the default settings for this view. Type **VP** to enter the View Properties dialog. In the Element Properties dialog for the current view, choose the Edit button for the View Range (see Figure 3–54).

| Phase | New Construction |
|---|---|
| View Range | [Edit...] |
| Underlay | 3RD FLOOR MILLER HALL |
| Wall Join Display | Clean all wall joins |

**Figure 3–54** *The View Range control*

7.17   In the View Range dialog, change the Offset value to the right of the Top parameter to 25'. Do not change the Top parameter. Change the Cut plane Offset value to 20' (see Figure 3–55). Choose Apply to see the changes in the roof display under the dialog boxes. Choose OK twice to exit the View Properties dialog.

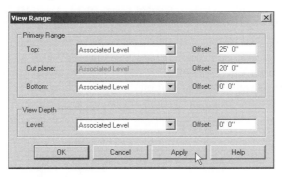

**Figure 3–55** *Edit the View Range values*

7.18   Type **ZF** to fill the View window with the roof plan. Choose the 3D (Default 3D view) button on the Toolbar (see Figure 3–56). Revit will open an isometric view showing the model and topography, and create a view named {3D} under 3D Views in the Project Browser. Type **Z**, then hit SPACE BAR to activate Zoom In Region, then select two points around the model.

**Figure 3–56** *The 3D View control*

7.19   From the View menu, choose View>Shading. Revit will display the generic walls and roofs in shades of gray and blue (see Figure 3–57). The parts designated as demolished in the New Construction Phase, which is the current (default) phase setting, are red and transparent.

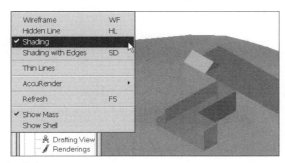

**Figure 3–57** *The shaded isometric view*

7.20    Type **HL** to return the view to Hidden Line mode. Choose Window>Close Hidden Windows. Save the file.

For the purposes of our hypothetical project we now need to put wall, door, and window information on certain walls of the building so it is an accurate model of parts of the building represented in the CAD files we have been studying in Revit. The first stage of the design project will involve transforming the North–South wing of this building. The ground level will be removed for a pedestrian concourse, and the upper two stories will become walkways between the existing East–West wing and a new building to be located at the south end of the new concourse. The next chapter's exercises will locate the shell of this new building and those of later phases in the ambitious master plan of this little college.

Once the walls of the existing North–South wing have been accurately represented on the parts that will remain, we will design the faces of support pillars for the concourse and add a glazed roof over the top floor. The final task for the job setup will be to export an elevation of this wing in AutoCAD, to represent electronic files sent to consultants or contractors.

## Exercise 8. Define exterior walls using information from imports

8.1    Continue with or open the file *Chapter 3 Phase 1.dwg*. Open the Floor Plan View 1ST FLOOR MILLER HALL. Select Ref Plane from the Basics tab of the Design Bar. For the first defining point of the new Reference Plane, select the upper-left (northwest exterior) corner of the building—Revit will show the Endpoint snap to make the selection easier. Pull the cursor to the right, past the right side of the building, to establish a horizontal plan at the exterior wall's exterior surface. Left-click to finish creating the plane.

Choose Modify from the Basics tab of the Design Bar. Select the new Reference Plane. Choose the Properties icon from the Options Bar. In the Name value field of the Element Properties dialog, type **NORTH FACE MILLER HALL**. Choose OK.

8.2    Open the West Elevation View. It will open with the NORTH FACE MILLER HALL Reference Plane still highlighted red, and its blue control dots showing. Choose the upper one and drag it past the imported elevation drawing, so we will be able to align an edge of the CAD file to an edge of our building.

8.3    Use Zoom in Region to the lower-left side of the building elevation in the CAD file. Choose the Align tool from the Toolbar. The cursor will change to the double arrow Align tool shape. Select the new Reference Plane first. It will highlight red. Put the cursor over the line representing the building edge. Revit will snap to the linework and the line will highlight green. The Status Bar and tooltip will identify the import symbol (see Figure 3–58). Left-click to accept the selection and the import file will move over to the Reference Plane.

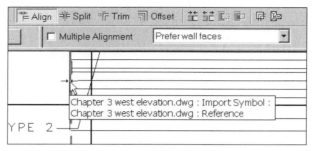

**Figure 3–58** *Align the import to the Reference Plane*

8.4 Choose Modify from the Basics tab of the Design Bar. Zoom to Fit. If the Level Lines for the Revit model terminate far to the right of the model, and thus take up excessive space on the screen, choose any one of them to edit certain properties of the elevation bubbles as a group. The control dots will appear. Choose the right control dot, hold the left button down, and drag the bubbles to the left, closer to the building (see Figure 3–59). This will make the Zoom to Fit more useful. Hit ESC to leave drag mode.

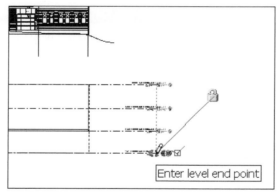

**Figure 3–59** *Dragging the Level Elevation bubbles*

8.5 Zoom in Region to the upper-right corner of the CAD file so you can see the bubble identifying Column 8 and the right edge of the building. Choose Ref Plane from the Basics tab of the Design Bar. Select the upper-right corner of the building in the CAD import. Type **ZF** to enlarge the view. Pull the cursor down at 90° past the Revit model to establish a Reference plane at the south side of the building wing.

8.6 Zoom in Region at the right side of the gable roof. The Reference Plane tool will still be active. Select a point on the line representing the edge of the wall under the right side of the roof—this is also the south edge of the East–West wing of the building. Type **ZF** to zoom back out. Pull the cursor down to a point below the Revit model at 90° (vertical), and click to establish a vertical Reference Plane at the building edge.

8.7    Choose Modify from the Basics tab of the Design Bar. Select the first new Reference Plane (on the right), choose Properties, and give it the Name value **SOUTH SIDE WALKWAY**. Choose OK. Use the same technique to name the other new Reference Plane **SOUTH SIDE MILLER HALL**. Choose OK.

8.8    Open the 1ST FLOOR MILLER HALL Floor Plan view. Zoom in Region to the south (bottom) side of the Revit model. Use the Align tool to align the south exterior face of the model with the SOUTH SIDE WALKWAY Reference Plane. Zoom Previous and repeat the alignment—make the two south side walls of the main wing align with the SOUTH SIDE MILLER HALL Reference Plane (see Figure 3–60). There are two unconnected walls that must be aligned separately.

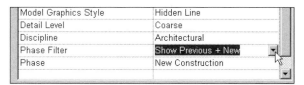

**Figure 3–60** *Align the walls to the Reference Plane*

8.9    Zoom to Fit in the Floor Plan view. Choose Modify from the Basics tab of the Design Bar to terminate the Align command. Open the West Elevation view. Zoom to Fit. Type **VP** to open the View Properties dialog. Change the value of the view Phase Filter to Show Previous + New (see Figure 3–61). This will turn off the display of components marked demolished in the current phase (New Construction), while still showing components identified as Existing or New Construction.

| Model Graphics Style | Hidden Line |
|---|---|
| Detail Level | Coarse |
| Discipline | Architectural |
| Phase Filter | Show Previous + New |
| Phase | New Construction |

**Figure 3–61** *Phase Filter for a View*

The walls will appear slightly different than before—there will be no dashed lines evident. We previously defined two west walls for the North–South extension—one from the first floor level to the second, and one from the second to the roof. The first floor wall was designated as demolished, so it has disappeared from this view. The wall section visible between the first and second floors is at the east side of the extension.

8.10   Move the cursor over the right side of the model until the second/third story wall section highlights. Left-click to select it. The wall type will appear in the Type Selector on the Options Bar. Choose the Properties icon. In the Element Properties dialog, change the Base Offset value to –1'4" to allow for a wall outside the floor structure at the second floor level. Choose OK. The bottom of the wall will now appear below the 2ND FLOOR MILLER HALL Level line.

## Create Reference Planes for alignment between views

8.11 Zoom in Region to the right side of the CAD import. Create a series of six vertical Reference Planes at the sides of the areas identified as the MP type wall (these areas contain windows type E). Name the Reference Planes **WALL DIVISION 1** through **WALL DIVISION 6**, from right to left (see Figure 3–62).

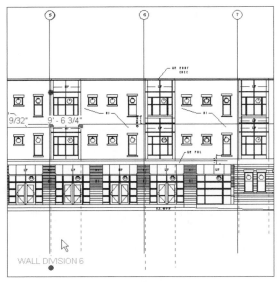

**Figure 3–62** *Reference Planes along the wall face*

8.12 Open the 2ND FLOOR MILLER HALL Floor Plan View. Open the View Properties dialog. Change the Phase Filter to Show Previous + New. Choose OK to exit the dialog. Choose the Split Tool from the Toolbar. The cursor becomes a scalpel shape. Select the west (left) wall of the extension at the intersection points of the new Reference Planes to split it into seven separate sections.

Revit will snap to the Reference Planes, but not very strongly. You may have to Zoom in Region closely to make sure your split point is on or reasonably close to the Reference Plane. Revit will show temporary dimensions to allow you to define the points by distance.

**Tip:** For repeated Zoom In and Zoom Out, you can use the wheel on a scroll mouse if your computer is so equipped. Rolling the wheel forward or back will Zoom In and Out, and holding the wheel down while moving the mouse will Pan back and forth. If this behavior does not take place and you have a Microsoft Intellimouse on a Windows 2000/XP system, open the Mouse Properties dialog in the Windows Control Panel. There should be a Wheel tab with a Troubleshooter section containing an Advanced button. Selecting the Advanced button will let you turn off the default Intellimouse behavior (scroll up and down rather than in and out) for Revit. Choose the Add button and navigate to the folder containing *Revit.exe*. Select the application file, and back your way out of the dialog.

8.13    Open the West Elevation. Zoom in Region to the left side of the CAD import. Create a horizontal Reference Plane. Place the first point at the left end of the heavy line identified as the 2ND FLOOR TOP OF SLAB. Zoom to Fit and drag the right end of the Reference Plane beyond the CAD import. Name the new Plane **IMPORT 2ND FLOOR T.O.S.**

## Identify wall materials using import files

8.14    Zoom in Region to the right side of the import. Study the structure of the wall as identified by the text and leaders. Use Zoom and Pan as necessary to find identifying text.

Below the Reference Plane we just created that marks the top of the floor slab behind the wall, there is a surface marked Type 1 at the right side of the imported elevation, with metal panel/storefront window combinations between pillars dressed with brick and CMU. All of that area is to be demolished, but study its structure so that we can match it with new elements. Above the Reference Plane is a flat roof with a metal edge, and a metal panel flashing against the wall above. Above the panel, surface B1 runs to the Metal Roof Edge at the top of the building. In between the B1 sections there are areas (we identified their edges with Reference Planes) marked MP. Window assembly type E appears in the MP sections.

8.15    From the View menu, choose View>New>Drafting View to create a separate View that will not be connected to the building model (see Figure 3–63). We will use it to hold information that comes to us from an outside source in dwg format, which may be updated during the project. In the New Drafting View dialog, name the View **WALL MATERIALS**. In the View Scale field, choose Custom from the dropdown list and make the Scale value 1.

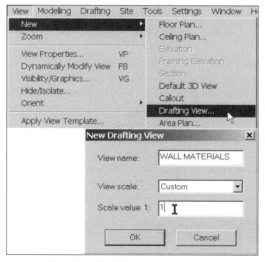

**Figure 3–63** *Open and Specify a Drafting View*

8.16 Click the + symbol that appears next to Drafting Views in the Project Browser. The new view WALL MATERIALS appears below. Double-click the name of the new view to open it. From the File menu, choose File>Import/Link>DWG, DWF, DGN. In the Import dialog, navigate to the folder that holds the file *materials notes.dwg* and select that file for import. In the Import or Link section of the dialog, check Link (instead of import). Select Black and white in Layer/Level colors, and Center-to-center and Automatically place in the Positioning control, as shown in Figure 3–64.

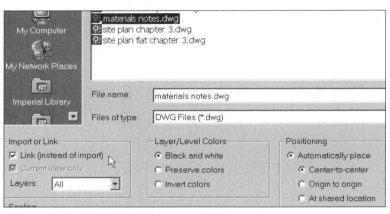

**Figure 3–64** *Link settings for the materials.dwg file*

8.17 Zoom to Fit in the new view. Study the materials as currently specified:

Brick Type B1 is brown Brick. Concrete Masonry Unit (block) type 1 is 4" x 8" Split-faced Concrete Masonry Unit. MP is a Metal Panel wall system. We linked this file into Revit rather than import it, so that there is a place in Revit to check if finish specifications for this project change. The new information is updated in the *materials notes.dwg* file.

8.18 From the Window menu, select Window>Close Hidden Windows. Double-click the West Elevation view in the Project Browser. From the Window menu, select Window>Tile. The WALL MATERIALS view and West Elevation view will appear side by side in the View window to allow you to learn the materials arrangement in the walls. Zoom and Pan as necessary to make yourself familiar with the wall surface designations in the import.

**Tip:** Revit does not currently have a mechanism for splitting the View Window (comparable to AutoCAD's Viewports) to allow for viewing or working in two parts of the model from the same view orientation. You can readily duplicate Floor Plan Views and tile them, as we have just done. In order to duplicate Elevation Views, you must copy an Elevation in a Floor Plan View to create a new Elevation View from an existing one. If the existing Elevation contains an imported or linked CAD file that you need to see in the new one, you will have to import the file into the new Elevation View as well.

8.19 Save the file.

## Exercise 9. Working with the walls—change wall types

9.1    Open or continue working in the file from the previous exercise. Maximize the West Elevation view. Move the cursor over the Revit model along the North–South wing. As the cursor moves, the sections of wall you created by using the Split tool on the original wall will highlight under the cursor. Select the new wall segment on the far right of the model. It will highlight red, and the cursor will change shape to a double-headed Move arrow. The Type Selector will show Basic Wall: Generic – 8", and wall editing commands will appear along the Options Bar.

9.2    Use the Type Selector drop-down list to change the type of the highlighted Generic wall to Exterior – Brick on Mtl. Stud, as shown in Figure 3–65. Repeat on the other walls to define a total of four brick walls in the building wing.

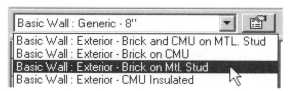

**Figure 3–65** *Change the wall type on one wall*

    **Tip:** You can select more than one wall by holding down the CTRL key while selecting wall sections as they highlight under the cursor (see Figure 3–66). Then all sections will change properties at the same time.

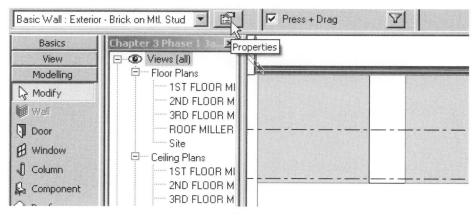

**Figure 3–66** *Change the wall type on more than one wall*

9.3    Hit the ESC key twice to clear the walls you have selected. Use the same technique to select the three walls corresponding to the MP wall type in the imported file. When all three wall sections are highlighted red, choose the Properties icon from the Options Bar.

9.4 In the Element Properties dialog, choose the Edit/New button. In the Type Properties dialog that opens, change the Type to Exterior – EIFS on Mtl. Stud. Choose the Duplicate button to use that existing type as the basis for a new one you are about to name and define.

In the Name dialog that opens, type **Exterior – MP on Mtl. Stud,** as shown in Figure 3–67. Choose OK. The Type Properties dialog will display the new name. Choose the Edit button in the Value field of the Structure Parameter.

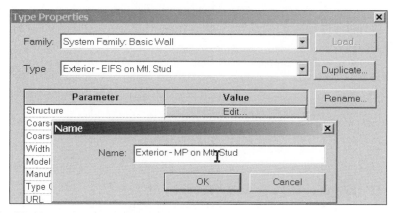

**Figure 3–67** *Name the new wall type*

9.5 The Edit Assembly dialog will open. It will display the Layers of the wall structure from the Exterior side to the Interior. Study the wall structure for a second. Select Layer 1 to highlight it. Change the Material field to Finishes – Exterior – Metal Panel, as shown in Figure 3–68. Choose OK three times to exit the dialog. The appearance of the walls will not change.

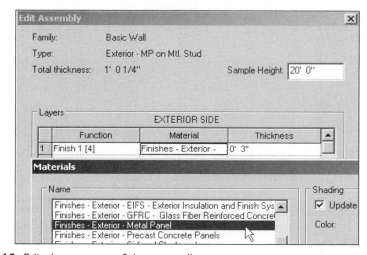

**Figure 3–68** *Edit the structure of the new wall type*

## Add windows using the CAD file for placement

9.6    Zoom in Region to the upper-right area of the CAD file. Choose the Tape Measure tool from the Toolbar (see Figure 3–69). It will read the distance and angle between any two points you then select. Use it to determine the sizes of the Type A and Type B windows, and write down the results. Use the Tape Measure to determine the sill height of the Type A and Type B windows—the distance between the bottom of the window unit and the IMPORT 2ND FLOOR T.O.S. Reference Plane—and write down that information. Type A will be 2' 8" x 6' 8" with a 3' 0" sill. Type B will be 2' 8" x 2' 8" with a 6' 8" sill height (see Figure 3–70).

**Figure 3–69** *The Tape Measure*

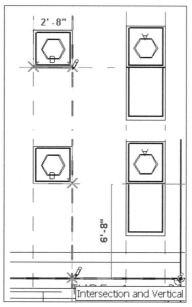

**Figure 3–70** *Measuring the windows*

9.7    Select Window from the Basics tab of the Design Bar. Since this file was started in a generic template, Revit will not have any special content loaded into it. The default window it will list for installation is Fixed: 36" x 48". Type B in our imported file is fixed 32" x 32"; choose the Properties button to begin creating a window type of the correct size.

9.8 In the Element Properties dialog, choose the Edit/New button, then the Duplicate button. In the Name dialog, title the new fixed window size **32" x 32"** and choose OK. Make the Height value of the new window **32"**. Make the Default Sill Height **80"**. Make the Width **32"**. Revit will convert inch values to foot-inch (2' 8", 6' 8"). Change the Type Mark value to **B** (see Figure 3–71). Choose OK twice to exit the dialog and get ready to locate a window. The cursor will change shape to show that you are inserting a component

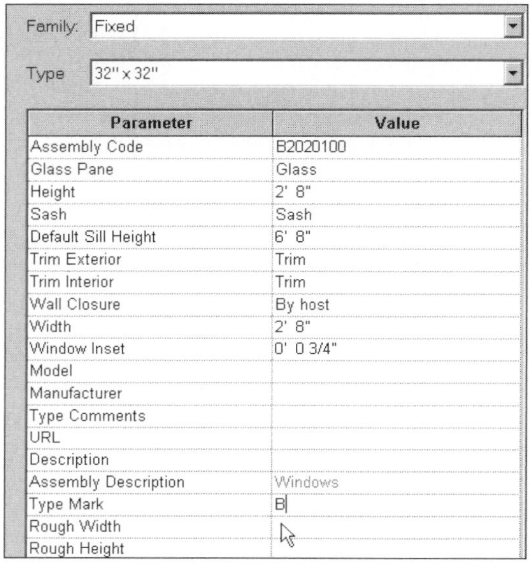

| Parameter | Value |
|---|---|
| Assembly Code | B2020100 |
| Glass Pane | Glass |
| Height | 2' 8" |
| Sash | Sash |
| Default Sill Height | 6' 8" |
| Trim Exterior | Trim |
| Trim Interior | Trim |
| Wall Closure | By host |
| Width | 2' 8" |
| Window Inset | 0' 0 3/4" |
| Model | |
| Manufacturer | |
| Type Comments | |
| URL | |
| Description | |
| Assembly Description | Windows |
| Type Mark | B |
| Rough Width | |
| Rough Height | |

Family: Fixed
Type: 32" x 32"

**Figure 3–71** *Properties for the Type B window*

9.9 Zoom or Pan to the lower-right corner of the model. Click to place an instance of the fixed window in the first floor section of the right-hand brick wall section. The exact location does not matter for the insertion. Continue to place instances of this window using the CAD file as a rough guide for locations, until you have nine instances.

Revit will attempt to locate the windows at the correct sill height for you, and will show temporary dimensions indicating the sill location and spacing from the center of the window to other building elements. If it's necessary to adjust the sill height of the first instance, select the dimension tag and adjust it.

9.10 Choose Load from Library from the Options Bar. Navigate to the *Imperial Library* folder on your system and double-click the *Windows* folder. Select the file *Double Hung.rfa* and choose Open.

9.11 Choose Window from the Basics tab of the Design Bar. Revit will default to the Double Hung: 36" x 48" size window. Choose the Properties button, then the Edit/New button, and then the Duplicate button, as before, to create and name a new size. In the Name dialog, type **32" x 80"** and choose OK. Change the Height value of this new size to **6' 8"**, change the Default

Sill Height to **3'**; and change the Width to **2' 8"**. Make the Type Mark value **A** (see Figure 3–72). Choose OK twice to exit the dialog.

| Family: | Double Hung |
|---------|-------------|
| Type | 32" x 80" |

| Parameter | Value |
|-----------|-------|
| Assembly Code | B2020100 |
| Description | |
| Glass Pane Material | Glass |
| Height | 6' 8" |
| Sash Material | Sash |
| Default Sill Height | 3' 0" |
| Wall Closure | By host |
| Width | 2' 8" |
| Window Inset | 0' 0 3/4" |
| Model | |
| Manufacturer | |
| Type Comments | |
| URL | |
| Assembly Description | Windows |
| Type Mark | A |
| Rough Width | |

**Figure 3–72** *Properties for the Type A window*

9.12    Place four instances of the new window as for the first window. Make sure the Sill Height is 3' (see Figure 3–73 for a partial example).

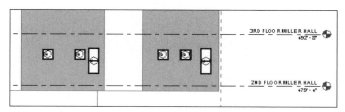

**Figure 3–73** *Windows inserted but not aligned*

## Temporary Reference Planes aid alignment

9.13    Choose Ref Plane from the Basics tab of the Design Bar. Zoom in Region to the upper part of the CAD file. Use the middle of a Type B window (head or sill) to start the Reference Plane. Zoom Previous. Pull the cursor straight down at 90° to below the model to create a vertical Reference Plane aligned with a window on the CAD import.

9.14    Choose Modify. Select the new Reference Plane. Choose Copy. Check the Multiple option from the Options Bar. Zoom in Region to the CAD file. Select the origin of the new Reference Plane as the first Copy point, then select the corresponding midpoints of the other Type A and Type B window symbols along the face of the CAD file, for a total of 13 temporary Reference Planes.

9.15 Zoom to Fit, then Zoom in Region to the new windows in the model. Use the Align tool to align the midpoints of the windows with the Reference Planes. Remember to select the Reference Plane first, then the appropriate window midpoint (see Figure 3–74).

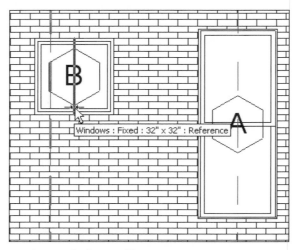

**Figure 3–74** *Align windows to Reference Planes*

9.16 Zoom to Fit. Choose Modify from the Basics tab of the Design Bar. Hold down the CTRL key and select the new Reference Planes you just used to align the windows. (You can also select them with a carefully located Crossing Window, by selecting two points right to left in empty space between the CAD import and the model.) Hit the DELETE key to remove them from the file.

9.17 Select two points to the left and right of the brick walls and above and below the windows so that your selection consists only of the windows and window tags. (If any walls or other objects highlight, choose the Filter button and uncheck all categories except windows and window tags, then Choose OK.) Choose Copy. Select the level line for 1ST FLOOR MILLER HALL as the Copy start point. Select the level line for 2ND FLOOR MILLER HALL to finish the copy. Hit the ESC key twice.

## Add a window type from the Revit Web site

Revit does not ship with combination window assemblies as in the Metal Panel wall sections of the CAD file. Autodesk maintain files of additional Revit component content on its Web site. You will need an Internet connection for this next step (or use an alternate file location specified by your instructor).

9.18 Choose Window from the Basics tab of the Design Bar. Choose Load from the Options Bar. Choose Web Library from the Open dialog. Revit will open an Internet browser to the Revit Library index page. Choose the link for Revit 6.1 Subscription Library. Choose the link for Windows. Right-click the Awning-Combo 1 link and choose the Save Target As option (see Figure 3–75). Choose Save on the File Download question box that appears. If necessary, navigate to the *Windows* folder in the *Imperial Library* folder on your system and choose Save. Close the browser.

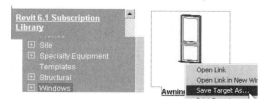

**Figure 3–75** *Finding the Awning-Combo1 window on the Revit site*

9.19    Re-open the Window tool if necessary. Revit will list Awning-Combo1: 36" x 48" in the Type Selector on the Options Bar. Choose Properties. Choose Edit/New. Choose Duplicate. In the Name Box, type **48" x 76"** and Choose OK.

9.20    In the Type Properties box, make the Height value **6' 4"**. Make the Operable Unit **3' 6"**. Make the Default Sill Height **3'.** Make the Width **4'**. Make the Type Mark **E**. Choose OK.

9.21    Place three instances of the new Window Type E in the Metal Panel walls— one in each section at the first floor level. Place them by eye—the left/right placement will be adjusted in the next step. Revit will locate them at the correct sill height (see Figure 3–76).

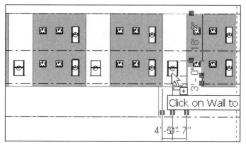

**Figure 3–76** *Placing windows*

9.22    Select Ref Plane from the Basics tab of the Design Bar. Zoom in Region to the CAD file. Select a point on the left side of one of the combination windows type E to start the Reference Plane. Zoom Previous. Pull the cursor down at 90° past the model to create a vertical Reference Plane and left-click (see Figure 3–77).

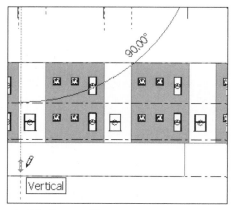

**Figure 3–77** *Make a vertical Reference Plane*

9.23 Choose Modify from the Basics tab of the Design Bar. Select the new Reference Plane. Choose Copy from the Toolbar. Check the Multiple option. Zoom in Region to the area of the CAD file that shows the combination windows. Select the origin point of the Reference Plane for the first Copy point. Select the corresponding points on the two other Type E windows assemblies in the CAD import to create copies.

You can just as easily select other points on the windows in the CAD file. Revit will find endpoints, midpoints, and intersections of lines. Just make sure that you can select corresponding points repeatedly. Revit's Align tool makes correcting locations later on easy.

9.24 Zoom to Fit. Choose Modify on the Basics tab of the Design Bar to terminate the Copy command. Choose the Align tool from the Toolbar. Align the new combination windows in the model with the new Reference Planes. Zoom and Pan as necessary (see Figure 3–78).

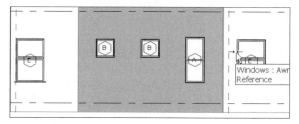

**Figure 3–78** *Align the windows*

9.25 Choose Modify from the Basics tab of the Design Bar. Hold the CTRL key down and select the sills of the new windows (but not the Tags), so that all three highlight red. Type **C** and then SPACE BAR to enter the Copy command. Zoom in Region around one of the highlighted windows. Select the upper-left corner of the window for the Copy start point. Select the upper-right corner of the same window to locate the copy edge-to-edge.

9.26   Zoom Previous to see the windows and the copies. Hit ESC to terminate the Copy. Hold down the CTRL key. Carefully select left to right around the new double window sets in turn, so that all six windows highlight red. Be careful not to select wall edges, or Revit will attempt to move any wall you select.

If you have trouble selecting left to right outside the windows to select them only, try selecting right to left (for a Crossing Window) inside the windows to avoid selecting walls.

9.27   Choose Copy from the Toolbar. Uncheck the Multiple option. Select a point on the 2ND FLOOR MILLER HALL Level Line for the Copy start point. Pull the cursor straight up to the 3RD FLOOR MILLER HALL Level Line and left-click to copy the six selected windows one level up.

9.28   Select one of the new Reference Planes you created to align the windows. Hold down the CTRL key and select the other two Reference Planes. Hit the DELETE key to remove them. Save the file.

### Exercise 10. Add a first floor wall concept sketch—create a new wall type from an existing one

10.1   Open or continue with the file from the previous exercise. Open the 1ST FLOOR MILLER HALL Floor Plan View. Choose Wall from the Basics tab of the Design Bar. Select Basic Wall Exterior – Brick on CMU in the Type Selector. Choose Properties. In the Element Properties dialog, Choose Edit/New. In the Type Properties dialog, choose Duplicate. In the Name box, type **Brick – CMU Column**, as shown in Figure 3–79. Choose OK.

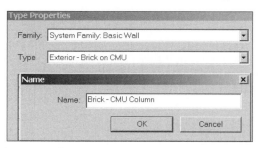

**Figure 3–79** *Name the new wall type*

10.2   In the Type Properties dialog, Choose Edit to modify the Structure. In the Edit Assembly dialog, choose <<Preview to open a view of the wall structure you are working with. In the Layers section, select the number designator for Layer 9. The entire row will highlight black. Choose Delete. Layer 8 will then be highlighted. Delete that layer as well. The view of the wall structure will adjust.

10.3 Select Layer 3 (insulation) so that it highlights. Choose Delete. Layer 3 will now be a Membrane Layer. Choose Delete. Place the cursor in the Thickness field for Layer 2. Change the value to 1", as shown in Figure 3–80, and choose OK to finish editing the wall assembly. Choose OK to leave the Type Properties dialog.

| Layers | | | |
|---|---|---|---|
| | | EXTERIOR SIDE | |
| | **Function** | **Material** | **Thickness** |
| 1 | Finish 1 [4] | Masonry - Brick | 0' 3 5/8" |
| 2 | Thermal/Air Layer [3] | Misc. Air Layers - Air | 0' 1" |
| 3 | **Core Boundary** | **Layers Above Wrap** | **0' 0"** |
| 4 | Structure [1] | Masonry - Concrete | 0' 7 5/8" |
| 5 | **Core Boundary** | **Layers Below Wrap** | **0' 0"** |

**Figure 3–80** *Working with the wall layers*

10.4 In the Element Properties dialog, make the Top Constraint of the new wall type Up to level: 2ND FLOOR MILLER HALL. Set the Top Offset to –1' 4". Make sure the Location Line value is Finish Face: Exterior (see Figure 3–81). Choose OK.

| Location Line | Finish Face: Exterior |
|---|---|
| Wall top is attached | ☐ |
| Wall bottom is attached | ☐ |
| Structural Usage | Non-bearing |
| Base Constraint | 1ST FLOOR MILLER HALL |
| Base Offset | 0' 0" |
| Top Constraint | Up to level: 2ND FLOOR ML |
| Top Offset | -1' 4" |
| Top extension distance | 0' 0" |

**Figure 3–81** *Assign a top offset value*

10.5 Select the intersection of the References Planes for WEST SIDE WALK-WAY and WALL DIVISION 1 to start the new wall (see Figure 3–82). Pull the cursor up at 90° until it snaps to the endpoint of the existing east–west wall and left-click.

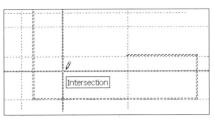

**Figure 3–82** *Start the wall here*

10.6 Select the six WALL DIVISION Reference Planes visible in this view. They are no longer needed, so delete them.

10.7    Open the West Elevation View. Select the new wall. Select Edit Elevation Profile from the Options Bar. The surface pattern will disappear, and the edges of the wall will highlight magenta. The Design Bar will switch to Sketch mode. Select Lines. Draw four lines from the edges of the Metal Panel wall sections (at the top of the new wall profile) down to the base of the new wall outline.

10.8    Choose the 3-point arc tool from the Options Bar, as shown in Figure 3–83. Create a 180° arc under the right-most brick wall section, using the lower-right corner of the wall outline and the end of one of the lines you just sketched. The 3-point arc tool works by selecting the two endpoints of the new arc, then dragging the cursor until the angle value is correct.

**Figure 3–83**  *The 3-point arc sketch tool*

10.9    Repeat the arc in the bay to the left of the one you just drew. Choose Modify from the Sketch tab of the Design Bar. Select the arc you just drew, then select Copy from the Toolbar. Copy the second arc to the left using the endpoints of the vertical alignment lines you drew earlier (a distance of 30'), as shown in Figure 3–84.

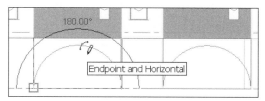

**Figure 3–84**  *Copy the new arc*

10.10   Choose the Split tool from the Toolbar. Use it to split the base line of the wall profile at the intersection points with the arcs. Delete the line segments under the arcs. Delete the vertical lines you drew earlier (see Figure 3–85). Choose Finish Sketch. If a dialog about constraints appears, choose Remove Constraints.

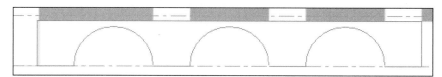

**Figure 3–85**  *The edited wall profile*

## Add a roof sketch

10.11    Open the ROOF MILLER HALL Floor Plan View. Select the gable roof section. Choose Copy. For the Copy start point, select the right endpoint of the ridge line. For the second point, select the midpoint of the wall at the north (upper) end of the walkway wing (see Figure 3–86).

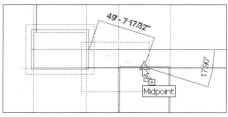

**Figure 3–86** *Copy the roof*

10.12    The new roof will be highlighted. Choose Rotate from the Toolbar. A Rotation arrow symbol will appear in the center of the copied roof. Put the cursor directly over that symbol and it will highlight black and indicate that you can now move it. Move the Rotation point to the right end of the ridge line (see Figure 3–87). Pull the cursor straight up to start a rotation ray at 90°. Select a point well away from the model. Pull the cursor down to the left to create a rotation angle of 90°; when Revit snaps to the 90° angle, left-click.

**Figure 3–87** *Shift the Rotation point*

10.13    The roof sketch will still be highlighted red. Choose Edit Sketch from the Toolbar. Select the lower edge of the roof. Dimensions will appear. Select the value of the one defining the roof length and type **90** in the edit field that appears, as shown in Figure 3–88. Choose Finish Roof from the Sketch tab on the Design Bar.

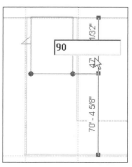

**Figure 3–88** *The length value for the roof sketch*

10.14    Open the 1ST FLOOR MILLER HALL Floor Plan View. Select the wall at the right side of the walkway so that it highlights. Choose the Properties icon. Change the Base Offset value to **12** (Revit will make it 12'), as shown in Figure 3–89. Choose OK. The wall will no longer be visible in the Floor Plan view.

| Structural Usage | Non-bearing |
| Base Constraint | 1ST FLOOR MILLER HALL |
| Base Offset | 12 |
| Top Constraint | Up to level: ROOF MILLER H |
| Top Offset | 0' 0" |

**Figure 3–89** *Adjust the wall top constraint*

10.15    Open the West Elevation view. Type **VG** to open the View Graphics dialog. On the Annotation Categories tab, uncheck Reference Planes. Choose OK. Zoom in Region around the model. Admire your work for a second, and then save the file.

### Exercise 11. Export your work into AutoCAD

For the purposes of our hypothetical project, as soon as we have the start on a redesign for the existing wing, we have to provide AutoCAD output to the site roadways engineer, who needs to approve the arches. File integration should be seamless in a well-run office. Revit makes export straightforward, with control over layers and line weights based on customizable standards.

11.1    Pan or Zoom so the right side of the CAD file is in the left half of the View window and the right half is blank. Type **VG** to open the Visibility/Graphics dialog. Place the cursor over the title bar of the dialog. Left-click and hold the button down. You can now move the dialog. Put the dialog window on the right side of the screen so you can see both the import linework and the dialog box Apply button, or part of it.

The bigger your screen, the easier this will be. Few of Revit's dialog boxes resize, but you can work with them as with any multi-window dialog sequence—the active window is available to move.

11.2    On the DWG/DXF/DGN Categories tab of the dialog, select the + symbol next to *Chapter 3 west elevation.dwg* to expand the category. A list of the layers in the file appears. All the layers are checked to make them visible in Revit. Uncheck the A-ANNO- and A-GRID layers. Choose Apply.

11.3    Uncheck each of the A-PATT layers in turn, and choose Apply after each one to see and note down the results (see Figure 3–90).

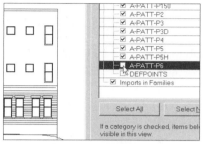

**Figure 3–90** *Stepping through the CAD file layers*

The layer organization in this CAD file is not perfect, as with the lines in the site plan file you imported in the exercises at the beginning of this chapter. These files were developed from actual project files that had gone through numerous hands in different firms, and present the often casual departures from standards that anyone familiar with CAD files has seen. The layer breakdown is as follows:

| | |
|---|---|
| A-PATT-PI | building edges |
| A-PATT-PI50 | CMU field pattern |
| A-PATT-P2 | grade |
| A-PATT-P3 | roof edges, wall edges, some window openings |
| A-PATT-P3D | foundations below grade |
| A-PATT-P4 | sills and wall sweeps, some window openings |
| A-PATT-P5 | window and door frames and mullions |
| A-PATT-P5H | door swings in elevation |
| A-PATT-P6 | brick field pattern |

11.4 Once you have examined the layers in the CAD file, Choose OK to close the Visibility dialog. Select the CAD file. Choose Delete to remove the file from the View so it will not be part of the export. Zoom to Fit. From the File menu, select File>Import/Export Settings>Export Layers DWG/DXF (see Figure 3–91).

**Figure 3–91** *Opening the Export Layers dialog*

11.5 Study the organization of this dialog for a minute. The title bar holds the location of an ASCII *.txt* file formatted to specify layer name and color ID. Revit comes with a dozen or more mapping files (read only by default) in a specific location. These files hold the values displayed in the dialog panels. The left side panel lists Revit object Categories. The middle panel covers Projection display (Elevations, or items not shown as cut in Plans and Sections); default layer names and color ID values come from the *.txt* file in use.

The right panel shows layers and color IDs for Cut display. Revit does not make fields available where a certain condition will not apply. Area Tags (the first Category listed) and other annotations will never have a Cut condition, for example, so the user cannot specify an inapplicable layer name.

11.6   Choose OK. Revit displays a notice that it cannot save the current file, as shown in Figure 3–92. Choose OK to exit this notice and return to the dialog.

This can be a little confusing. OK in this case means that you have made changes to the values in the dialog and that Revit will write the values back into the text file. The default files are read-only to prevent users from making changes to them unintentionally. It is possible to change the supplied text files to allow edits, but simpler and safer to use the Save As button and supply another file name and/or location.

**Figure 3–92** *Revit warns about trying to save the layer specification file*

11.7   Change the values for Revit object Categories in the Export Layers dialog according to the chart that follows. See Figure 3–93 for a view of the dialog.

| Category | Projection | Layer Name Color ID |
| --- | --- | --- |
| Doors | A-PATT-P5 | 5 |
| Door Elevation Swing | A-PATT-P5H | 5 |
| Door Frame/Mullion | A-PATT-P5 | 5 |
| Door Opening | A-PATT-P5 | 5 |
| Roofs | A-PATT-P5 | 5 |
| Roof Fascias | A-PATT-P3 | 3 |
| Roof Gutters | A-PATT-P3 | 3 |
| Walls | A-PATT-P1 | 1 |
| Walls Surface Pattern | A-PATT-P6 | 6 |
| Walls Wall Sweeps | A-PATT-P4 | 4 |
| Walls/Exterior | A-PATTP-1 | 1 |
| Windows | A-PATT-P5 | 5 |
| Window Frame/Mullion | A-PATT-P5 | 5 |
| Window Opening | A-PATT-P3 | 3 |
| Window Sill/Head | A-PATT-P4 | 4 |
| Window Trim | A-PATT-P5 | 5 |

| Walls | A-PATT-P1 | 1 |
|---|---|---|
| Surface Pattern | A-PATT-P6 | 6 |
| Wall Sweeps - Cornic | A-PATT-P4 | 4 |
| **Walls/Interior** | A-WALL | 2 |
| **Walls/Exterior** | A-PATT-P1 | 1 |
| **Walls/Foundation** | S-WALL-LOAD | 2 |
| **Walls/Retaining** | C-TOPO-RTWL | 7 |
| **Window Tags** | A-GLAZ-IDEN | 6 |
| **Windows** | A-PATT-P5 | 5 |
| Elevation Swing | {A-PATT-P5} | 5 |
| Frame/Mullion | A-PATT-P5 | 5 |
| Glass | A-GLAZ-IDEN | 6 |
| Opening | A-PATT-P3 | 3 |
| Plan Swing | {A-PATT-P5} | 5 |
| Sill/Head | A-PATT-P4 | 4 |
| Trim | {A-PATT-P5} | 5 |
| Trim Projection | {A-PATT-P5} | 5 |

**Figure 3–93** *Edits in the Export Layers dialog*

11.8 Notice that as you apply color values to layer names, Revit will supply the color for you. Once a layer name has a color specified, layer names in {} parentheses will pick up layer specifications from the parent category (Window Trim from Windows, for example). When you have made the changes outlined above, choose Save As. Save the file as *export-layers-dwg-CHAPTER 3.txt* in the *Program Files\Autodesk Revit 6.1\Data\* folder, or as specified by your instructor. Choose Save.

From the File menu, choose File>Export>DWG/DXF/DGN. In the Export Dialog, navigate to a folder where you will save an AutoCAD file. If you do not have AutoCAD 2000/2004 available but do have an earlier version, find your version in the Save as type field list. Make sure Current view is selected in the Export range section. Choose Save.

11.9 Save the Revit file.

## SUMMARY

Congratulations. You have created a new Revit file, imported files into it from outside of Revit, and used the CAD content of these files to create a site plan and building model. You have begun the process of detailing the shell of this sizable structure, and exported an exterior elevation view of one face of this building from Revit into AutoCAD. In the process you have worked with many of the basic design and editing tools within Revit, and some techniques that are a little more advanced than the basic level. In the next chapter we will use the model created in this chapter, plus an imported site plan sketch, to set up a multi-building, multi-file campus development project.

# REVIEW QUESTIONS

## Multiple Choice

1. To see text in an imported CAD file

    a) link instead of import

    b) invert the layer colors

    c) check Current view only

    d) you can't see the text in an import

2. Phases in Revit projects control

    a) the display of objects in views

    b) how many objects of a certain type the file can contain

    c) the voltage the computer uses while working in Revit

    d) the date display in titleblocks

3. Reference Planes

    a) can be named

    b) can be turned on and off in views

    c) are excellent alignment tools

    d) all of the above

4. Sketch mode on the Design Bar is active

    a) when creating or editing a roof

    b) when editing a wall elevation profile

    c) whenever no other command is active

    d) a and b, but not c

5. Revit's default Wall and Window families

    a) can't be edited

    b) are made to be edited

    c) can be edited if you pay a special fee

    d) fight all the time, like most families

## True/False

6. The View Properties dialog controls the appearance of nearly everything in Revit

7. Reference Planes can't be deleted

8. The Align tool will snap to walls, but not to windows

9. Plan views can be duplicated in the Project Browser, but Elevations have to be copied in Plan views to create new Elevaton views

10. An object's Phase Created and Phase Demolished properties work with View Phases and Phase Filters to control display of the object.

 Answers found on the accompanying CD.

# CHAPTER 4

# Sharing with Revit

## INTRODUCTION

This section continues a series of exercises using basic Revit techniques to create related, interconnected project files. The exercises in Chapter 3 showed how to use 2D imported CAD file information as the basis for a Revit model. For our hypothetical multi-building project, the model is an existing building which will be modified. In this chapter's exercises we will start modeling three structures for Phases 1, 2, and 3 of this campus development.

To get started we will import the file holding the site topography/building model we created in Chapter 3 and a raster image sketch of a proposed site plan into a project file, and then correlate these two sources. We will trace building footprints on top of the raster image and copy/paste those footprints into three additional empty project files. Each of those project files will become a building model, and we will work in each one using massing components to start the building. We will link those building model files back into the first file we created, which will hold the overall view of this extensive development.

## OBJECTIVES

- Create mass models using a variety of techniques in three separate files
- Link Revit files together; share locations and coordinates
- Create perspective views of mass models for client presentation

### REVIT COMMANDS AND SKILLS

*File save settings*

*File link*

*Shared levels*

*Coordinates—acquire; relocate project*

*File import attributes*

*Model lines—line, arc, fillet arc*

*Edit commands—move, copy, trim, rotate, copy to clipboard, paste*

*Add/Cut Mass—Extrusion, Sweep, Blend*

*View Camera*

## Exercise 1. Import project data—acquire coordinates

1.1 Launch Revit. It will open to an empty project file. Select File>Save As from the File menu. On the Save As dialog, choose the Option button. In the File Save Option dialog, make the Retain at most __ backup(s) value **1**, as shown in Figure 4–1.

 **Note:** This value is set in each Revit file. It is *not* global. Do this step for each file you create in this and other exercises.

If you are working in a class situation, set the value as directed by your instructor. If you are working on your own on a computer that may have limited storage space, you will want to make this a habit when first saving project files.

1.2 Save the file as *Chapter 4 Project Overview.rvt*, in a location specified by your instructor if you're in a class, or in a folder on a drive with adequate space and read-write permission.

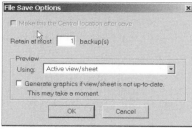

**Figure 4–1** *Set the backup quantity value*

1.3 Select File>Import/Link>RVT from the File menu, as shown in Figure 4–2. Navigate to the folder containing *Chapter 4 existing.rvt*; select that file name so it appears in the File name field. Accept the default option to Automatically place Center-to-center (see Figure 4–3). Choose Open. The file contents will not be visible in the Floor Plan: Level 1 view.

**Figure 4–2** *Link a Revit file*

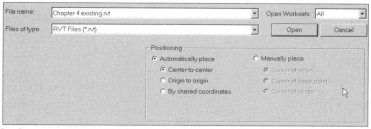

**Figure 4–3** *Center-to-center positioning for the link*

1.4 Open View Elevations: East. Type **VG** to open the Visibility/Graphics Overrides dialog for this view. Select the Linked RVT Categories tab. Choose the + symbol next to the name of the linked file to expand the list of content in that file. Uncheck Topography, as shown in Figure 4–4. Choose OK. The Toposurface under the building will disappear.

**Figure 4–4** *Turn off the topography for clarity*

1.5 From the Tools menu, select Tools>Locations and Coordinates>Acquire Coordinates, as shown in Figure 4–5. The Status Bar will read Select a linked project from which to acquire shared coordinate system. Select the elevation of the linked file. The View window will not alter its appearance.

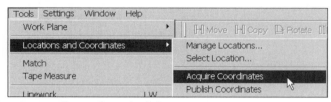

**Figure 4–5** *Acquire coordinates from the link*

Revit does not use an "exposed" coordinate system, in that there is no way to specify an absolute position for entities in a file independent of other entities. Revit does track and use the center point of each file, which changes as the building model footprint changes. Revit also provides a coordinate location point (unseen by the user) with each project file, so that linked files can synchronize positions. Generally the project file that contains site information becomes the basis for shared coordinates, and building models will be located according to the topography.

Since most projects are located some distance away from sea level and the absolute level will generally need to appear in each model file, the next step will be to tie the Level Elevations in this file to the shared coordinates. To do this, we will create a Shared Level type, then Move the project up or down the required distance to pick up the base level (above or below sea level) from the site file. This sounds more complicated than it is.

1.6 Zoom in Region to the Level lines at the bottom of the screen. These are the default Levels 1 and 2 for the current file. Select the Level line (not the bubble) for Level 2. Select the Elevation text so the field opens for editing. Change the Elevation of Level 2 to **13'4**, as shown in Figure 4–6. (Revit will supply the inch value.) Hit ENTER. The Level will adjust.

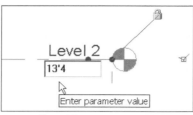

**Figure 4–6** *Adjust the Elevation of Level 2*

1.7 Select the Level line for Level 1. Select the Properties icon from the Options Bar. Choose Edit/New in the Element Properties dialog. Choose Duplicate in the Type Properties dialog.

1.8 In the Name box, type **Level – Shared**. Choose OK. In the Type Properties dialog, change the Base value to Shared, as shown in Figure 4–7. Choose OK.

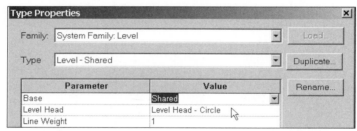

**Figure 4–7** *Add a Level type named Shared, and make its base value Shared*

1.9 From the Tools menu, choose Tools>Locations>Relocate this Project, as shown in Figure 4–8. Read the information box that appears. Choose OK. The cursor changes to the Move icon. Left-click anywhere off the Levels for the first Move point. Pull the cursor up at 90°. Type **466**. Hit ENTER. Type **ZF**. If necessary, drag the end of a Level line to align with the others.

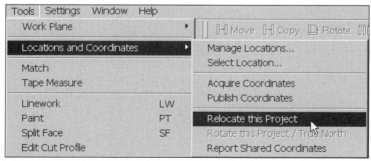

**Figure 4–8** *Relocate the project to line up with the link*

1.10 Open View Floor Plans: Level 1. The topography from the linked file now appears. Zoom to Fit to see the linked model. Save the file.

## Exercise 2. Sketch building footprints from an imported image

2.1 Open or continue working with the *Chapter 4 Project Overview.rvt* file from the previous exercise. Type **VV** to open the Visibility/Graphic Overrides dialog. On the Annotation Categories tab, uncheck Elevations. On the Linked RVT Categories tab, uncheck Topography. Choose OK. Only the building outline will now be visible in the View window.

2.2 From the File menu, select File>Import/Link>Image, as shown in Figure 4–9. Navigate to the folder that contains the file *proposed site plan.jpg*. Select the file name. Choose Open. Four blue control dots indicating the boundary of the image will appear around the cursor. Left-click anywhere in the View window. The image will appear.

**Figure 4–9** *Add a raster image to the file*

2.3 Choose the Properties icon on the Options Bar. In the Element Properties dialog, uncheck Maintain aspect ratio. Change the Width value to **13' 10"**. Change the Height value to **10' 11"** (see Figure 4–10). Choose OK.

| Instance Parameters - Control selected or to-be-created instance | |
| --- | --- |
| **Parameter** | **Value** |
| Width | 13' 10" |
| Height | 10' 11" |
| Maintain aspect ratio | ☐ |

**Figure 4–10** *Adjust the size of the image to fit the model*

**Note:** These values have been ascertained by trial and error. You will have to adjust the size of other images you may import or link into project files experimentally.

2.4 The image will change size. It will still be selected—the cursor is the double-headed double arrow, indicating that you can drag the selected item to a new position. Select Move from the Toolbar.

2.5 Zoom in Region around the image file where the hand lettering reads Existing Building 3 Story. Left-click at one of the corner points of the drawn building outline, then left-click at the corresponding corner of the Revit building model to move the image over the model (see Figures 4–11 and 4–12).

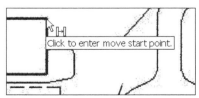

**Figure 4–11** *Start moving the image*

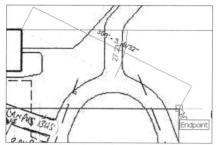

**Figure 4–12** *The second click for moving the image*

2.6  Zoom in Region to the right side of the building model. The walls of the model will show as white over the black raster image pixels. The image will still be selected. Move the image as necessary to get a careful alignment with the walls (see Figure 4–13). When you are satisfied, Zoom to Fit and study the site plan sketch.

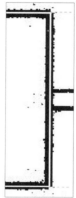

**Figure 4–13** *Careful alignment is possible*

2.7  Zoom in Region around the Phase 1 building outline. Select Lines from the Basics tab of the Design Bar. Select the Chain option from the Options Bar. Sketch three lines of the lengths and angles shown in Figure 4–14, starting from the left end of the wall of the linked model as shown. Do not add the dimensions; they are for guidance only.

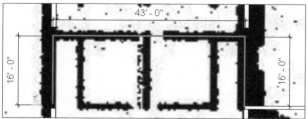

**Figure 4–14** *The first sketch lines for the new building outline*

2.8    From the File menu, choose File>Manage Links (see Figure 4–15). On the RVT tab, select the name of the *Chapter 4 existing.rvt* file so that it highlights (See Figure 4–16.). Select Unload. Choose OK. The building model will disappear.

**Figure 4–15** *Manage Links*

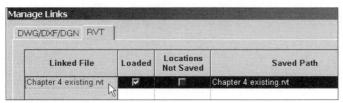

**Figure 4–16** *Highlight the linked Revit file so you can unload it*

2.9    Continue sketching five lines around the footprint of the Phase I building as shown in Figure 4–17. Do not add the dimensions; they are for guidance only.

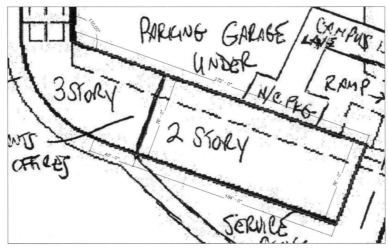

**Figure 4–17** *Draw the simple building outline*

2.10   Select the Fillet Arc icon on the Options Bar, as shown in Figure 4–18. Revit will prompt you to select two lines near the endpoints where you want the fillet to be applied. Select the two lines shown in Figure 4–19 at the approximate locations shown. Revit will start an arc and show a temporary radius dimension. Carefully move the cursor so the dimension reads as close to

100' as possible (see Figure 4–20), then left-click. Revit will trim back the two lines as necessary to create the fillet arc with a radius of 100' – 0" (see Figure 4–21).

**Figure 4–18** *The Fillet arc tool*

**Figure 4–19** *The selections for the fillet arc*

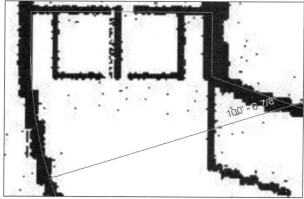

**Figure 4–20** *Pull the cursor to get a radius reading near 100'*

**Figure 4–21** *Revit will snap to the desired radius*

2.11   With the Lines command still active, change the Type Selector to <Over-head>, as shown in Figure 4–22. Draw the two lines shown in Figure 4–23. Do not draw the dimension.

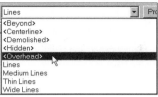

**Figure 4–22** *Change the line type to Overhead*

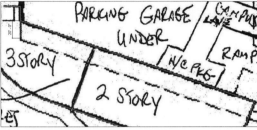

**Figure 4–23** *The new Overhead lines*

2.12   Continue adding sketch lines as shown in Figure 4–24. The raster image has been removed in this figure for clarity. Do not draw the dimensions. Use a combination of Lines and <Overhead> lines as shown. Use the 3-Point Arc tool for the top and bottom lines of the outline.

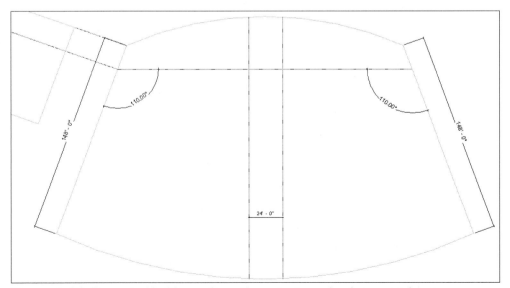

**Figure 4–24** *The central building outline—the raster image has been moved away.*

2.13   Zoom to Fit. Select Modify from the Basics tab of the Design Bar to terminate the Lines command. Left-click over the raster image to select it. The cursor will change to the double arrow. Choose Move from the Toolbar. Left-click anywhere in the View window. Pull the cursor vertically at 90°, type **1000**, and hit ENTER (see Figure 4–25). The raster image will move off the screen.

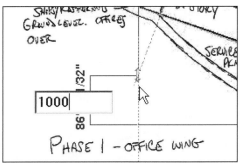

**Figure 4–25** *Move the raster image 1,000 feet up*

 **Note:** Revit does not have a visibility or load/unload control for raster images at this time. Nor can they be rotated by any angle other than a multiple of 90°. However, by moving an image an easily remembered distance in an orthogonal direction, images can be managed without deleting and re-inserting them.

2.14   Zoom in Region around the sketch lines. Choose Ref Plane from the Basics tab of the Design Bar. Create a vertical Reference Plane at the midpoint of the arcs. Select Modify to terminate the Reference Plane command.

2.15   Select the lines as shown in Figure 4–26. Select Mirror from the Toolbar. Select the new Reference Plane. The selected lines will be duplicated in reverse at the right side of the Reference Plane. Hit ESC twice to terminate the selection and command.

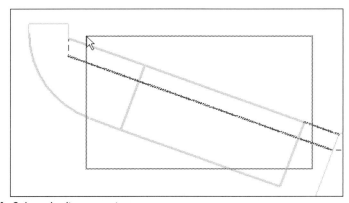

**Figure 4–26** *Select the lines to mirror*

2.16    Select the Reference Plane and delete it. Trim the top and curved side of the right-hand building outline sketch to eliminate the gap. Trim other lines as necessary to make the sketch lines look like Figure 4–27.

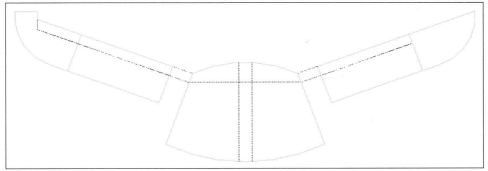

**Figure 4–27** *The linework for three building footprint outlines is complete*

2.17    Zoom to Fit. Select and Move the raster image 1000' at 90° down to locate it over the sketch lines. Check your sketch for accuracy and make any necessary adjustments. Select the raster image and delete it. Zoom to Fit. Save the file.

Now that we have used the existing model file and a sketch of the proposed site design to capture information accurately, we are ready to set up files for the individual designs that will constitute this campus development. First we will move the sketched lines into separate files for each building model, then link those model files back into the Overview file. The building models will be developed and revised extensively during the course of this (or any) project. The Overview file provides a central place for designers or project managers to check progress and design correlation.

### Exercise 3. Export sketch lines into new project files.

3.1    Open a new empty project file. In the new file, open any Elevation view. Set Level 2 to **13' – 4"**. Create a Shared Level type and Relocate the new project to **466'**, as you did in steps 6–9 of Exercise 1.

3.2    From the Window menu, choose *Chapter 4 Project Overview.rvt – Floor Plan: Level 1* from the list of available views in the bottom section of the menu to return to that view (see Figure 4–28). Your Project list may look different from the figure.

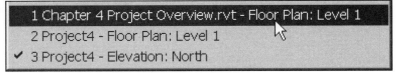

**Figure 4–28** *Return the view to the Overview file*

3.3 With no command active, select two points around the building outline on the left side of the View window, as shown in Figure 4–29.

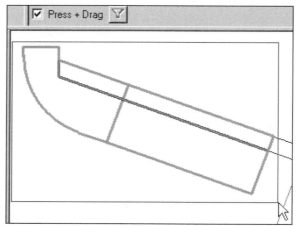

**Figure 4–29** *Select the Phase 1 building outline linework*

3.4 Choose Edit>Copy to Clipboard from the Edit menu, as shown in Figure 4–30.

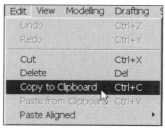

**Figure 4–30** *Copy the building outline lines to the Windows Clipboard*

3.5 From the Window menu, select *Floor Plan: Level 1* for your new project to return to that view. From the Edit menu, choose Edit>Paste from Clipboard, as shown in Figure 4–31. The linework will appear highlighted to one side of the View window.

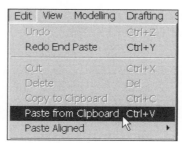

**Figure 4–31** *Edit Paste from the Clipboard*

3.6 Right-click and choose Zoom Out (2X) until you can see the pasted outline. The pasted elements will remain a Group so that you can manipulate them as an object if you desire (see Figure 4–32). Pass the cursor over the paste Group and the cursor will change shape. Select inside the paste border of the outline and hold down the left button for drag positioning. Locate the pasted contents to the right of the Elevations. Until you select off the paste object, it will be available to edit as a single object.

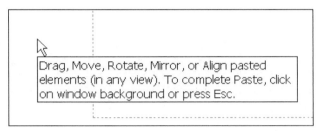

**Figure 4–32** *Editing Option available on pasted content*

3.7 Select Rotate from the Toolbar. Select a point to the right of the Rotate icon that appears in the center of the pasted content, pull the cursor up until the temporary angle dimension reads 20°, and then left-click to rotate the highlighted elements counterclockwise together (see Figure 4–33). Left-click off the linework to terminate the paste command.

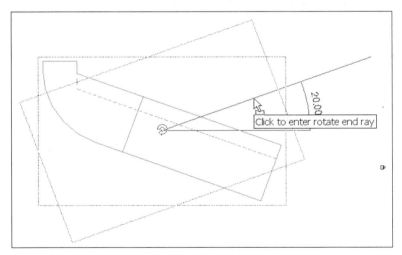

**Figure 4–33** *Rotate the new lines to horizontal orientation*

3.8 Select two points around the North and South elevations to select them, as in Figure 4–34. Select Move from the Toolbar. Select Move start and end points to relocate the Elevations above and below the outline.

**Figure 4–34** *Select two Elevations to move them*

3.9    Repeat the Move command for the East and West Elevations. Zoom to Fit (see Figure 4–35).

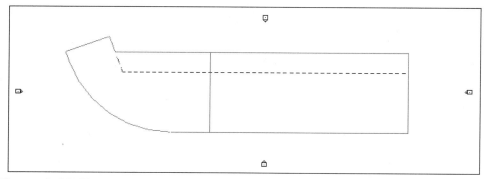

**Figure 4–35** *The new project with the building outline ready to work on*

3.10    From the File menu, select File>Save As. Save the new file as *Chapter 4 Phase 1.rvt* in a location specified by your instructor. Use the Options button in the Save As dialog to set the backup count for this new file to **1**, or as specified by your instructor.

3.11    From the File menu, select File>Close.

3.12    Open a new empty project file. In the new file, open any Elevation View. Set Level 2 to **13' – 4"**. Create a Shared Level type and Relocate the new project as you did in Exercise Steps 1.6–1.9. Make the offset distance **474**, as shown in Figure 4–36.

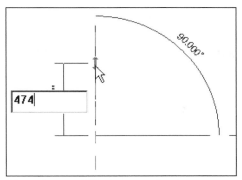

**Figure 4–36** *Relocate this project to 474' above sea level*

3.13 Return to the Floor Plan: Level 1 view of the Project Overview file, as you did before. Select the right-hand building outline, as shown in Figure 4–37. Copy the selection to the Clipboard.

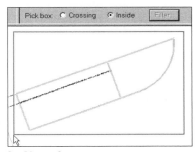

**Figure 4–37** *The selection for Phase 2*

3.14 Return to the Floor Plan: Level 1 view of the new file. Paste in the contents of the Clipboard as you did before. Rotate the contents 20° counterclockwise this time. Move Elevations as necessary. Zoom to Fit (see Figure 4–38).

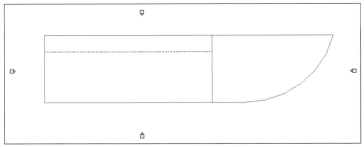

**Figure 4–38** *The building outline placed horizontally*

3.15 Save the new file as *Chapter 4 Phase 2.rvt*, using the Save As, Option tool to set the backup count as before. Close the file.

3.16 Open a new empty project file. In the new file, open any Elevation View. Set Level 2 to **12'** (see Figure 4–39). Create a Shared Level type, and Relocate the new project as you did in Exercise Steps 1.6–1.9. Make the offset **470'**, as shown in Figure 4–40.

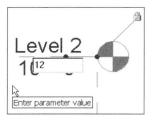

**Figure 4–39** *Set Level 2 to 12' for this building model*

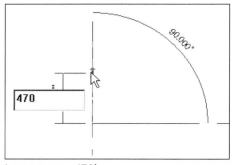

**Figure 4–40** *Relocate this project to 470'*

3.17 Return to the Floor Plan: Level 1 view of the Project Overview file, as you did before. Select the middle building outline, as shown in Figure 4–41. Copy the selection to the Clipboard.

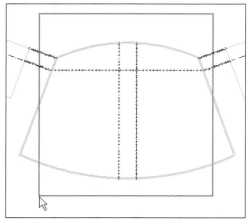

**Figure 4–41** *Copy the middle outline to the Clipboard*

3.18    Return to the Floor Plan: Level 1 view of the new file. Paste in the contents of the Clipboard using the Edit>Paste Aligned>Current View option (see Figure 4–42). Move Elevations as necessary. Zoom to Fit (see Figure 4–43).

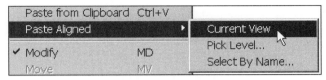

**Figure 4–42** *Paste the outline using an alignment option*

Revit allows the user to copy and paste from file to file, as we have just done, and also allows copy/paste operations within a project, with alignment as an option. A particular arrangement of walls or furniture components can easily be copied from floor to floor of a multi-level project in exact alignment. Revit's capacity to align objects with each other or with Reference Planes is sophisticated, and extremely useful for the busy designer.

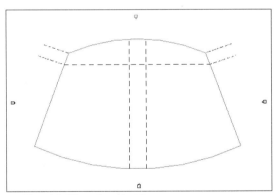

**Figure 4–43** *The new file with Elevations in place*

3.19    Save the new file as *Chapter 4 Phase 3.rvt*, using the Save As Option to set the backup count to 1. Close the file.

## Exercise 4. Link the new project Phase files to the Overview file

4.1    Open or continue working with the *Chapter 4 Project Overview* file. Zoom to Fit in the Floor Plan: Level 1 view. Zoom Out (2X) to make screen space open up around the linework.

4.2    Open Elevations: East view. Use Zoom in Region around the Elevation bubbles at the right side of the View window. Select the text label identifying Level 1. Select the same text again to open the label for editing. Rename the Level **1ST FLOOR EXISTING**. Choose Yes to rename the corresponding Views.

4.3    Repeat for Level 2, naming it **2ND FLOOR EXISTING** (see Figure 4–44).

**Figure 4–44** *Levels in the current file renamed*

4.4    Choose Level from the Basics tab of the Design Bar. Create a level **4' –0"** above 1ST FLOOR EXISTING. Select to the left of the screen, then pull the cursor to the right. Revit will snap to an alignment above 1ST FLOOR EXISTING.

4.5    Create a Level **4' – 0"** above the new one (8' – 0" above 1ST FLOOR EXISTING).

4.6    Rename the first new Level (named Level 3 by default) at the 4' offset **1ST FLOOR PHASE 3**. Choose Yes to change View names. Rename the second new Level **1ST FLOOR PHASE 2**. Choose Yes to change View names.

4.7    Select the Level line for 1ST FLOOR PHASE 2. Use the Type Selector on the Options Bar to change the Level type to Level – Shared. The Elevation will read 470' – 0". Repeat for 1ST FLOOR PHASE 2 (see Figure 4–45).

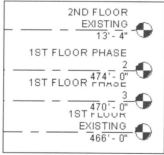

**Figure 4–45** *New Levels shared*

4.8    Open the (renamed) View Floor Plan: 1ST FLOOR EXISTING. From the File menu, select File>Import/Link>RVT, as shown in Figure 4–46. In the Look in: section of the Add Link dialog, navigate to the folder that contains the file *Chapter 4 Phase 1.rvt* that you created earlier. Select that file name so that it appears in the File Name field in the Add Link dialog box. In the Positioning section of the dialog, select Manually Place>Cursor at origin. Choose Open.

**Figure 4–46** *Link a Revit file into the current file*

4.9  Linework from the linked file will appear in the View window. The cursor tooltip and Status Bar will read Left-Click to place Import Instance. Place the linked instance to the left of the View window near the linework you pasted into the link file.

4.10  Repeat the Add Link process for files *Chapter 4 Phase 2.rvt* and *Chapter 4 Phase 3.rvt*. In the Add Link dialog for the Phase 3 file, select Automatically Place/Center-to-center in the Positioning section of the dialog. Zoom to Fit.

4.11  Select the Phase 3 link instance. It will highlight red and a boundary will appear around it. The cursor will become the double-headed edit arrow. Move the linked file using appropriate selected points so that it sits directly on top of the linework in the current file.

4.12  Select the Phase 1 link instance. Move it so that the upper-left corner of the link sits directly on the corresponding corner of the linework in the current file (see Figure 4–47). Choose Rotate from the Toolbar. The Rotation icon will appear in the center of the link instance.

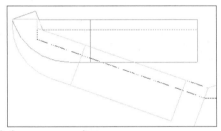

**Figure 4–47** *Move the link so corners align*

4.13  Place the cursor over the icon so that it turns black. Left-click and hold the button down. Drag the Rotation icon to the corner you used for your alignment point. Pull the cursor vertically and select to start a rotation angle vector. Pull the cursor down to the right (clockwise), so that the angular listening dimension reads 20° (see Figure 4–48). Left-click to enter the angle.

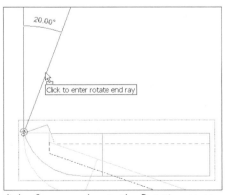

**Figure 4–48** *Rotate the link after you relocate the Rotation center*

4.14 Select the Phase 2 import instance. Repeat Move and Rotate (20° counter-clockwise) so that the import sits directly on top of the linework. Zoom to Fit.

4.15 Choose two points around the edges of the View window to select every-thing visible. Use the Filter button on the Toolbar to deselect the imports as shown in Figure 4–49. Choose OK. Delete the lines from the project. The screen will not change in appearance, but as you move your cursor over the visible entities, it will only read the import instances.

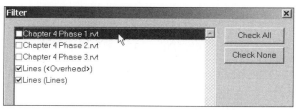

**Figure 4–49** *Filter out links from the selection before deleting lines*

4.16 Type **VV** to activate the Visibility/Graphic Overrides dialog. On the Anno-tations tab, check Elevations, as shown in Figure 4-50. Choose OK. Move the now visible Elevations out around the imported file instances as you have done previously.

| | | |
|---|---|---|
| ⊞ ☑ Dimensions | | ☐ |
| ☑ Door Tags | By Category | ☐ |
| ☑ Electrical Equipment Tags | By Category | ☐ |
| ☑ Electrical Fixture Tags | By Category | ☐ |
| ☑ Elevations | | ■ |
| ☑ Furniture System Tags | By Category | ☐ |

**Figure 4–50** *Select Elevations in the Visibility control dialog*

4.17 Open the View Elevations: South. Zoom in Region to the Level Elevation bubbles in the middle of the screen. As you move the cursor over Level lines, Revit will let you know that some belong in the current file and some are in linked files. Select the Align tool from the Toolbar. Select 1ST FLOOR EXISTING as the reference to align to. Carefully select Level 1 (470' – 0") in the *Chapter 4 Phase 3.rvt* Linked Revit Model with the second click, so that the Phase 3 file moves to the correct Elevation relative to the current file. (See Figure 4-51.)

**Figure 4–51** *Align the link Level to the host file Level*

4.18 The Align tool will still be active. Select 1ST FLOOR PHASE 2 as the Refer-ence. Select Level 1 (474' –0") in the *Chapter 4 Phase 2.rvt* linked model to move that file to the correct Elevation.

4.19 Open the View Floor Plans: 1ST FLOOR EXISTING. The Phase 2 file is no longer visible. That instance now sits above the default cut plane for this Floor Plan View.

4.20 Open the 1ST FLOOR PHASE 2 and 1ST FLOOR PHASE 3 Floor Plan Views in turn to note the differences in default visibility.

4.21 Open the View Floor Plans: 1ST FLOOR EXISTING. Type **VP** to open the View Properties dialog. Choose the Edit button in the View Range Value field. Use the drop-down list in the Top field to select 2ND FLOOR EXISTING. Make the Offset from that Level **0' - 0"**, and make the Cut Plane offset **9' - 0"**. Choose Apply. Phase 2 will now appear on the right side of the screen. Choose OK. Choose OK again to exit the View Properties dialog.

4.22 From the File menu, select File>Manage Links. On the RVT tab in the Manage Links dialog, choose *Chapter 4 existing.rvt* so that it highlights. Select Reload. Choose OK. Zoom to Fit. The building outline will appear at the upper left. (See Figure 4–52.)

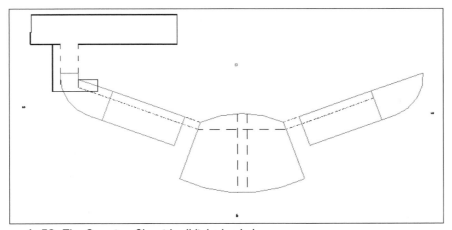

**Figure 4–52** *The Overview file with all links loaded*

4.23 Save the file. Close the file.

This file is now set up. As the designs of the other project building models develop and evolve, a project manager or senior designer can monitor overall progress. In the next exercise we will start mass models in the three associated project files, then return to this file to create a perspective view of the combined facades for a hypothetical client presentation.

## Exercise 5. Start building models using Massing—Extrusions

5.1 Launch Revit if it is not running. Open file *Chapter 4 Phase 1.rvt*. Open View Elevations: East. Select Level from the Basics tab of the Design Bar. Add a Level at **13' – 4"** above level 2. Add a level at **14' – 4"** above the new Level 3.

5.2 Rename Level 1, **1ST FLOOR PHASE 1**. Rename Level 2, **2ND FLOOR PHASE 1**. Rename Level 3, **3RD FLOOR PHASE 1**. Rename Level 4, **ROOF PHASE 1**. Choose Yes to rename the corresponding Views in each case (see Figure 4–53).

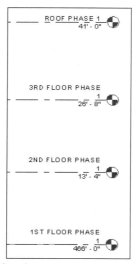

**Figure 4–53** *New and renamed Levels*

5.3 Open the 1ST FLOOR PHASE 1 View. If the Design Bar does not display a Massing tab head in the stack at its bottom, put the cursor over the Design Bar and right-click to bring up the list of available tabs. Check Massing, as shown in Figure 4–54. That tab will open up in the Design Bar.

**Figure 4–54** *Toggle on the Massing tab*

5.4 Choose Add Mass from the Massing tab of the Design Bar, as shown in Figure 4–55. In the Form dialog, select Extrude and choose OK. The Design Bar will switch to Sketch mode with the Lines tool active. The Options Bar will display option icons for Lines tools (straight line, arc, etc.). The cursor will show a pencil icon. The linework will display as halftone gray.

**Figure 4–55** *The Add Mass tool*

5.5 Set the Depth value for this extrusion to **28'**. Select the Pick icon (arrow-head) on the Options Bar. Line Option will disappear from the Options Bar once you make that selection. The cursor shows the selection arrow icon. Select the green lines and fillet arc that make up the footprint of the building. Each will highlight magenta as it is selected, and will display editable dimension information so that you can start to modify the profile sketch you are creating (see Figure 4–56).

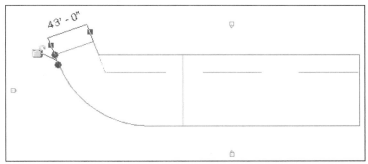

**Figure 4–56** *Selecting a line for the Extrusion sketch displays dimension information you can edit*

5.6 When you have a complete outline selected and highlighted, choose Finish Sketch on the Sketch tab of the Design Bar. The linework will disappear under the new extrusion. Type **WF** to change the View mode to Wireframe.

5.7 Repeat the Add Mass>Extrusion command. Set the Depth value for this extrusion to **41'**. Select Pick on the Options Bar. Start by selecting the green line in the middle of the outline, then the lines and arc to its left (see Figure 4–57).

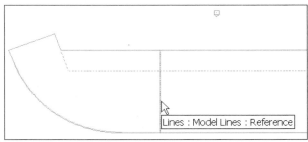

**Figure 4–57** *The first selection for the second extrusion*

5.8    Use the Trim command to trim back the upper and lower outline lines as necessary to make a complete outline with no overlapping lines (see Figure 4–58). Select Finish Sketch from the Sketch tab on the Design Bar.

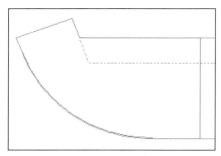

**Figure 4–58** *The complete, trimmed outline for the second extrusion*

5.9    Choose Cut Mass from the Massing tab of the Design Bar. Select Extrusion in the Form dialog. Set the Depth value of the Cut to **12'**. Select the Pick icon on the Options Bar.

5.10    Select the two Overhead lines inside the building outline as shown in Figure 4–59. Select the top and right edge lines of the outline. Trim as necessary to make a complete outline. Select Finish Sketch from the Sketch tab on the Design Bar.

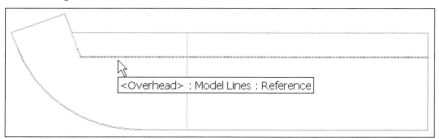

**Figure 4–59** *The first Overhead line selection for the Cut*

5.11    Type **HL** to return the current View to Hidden Line mode. Select the Default 3D View icon on the Toolbar, as shown in Figure 4–60. Revit will open an SE isometric view of the model and add a View named {3D} under 3D Views in the Project Browser.

**Figure 4–60** *The Default 3D View icon*

5.12 Left-click anywhere off the model in the View window to make sure that view is active. Select View>Orient>Northeast from the View menu, as shown in Figure 4–61.

**Figure 4–61** *Orient the 3D View to the Northeast*

 **Caution:** At this time, Revit's 3D Views do not hold their orientations if a file is closed and reopened. Be aware that if you create numerous 3D views of your model, they will revert to the default Southeast orientation when the file is reopened.

5.13 Choose Show Shell from the Massing tab of the Design Bar. Revit will display the faces of the extrusion as walls, roofs, and floors. Move the cursor over the model—these components will highlight and read their category information at the Status Bar and Tooltip.

The Show Mass/Show Shell toggle is view-specific, not a global value. You can switch views of Mass components to Show Shell at any time to make adjustments in the Mass component properties, or Show Shell to see display of walls as walls, and so on.

5.14 Select Settings>Mass/Shell from the Settings menu, as shown in Figure 4–62. The Mass/Shell settings dialog will show the default types for Walls, Floors, and Roofs (see Figure 4–63). These types will be applied to vertical and horizontal Mass components when the Show Shell toggle is activated. Select Cancel to close this dialog without making changes.

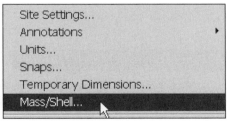

**Figure 4–62** *Reaching the Mass/Shell Settings*

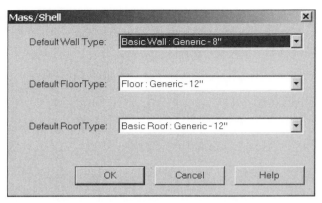

**Figure 4–63** *Default types in the Mass/Shell Settings dialog*

5.15 Select View>Shading from the View menu. Revit will display the default Roof and Wall styles in Shade mode, as shown in Figure 4–64. Type **HL** to return to Hidden Line mode. Save the file. Close the file.

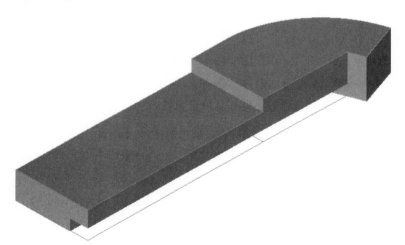

**Figure 4–64** *Shaded view of the Phase 1 extrusions*

## Exercise 6. Edit an Extruded model with Sweeps

6.1 Open the file *Chapter 4 Phase 2.rvt*. Make the Basics tab on the Design Bar active. Open the East Elevation View. Add 5 Levels (3 through 7) at a standard offset distance of **13' – 4"** from the Level below. Add a Level **14' – 4"** above Level 7. Rename Levels 1 through 7 to **1ST FLOOR PHASE 2** through **7TH FLOOR PHASE 2**. Rename level 8 to **ROOF PHASE 2**. Choose Yes each time to rename the corresponding Views (see Figure 4–65).

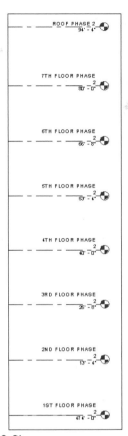

**Figure 4–65** *Levels in the Phase 2 file*

6.2 Open the view IST FLOOR PHASE 2. Select the Massing tab head on the Design Bar to activate that tab. Select Add Mass from the Massing tab of the Design Bar. Select Extrude. Choose OK.

6.3 Set the Depth Value for the first extrusion to **28'.** Select the Pick icon on the Options Bar. Select the outer lines of the footprint, as shown in Figure 4–66. When the outline is highlighted magenta, select Finish Sketch.

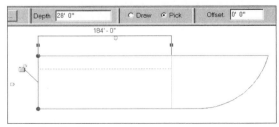

**Figure 4–66** *The outline for the first extrusion*

6.4    Type **WF** to set the View mode to Wireframe. Select Add Mass on the Massing tab of the Design Bar. Choose Extrude in the Form dialog. Choose OK. Set the Depth to **94' 4"** (the Elevation of the Roof Level). Select Pick on the Options Bar. Select the vertical green line in the middle of the footprint as the first click, as shown in Figure 4–67.

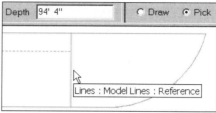

**Figure 4–67** *The first click for the second extrusion*

6.5    Choose the lines and fillet arc to the right of that line to complete the outline. Trim lines as necessary (see Figure 4–68). Select Finish Sketch.

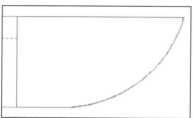

**Figure 4–68** *The outline for the second extrusion*

6.6    Select Cut Mass on the Massing tab of the Design Bar. Choose Extrude in the Form dialog. Choose OK. Set the depth to **12** (Revit will convert it to 12' 0"). Select the Pick icon on the Options Bar. For the first line, choose the horizontal Overhead line as shown in Figure 4–69. Select other lines and trim as necessary to create the outline for the Cut (see Figure 4–70). Choose Finish Sketch.

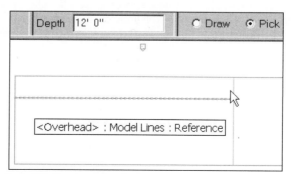

**Figure 4–69** *The first click for the extrusion cut*

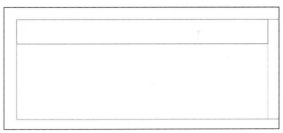

**Figure 4–70** *Completed outline for the extrusion cut*

6.7 Type **HL** to reset the view to Hidden Line. Select the Default 3D View icon from the Toolbar. Choose View>Orient>Northwest from the View menu. Type **WF** to set this view to Wireframe (see Figure 4–71).

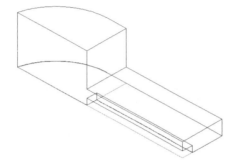

**Figure 4–71** *Wireframe view of the extrusions*

6.8 Select Cut Mass on the Massing tab of the Design Bar. Choose Sweep in the Form dialog, as shown in Figure 4–72. Choose OK.

**Figure 4–72** *Sweep*

6.9 Choose Sketch Path on the Sketch tab of the Design Bar. The Sketch tab will change its Option. Select Pick Path, as shown in Figure 4–73.

**Figure 4–73** *Select lines to define a path*

6.10 Carefully select the line, as shown in Figure 4–74. The profile locator will appear. Choose the fillet radius, as shown in Figure 4–75. Select Finish Path on the Sketch tab.

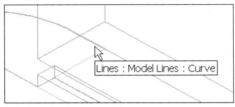

**Figure 4–74** *The first click for the Sweep path*

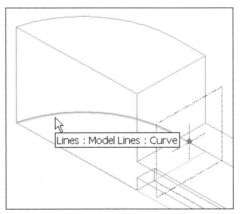

**Figure 4–75** *The second click to create the path—note the profile icon*

6.11 Select Sketch Profile on the Sketch tab, as shown in Figure 4–76. Open the West Elevation View. Choose Lines from the Sketch tab. Select the Chain option. Start a profile sketch 12' from the right side of the massing, as shown in Figure 4–77. Create a profile outline, as shown in Figure 4–78. Do not draw the dimensions; they are for your guidance only.

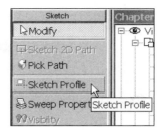

**Figure 4–76** *Sketch profile*

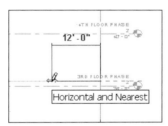

**Figure 4–77** *Start the Profile*

**Figure 4–78** *The finished profile*

You will see by a study of the figure that the profile consists of a series of 12' setbacks moving from right to left, starting at the height of the lower extrusion and using the project Levels above that.

6.12 When the Sweep Profile is complete, select Finish Profile. Choose Finish Sweep. Choose Unjoin Elements if an error message appears, as shown in Figure 4–79.

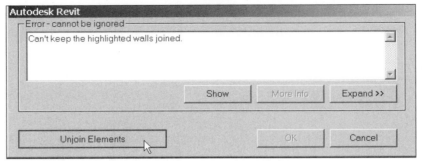

**Figure 4–79** *Unjoin Elements to prevent an error*

6.13 Open the {3D} View. Type **HL** to examine the results of your Sweep Cut. Select View>Orient>Southeast from the View menu.

6.14 Select Show Shell from the Massing tab. Choose View>Shading from the View menu (see Figure 4–80). Type **HL** to return to Hidden Line view mode.

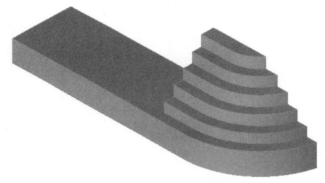

**Figure 4–80** *Shaded view of the stepped extrusion*

6.15 Save the file. Close the file.

## Exercise 7. Create a mass model using blends

7.1 Open the file *Chapter 4 Phase 3.rvt*. Open View Elevations: East. Choose Level from the Massing tab. Add a Level at **12' – 0"** above level 2. Add a level at **16' – 0"** above the new Level 3.

7.2 Rename the Levels **1ST FLOOR PHASE 3**, **2ND FLOOR PHASE 3**, **3RD FLOOR PHASE 3**, and **ROOF PHASE 3**. Choose Yes to accept View name changes (see Figure 4–81).

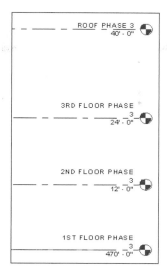

**Figure 4–81** *Levels in the Phase 3 file*

7.3 Open the IST FLOOR PHASE 3 Floor Plan View. Select Add Mass from the Massing tab. In the Form dialog, select Blend (see Figure 4–82). Choose OK. The Sketch Bar will appear.

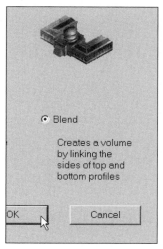

**Figure 4–82** *The blend choice for a Mass component*

A blended Mass in Revit uses two profiles, for the top and bottom, and creates a volume between the two. You will start by sketching the bottom profile, then the top.

7.4 Set the Depth value to **40'**. Select the Pick icon and select the sides of the building outline. Choose the Draw icon and draw horizontal lines between the endpoints of the magenta lines you created by your clicks (see Figure 4–83).

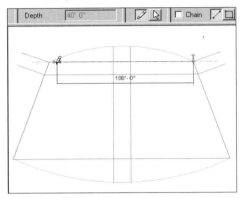

**Figure 4–83** *The outline for the blend bottom*

7.5 Choose Edit Top in the Sketch tab, as shown in Figure 4–84. Select the Pick icon and select the outline of the building footprint (see Figure 4–85).

**Figure 4–84** *The Edit Top icon*

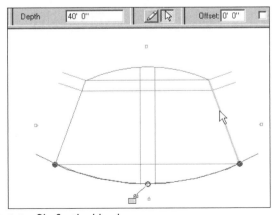

**Figure 4–85** *The top profile for the blend*

7.6 Select Finish Sketch. Choose the Default 3D View icon. Select the Dynamic View icon (eyeball) from the toolbar, as shown in Figure 4–86. (Function key F8 is the typed shortcut.) In the Dynamic View dialog, select Spin (Shift), as shown in Figure 4–87. The cursor will change shape to a curved double-headed arrow. Hold down the left button and move the mouse in the View window—the model will appear to rotate, as shown in Figure 4–88. Study the effect of the blend.

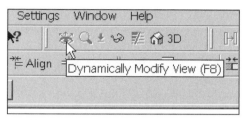

**Figure 4–86** *The Dynamic View icon*

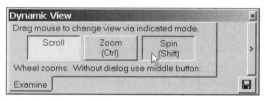

**Figure 4–87** *Controls for the Dynamic View*

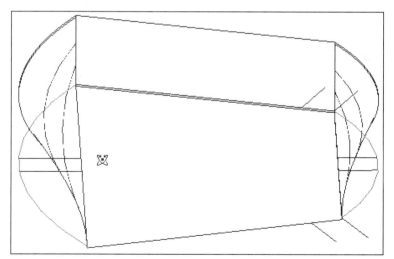

**Figure 4–88** *Spin the View*

7.7 Open the 1ST FLOOR PHASE 3 Floor plan View. Type **WF** to see the model lines. Select Cut Mass on the Massing tab. Choose Extrude on the Form dialog. Choose OK.

7.8 Set the Cut Depth to **20'**. Select the Pick icon on the Options Bar. Select the Overhead line near the top of the figure. Select the arc at the top of the building outline. Choose the Draw icon and trace lines connecting the end-points of the magenta lines created by your previous clicks (see Figure 4–89).

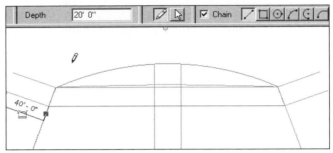

**Figure 4–89**  *Outline for the Extrusion cut*

7.9 Choose Finish Sketch. Examine the results in the 3D View using the Dynamic View. Return to the 1ST FLOOR PHASE 3 Floor Plan View.

7.10 Select Cut Mass from the Massing tab. Choose Blend in the Form dialog. Choose OK.

7.11 Set the Depth to **40'**. Select the Pick icon from the Options Bar. Select the two Overhead lines running vertically down the middle of the outline. Select the upper and lower arcs. Trim the arcs back to the vertical lines to complete the outline of the base (see Figure 4–90).

**Figure 4–90**  *Outline for the blend cut bottom*

7.12 Select Edit Top. Select Pick. Make the Offset value **8'**. Select the two Overhead lines so that the offset lines appear to the outside of each (wider).

7.13 Make the Offset value **0**. Select the upper and lower arcs. Trim the arcs back to the vertical lines (see Figure 4–91). Select Finish Sketch. Examine the results in the 3D view, as shown in Figure 4–92.

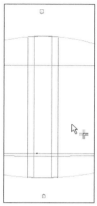

**Figure 4–91** *Outline with offsets for the blend top*

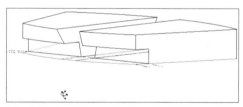

**Figure 4–92** *Results of the blend cut*

7.14 Open the ROOF PHASE 3 Floor Plan View. Make the Basics tab active. Select Roof>>Roof by Footprint.

7.15 Choose Lines on the Sketch tab. Select Pick from the Options Bar. Check Defines Slope. Carefully select the outer (upper) sides of the blend cut. Select Draw from the Options Bar. Uncheck Defines Slope. Draw lines connecting the endpoints of the lines you just created (see Figure 4–93). Select Finish Roof.

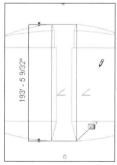

**Figure 4–93** *Profile for the atrium roof*

7.16 Open the 3D View to see the roof you just created.

7.17 Open the 3RD FLOOR PHASE 3 View. Type **VP** to open the View Properties dialog. Set the Model Graphics Style to Wireframe. Set the Underlay to 1ST Floor Phase 3 (see Figure 4–94).

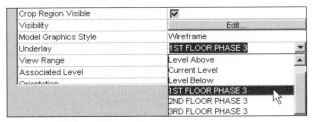

**Figure 4–94** *Setting the View Properties to show model lines under the Massing Objects*

7.18 Select Roof>>Roof by Footprint. Choose Lines from the Sketch tab. Select the Pick icon from the Options Bar. Put a check next to Defines Slope. Select the two Overhead lines at the upper-left of the building outline.

7.19 Uncheck Defines Slope. Select Draw. Draw two lines between the endpoints of the lines you just created (see Figure 4–93). Select Finish Roof.

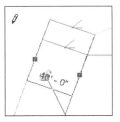

**Figure 4–95** *Profile for the passageway roof*

7.20 Repeat the Roof on the right side of the building to cover the other walkway between buildings. Open the 3D View. Examine the model using Dynamic View (see Figure 4–96)

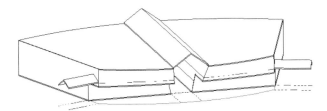

**Figure 4–96** *Dynamic view of the building roofs*

7.21 Select View>Orient>Northeast. Make the Massing tab active in the Design Bar. Select Show Shell. Type **SD** to turn on Shading with Edges display (see Figure 4–97). Type **HL** to return to Hidden Line display.

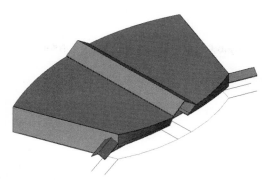

**Figure 4–97** *The building with shading applied*

7.22    Save the file. Close the file.

## View the new models in the Overview file

7.23    Open the file *Chapter 4 Project Overview.rvt*. Open the Default 3D View. Select View>Shading from the View menu. Activate the Dynamic View and examine the model.

7.24    Open the IST FLOOR EXISTING Floor Plan View. Activate the View tab on the Design Bar. If the View tab header does not appear, right-click over the Design Bar and place a check next to View in the tab header list. Select the Camera tool, as shown in Figure 4–98.

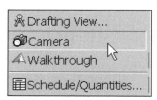

**Figure 4–98** *The camera tool on the View tab*

7.25    The cursor will show a camera icon. On the Options Bar, set the Offset value to **8'** From IST FLOOR PHASE 3, as shown in Figure 4–99. Place the camera (with a left-click) to the upper right of the model. Pull the cursor down and to the left, so that the camera vectors point towards the linked models (see Figure 4–100).

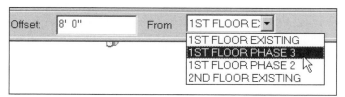

**Figure 4–99** *The camera height setting*

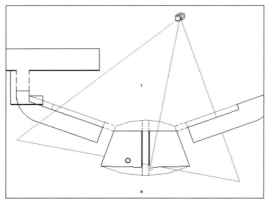

**Figure 4–100** *The camera in plan view*

7.26 Revit will create and open a view named 3D View I. The boundary for the View will be highlighted red, and blue control dots become visible. Zoom Out (2X). Drag the sides and top of the view to emphasize the view of the linked models and de-emphasize the foreground and sky. You can also adjust the location and angle of the camera in the Floor Plan View.

7.27 While in the 3D View I, select View>Shading with Edges from the View menu. Zoom to Fit (see Figure 4–101). Your view will probably look slightly different.

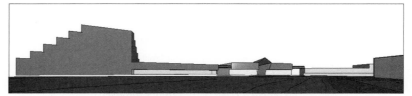

**Figure 4–101** *Shaded wide-angle 3D perspective view of the linked models*

7.28 Save the file. Close the file. Exit Revit.

## SUMMARY

We have seen in this series of exercises how Revit accepts information from other sources, including CAD files, images, and Revit files. We used simple modeling techniques based on Massing to create building shells for a multi-building project, and linked them into a central file so as to view them all at once.

# REVIEW QUESTIONS

## Multiple Choice

1. Two-letter typed shortcuts that deal with Views include

    a) VV, VG, VP

    b) WF, HL, SD

    c) F8

    d) all of the above

2. To create a Blended Massing Form you define

    a) a Top Profile

    b) a Bottom Profile

    c) both a and b

    d) both a and b plus a Profile Angle

3. Sharing Coordinates by Acquire or Publish between files

    a) allows synchronized elevations

    b) requires Administrator privileges on the network

    c) takes too much system memory to be useful

    d) all of the above

4. A Shared Level

    a) is easily created from a default Level by changing the Project Base for the Level

    b) takes its elevation information from a file with Shared coordinates

    c) always has to be a different color from regular Levels

    d) a and b, but not c

5. Edit/Copy and Edit/Paste

    a) are only for use between files, not within a project file

    b) can only be used on door and window objects, not model lines

    c) can take advantage of Paste Aligned

    d) can only be used once per file

## True/False

6. Erasing a linked Revit file does not remove the link from the host

7. Linked images can't be deleted

8. The Add Mass and Cut Mass Form types are Extrude, Revolve, Sweep, and Blend

9. Sketches for Roofs and Massing always have to be drawn—there is no option to select lines or edges

10. Model Line types include Medium, Thin, and Wide lines, plus Overhead, Demolished, and Hidden

 Answers found on the accompanying CD.

# CHAPTER 5

# Multiple Users

## INTRODUCTION

We have repeatedly mentioned and concentrated on the teamwork aspects of design projects. The ability to share information directly from a building design model with collaborators, consultants, clients, and officials will only grow in importance for designers of buildings. In Chapter 3 we examined how Revit imports from and exports to CAD files. In Chapter 4 we explored how to link Revit model files together in order to coordinate multi-building site developments.

In this chapter we shall look at Revit's Worksets and Phasing mechanisms. Worksets are designed to allow real-time collaborative team design. A Workset is simply a named, segregated collection of elements in a model file. When a project model is divided according to Worksets, more than one user can work simultaneously on different parts of the model. Worksets function similarly to the XREF (external reference) capability in AutoCAD, but within a single file rather than a collection of files linked together. All Worksets are part of the same building model file, so when design developments are published from a local copy to the Central File and reloaded by other users, changes to the model are propagated and coordinated tightly.

Worksets are a professional toolset, and their proper and efficient use takes planning and adherence to common-sense best practices. There isn't enough room in this book to discuss advanced uses of Worksets (editing at risk, rolling back changes, or restricting references). Worksets provide the ability to open only part of a building file's contents. Many Revit users divide nearly all their projects—even those that only one person will work on—into Worksets according to a standard setup, to take advantage of the reduced load on their computer system resources.

Phases are Revit's way of putting time value into a building model. All building projects have at least two implicit phases (existing conditions and new construction); all remodeling projects have at least three phases (existing, demolition, and new); many projects have four or more phases (existing, demolition, temporary construction, removal of temporary work, new construction phase 1, and so forth). Revit allows the user to define any number of phases for purposes of viewing a project at any stage along its projected timeline, which can be of significant value in a collaborative setting.

Phases and Worksets are not designed explicitly for exploring alternative options or variations on a design. The Design Options feature in Revit 6.1 will be explored in Chapter 10. Both phases and Worksets can be utilized in linked files, allowing for specific views across many models in a multi-building project.

## OBJECTIVES

- Activate, create, and use Worksets in a model file
- Examine Worksets in linked files
- Set-up and use Phases in a model file
- Examine phases in linked files

### REVIT COMMANDS AND SKILLS

| | |
|---|---|
| *Worksets—rename, new* | *Convert to Curtain wall* |
| *Save to Central* | *Add and Edit Curtain Grid* |
| *Save Local* | *Manage Links* |
| *Create walls from mass elements* | *Phasing—create Phases* |
| *Create sloping floor* | *View Properties for Phases* |
| *Split walls* | *Add Components in Phases* |

### Exercise 1. Activate and create Worksets in a model file

1.1 Launch Revit. Open the file *Chapter 5 start.rvt*. It will open to a floor plan view that shows exterior walls, interior walls, and a collection of toposurfaces. In our hypothetical project the design for the building and site has thus far been prepared by a single person working in the file. We will now prepare the file so that a team will be able to work in the file simultaneously.

1.2 From the File menu, choose File>Worksets, as shown in Figure 5–1. An alert box recommending training will appear. Choose OK. The Project Sharing – Worksets dialog will appear, which names the default Worksets that will be created automatically (see Figure 5–2). Choose OK to accept the default Workset names.

**Figure 5–1** *Activating Worksets from the file menu*

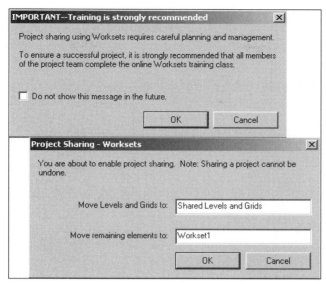

**Figure 5–2** *The Project Sharing – Worksets dialog*

1.3 The Worksets dialog appears. It shows the active Workset (Workset1 by default) and the properties of the User-Created Worksets now in the file. They are all Opened and Editable. Your Revit username will appear next to the Workset name in the Owner column (see Figure 5–3).

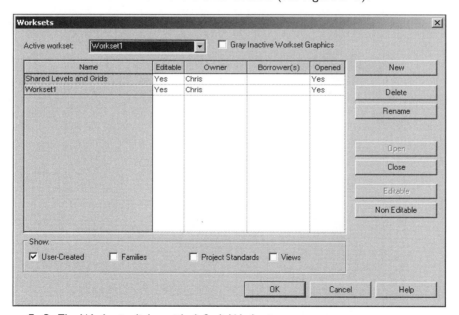

**Figure 5–3** *The Worksets dialog with default Worksets*

1.4 Take a moment to study the Worksets that Revit creates, other than the User-Created ones. Uncheck User-Created in the Show section of the Worksets dialog, and check Families. Note the Revit object categories that have been moved into Worksets, as shown in Figure 5–4.

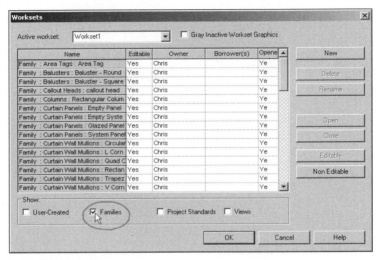

**Figure 5–4** *Family Worksets*

1.5 Uncheck Families, and check Project Standards and Views, in turn, to display the relevant Worksets, as shown in figures 5–5 and 5–6. Uncheck Views and check User-Created. Select Workset1 in the Name field of the dialog, and choose Rename.

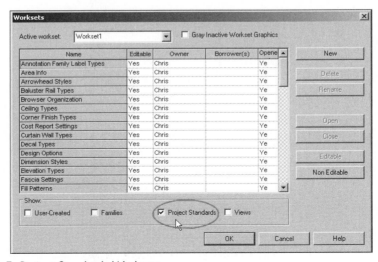

**Figure 5–5** *Project Standards Worksets*

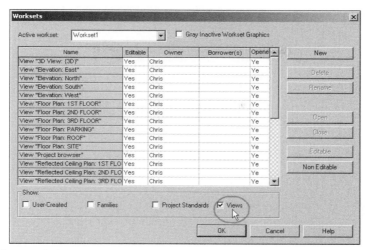

**Figure 5–6** *Views Worksets*

1.6    In the Rename dialog, type **Exterior** in the New name field (see Figure 5–7). Choose OK.

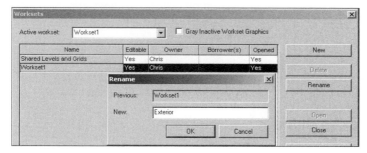

**Figure 5–7** *Rename Workset1 to Exterior*

1.7    Select New. Enter the name **Interior** (see Figure 5–8). Note that Visible by default in all views is checked—this is appropriate for interior walls. Choose OK.

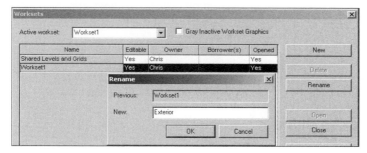

**Figure 5–8** *A new Workset named Interior*

1.8    Select New. Enter the name **Site**. Uncheck Visible by default in all views, as shown in Figure 5–9. Choose OK.

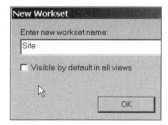

**Figure 5–9** *The Site Workset will not be visible in all views*

1.9    Select New. Enter the name **Furniture**. Uncheck Visible by default in all views. Choose OK. Choose OK in the Worksets dialog to close it.

1.10   Place the cursor anywhere over the Toolbar. Right-click to bring up the list of available toolbars. Check Worksets. The Worksets toolbar will appear, with Exterior in the drop-down field, indicating that Exterior is the current Workset. Any elements created now will appear in the Exterior Workset (see Figure 5–10).

The Pick selection on the Options Bar now shows Editable Only, which is checked by default. With this option checked, you will not be able to select elements that are in a non-editable Workset.

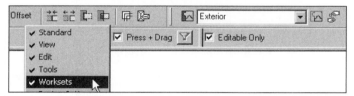

**Figure 5–10** *Turn on the Worksets toolbar*

## Place the model contents into Worksets

1.11   Zoom to Fit. Select two points to select the entire visible model. Select the Filter icon. In the Filter dialog, choose Check None, then check Topography, as shown in Figure 5–11. Choose OK.

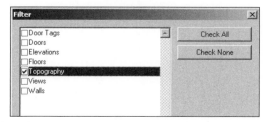

**Figure 5–11** *Filter the selection*

1.12    Choose the Properties icon. In the Element Properties dialog, change the Workset Value to Site, as shown in Figure 5–12. Choose OK. The highlighted topography will disappear, as the Site Workset is not visible in all views by default.

| Instance Parameters - Control selected or to-be-created instance | |
|---|---|
| **Parameter** | **Value** |
| Comments | |
| Mark | |
| Workset | Site |
| Phase Created | New Construction |
| Phase Demolished | None |

**Figure 5–12** *Change the selected elements to the Site Workset*

1.13    Zoom to Fit again. Select all interior doors and walls, as shown in Figure 5–13. Select the filter icon and uncheck Door Tags. Select the Properties icon. Change the Workset Value to Interior, as shown in Figure 5–14. The doors and walls will not disappear, as the Interior Workset is visible in all views by default.

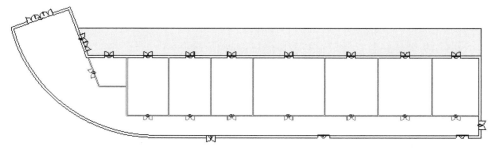

**Figure 5–13** *Select interior doors and walls in the 1ST FLOOR view*

| Instance Parameters - Control selected or to-be-created instance | |
|---|---|
| **Parameter** | **Value** |
| Comments | |
| Mark | |
| Workset | Interior |
| Phase Created | New Construction |
| Phase Demolished | None |

**Figure 5–14** *Place the selection in the Interior Workset*

1.14    Left-click to the right (interior) side of the curved wall, hold down the left button, pull the cursor to the left (exterior) side of the wall, and release the button, to select the wall and also the floor that touches it. Use the Filter to deselect Walls. Choose the Properties icon and place the floor into the Interior Workset.

 **Note:** The floor in the current floor plan view is selectable, but since no pattern has been applied it's difficult to know if a floor has been created as an object in this view. Depending on the complexity of a plan, it can also be tricky to select the desired object. Using right to left selections to create a Crossing selection window at wall-floor joins is a quick way to "fish" for floor edges and other hard-to-select objects.

1.15   Open the 2ND FLOOR plan view. Select all interior doors and windows, as shown in Figure 5–15 (filter out the Door Tags, as in Step 1.13), and place them in the Interior Workset.

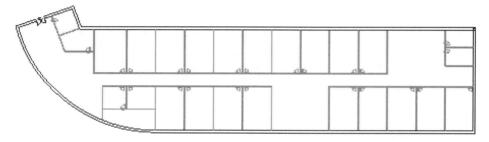

**Figure 5–15** *The 2ND FLOOR interior walls and doors selected*

1.16   Select the floor. Select the Properties icon and place the floor into the Interior Workset.

We are removing Door Tags from our selections for the Interior Workset because Tags are automatically placed in a Workset by Family type (as are all annotations).

1.17   Open the PARKING plan view. Select two points around the model to select everything visible. Only the Elevations and the green trapezoidal shape (sketch lines for the Parking Garage) will highlight red. Use the Filter to deselect Elevations and Views, leaving Lines (Lines) checked. Choose OK to exit the Filter. Hold down the CTRL key and select the two lines to the right of the trapezoid. These are also sketch lines to use in creating a ramp (see Figure 5–16).

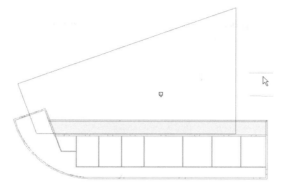

**Figure 5–16** *Select the two short lines one by one with the* CTRL *key held down*

1.18 Select the Properties icon. Place the lines in the Site Workset. Choose OK. The lines will disappear, as the Site Workset is not visible in all views by default.

1.19 Open the SITE plan view. Type **VV** to enter the Visibility/Graphic Overrides dialog. There is now a Worksets tab. Make that tab active. Uncheck Furniture and Interior. Check Site (see Figure 5–17). Choose OK. Topography is now visible around the building outline.

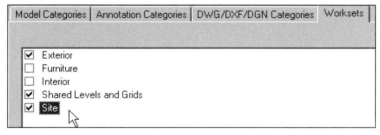

**Figure 5–17** *Change the Workset visibility in the SITE plan view.*

1.20 Open the PARKING view. Type **VV** to enter the Visibility/Graphic Overrides dialog. Repeat the previous checks on the Worksets tab to make the Site Workset visible, and Interior not visible, in this view.

1.21 Open the 1ST FLOOR view.

### Exercise 2. Create a Central File

2.1. Choose File>Save As to save the project as a Central File.

The Central File is created automatically the first time a file is saved after Worksets are enabled. Location of this file is important—all team members who will use the file should have access to it, using the same network path.

2.2.    In the Save As dialog, select Options. In the File Save Options dialog, notice that the option to Make this the Central location after save is checked and grayed out—this will automatically happen the first time, and then become optional for later saves. Change the number of backups to 1, or as specified by your instructor (see Figure 5–18). Choose OK.

**Figure 5–18** *The backup and Make this Central options*

2.3.    In the Save As dialog, give the file you are about to save the name *Chapter 5 central.rvt,* and save it in a location specified by your instructor. If you are working this exercise on your own, or outside a network environment, save the Central File on your hard drive. Choose Save.

2.4.    Open the File Manager or Windows Explorer (My Computer) on your computer and navigate to the folder where you just saved the Central File.

Note that Revit has created a folder named *Chapter 5 central_backup.* Open that folder. There will be over 150 folders cryptically named, and one file named *Chapter 5 central 0001.bak.* More *.bak* files will be created as the Central File is updated, up to the number of copies you entered in the Save As Options dialog earlier (see Figure 5–19). Revit will not create *Chapter 5 central.000x.rvt* files for the Central File as it does for non–Central Files.

 **Caution:** If you ever move a Central File after creating it, the backup folder and its contents should be moved with it. Make no changes to the contents of the backup folder at any time.

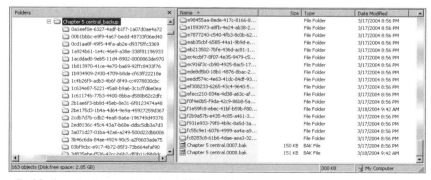

**Figure 5–19** *A typical central_backup folder and contents*

2.5. Close the File Manager without making any changes to it or in the *Chapter 5 central_backup* folder.

2.6. Before closing the new Central File, release the Worksets so they will be accessible to users who will use this file to create local files to develop the design. Choose the Worksets icon on the toolbar you made visible earlier, as shown in Figure 5–20. Choose OK in the alert box if it appears again. You can check the option to suppress the alert.

**Figure 5–20** *The Worksets icon*

2.7. The Worksets dialog will open. Select the top Workset name field, hold down the left mouse button, and drag down to select all the Worksets. Select Not Editable. Your name will be removed from the Owner fields and the Editable values become No, as shown in Figure 5–21. Choose OK.

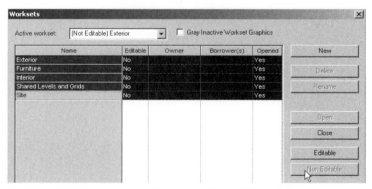

**Figure 5–21** *Relinquish Workset editability in the Central File*

2.8. Choose File>Close from the File menu.

## Creating and using local files

At this point you are about to create a copy of the Central File for your local use. You will need to check out at least one Workset to make edits. To publish your design changes so others can see them, you should save work frequently (every 30 minutes is recommended) and save centrally—a different step—every one to two hours. After your last central save, remember to release Worksets as you just did, then save locally and close the file.

In the next part of the exercise we will create three local files and work in each as a separate user, accessing the same project Central File. If you are working individually without a network you can open a separate session of Revit for each file (depending on system resources you may want to limit yourself to two open sessions) to see the changes propagate.

    2.9.    Choose File>Open from the File menu. In the File Open dialog, use the Open Worksets drop-down list and select Specify. Select the *Chapter 5 central.rvt* file you created (see Figure 5–22).

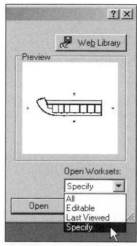

**Figure 5–22** *The choices for opening Worksets*

    2.10.    In the Opening Worksets dialog, you will choose which Worksets to open (make visible) in the copy of the Central File you are about to create. Any Worksets which you had left open would open automatically—all in this case. In this file you will work on the site, so select the Furniture and Interior Worksets (as shown in Figure 5–23) and pick Close. Choose OK.

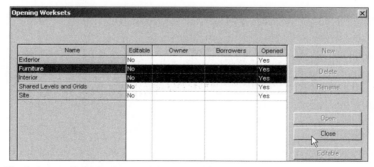

**Figure 5–23** *Select and close the Worksets you do not need to see*

    2.11.    The *chapter 5 central.rvt* file opens. No interior walls or doors are visible. Select File>Save As from the File menu. Choose Options. Note that the option to Make this the Central location after save is now available. Make sure it is not checked (see Figure 5–24). Choose OK to close the options dialog. Save the file as *Chapter 5 site.rvt* on your local drive. Choose Save.

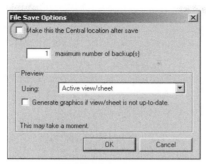

**Figure 5–24** *Do not make this the Central File*

 **Note:** Once you have saved the file as a central copy, Revit activates the icon for the Save to Central function on the Toolbar (see Figure 5–25). Set the save reminder as appropriate so you can keep your work safe and your design changes published to the Central File regularly. Remember—Revit *does not* have an autosave.

This file is now your local copy to work in. It will maintain a connection to the Central File so that certain functions and conditions (Workset editability) are constantly monitored, while other changes update when local files are saved and Worksets reloaded.

**Figure 5–25** *The Save to Central icon appears in Workset-activated files*

2.12.  Choose the Worksets icon on the Toolbar. Certain Worksets are open (visible), but none have been made editable (checked out of the Central File). Select the Site Workset. Choose Editable. Your name appears as the Owner, as shown in Figure 5–26. Use the Active Workset drop-down list to make Site the active Workset. Choose OK.

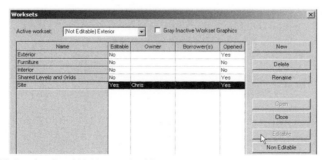

**Figure 5–26** *Make the Site Workset editable*

2.13. Uncheck Editable Only on the Options Bar to enable you to select elements not in the editable Worksets. Otherwise walls will not pre-highlight under the cursor or allow you to select/identify them (see Figure 5–27).

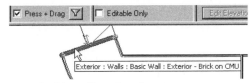

**Figure 5–27** *Select a non-editable wall to identify it*

2.14. Select a wall. The tooltip will identify its type. Choose the Properties icon. All the properties are visible, but if you try to modify any of them an error message appears (see Figure 5–28). Change the wall type to Exterior – CMU Insulated. Select Cancel to close the warning. Select Cancel to close the Element Properties dialog.

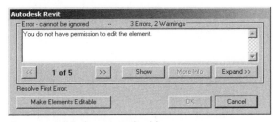

**Figure 5–28** *Properties are visible but not editable*

## Exercise 3. Working simultaneously

If you are working this exercise on your own, at this point you will open a second session of Revit and run the second session under a different username. If you are working in a class situation or with a partner, skip to Step 3.3.

3.1. Minimize the current session of Revit. Start another session using the desktop icon or Start menu. Close the file that opens. From the Settings menu, choose Settings>Options, as shown in Figure 5–29.

**Figure 5–29** *Reaching Revit's Options with no project open*

3.2. On the General tab of the Options dialog, enter **User2** in the Worksets Username value field, as shown in Figure 5–30. Choose OK.

**Figure 5–30** *Changing your username in Revit*

The Workset Username value field will hold the last entered value. After completing this exercise and closing all open Revit sessions, change the Worksets Username back to your Windows Login name to coordinate Revit with your usual login for future work.

### Create a second local file

If you are working with a partner in this part of the exercise, you will both have created a Central File on the network. Choose one of those Central Files to use for the rest of this exercise. The open local file from the selected Central File will stay open—that user will now be called User1, and the user who creates a second local file in the next steps will be called User2.

### User2:

3.3. Close any open Revit files. Open the Central File *Chapter 5 central.rvt* (open the file created by your partner for this exercise if you're working with a partner). Accept the Last Viewed option for Visible Worksets. Select File>Save As from the File menu.

3.4. In the Save As dialog, select Options and make sure that Make this the Central Location after save is **not** checked. Save the file on your local hard drive as *Chapter 5 exterior.rvt.*

You are now working in a local file. Revit places a Save to Central icon on the Toolbar so design changes can be published to the Central File and shared with your partner.

3.5. Place the Worksets icon on the Toolbar as before. Open the Worksets dialog. Select Exterior in the Name field to highlight it. Left-click in the Editable column; change the value from No to Yes. Your username (which will be different from User2 if you are working the exercise with a partner) will appear in the Owner column (see Figure 5–31). Choose OK.

| Name | Editable | Owner | Borrower(s) | Opened |
|---|---|---|---|---|
| Exterior | Yes | User2 | | Yes |
| Furniture | No | | | Yes |
| Interior | No | | | Yes |
| Shared Levels and Grids | No | | | Yes |
| Site | No | Chris | | Yes |

**Figure 5–31** *Making the Exterior Workset Editable*

## User1:

3.6. You are still working in the file *Chapter 5 site.rvt*. Select File>Worksets from the File menu to open the Worksets dialog. The Exterior Workset will show that User2 has checked it out. The Editable status of that Workset in your file is No. Try to change the status of the Exterior Workset to Yes—Revit will display a warning that the Workset is being edited by another user, as shown in Figure 5–32. Choose OK to close the warning. Choose Cancel to exit the Worksets dialog.

**Figure 5–32** *Revit knows this Workset is checked out*

 **Instructor's Note:** The next steps in this exercise for both User1 and User2 introduce new skills other than Worksets. We recommend that you have users complete both sections of this exercise. Once both users have completed their relative exercises 4 or 5, have them return to step 3.3, open the copy of *Chapter 5 central.rvt* NOT used the first time through, and work the exercise steps (site or exterior) that they did not complete the first time.

## User2:

3.7. Move to Step 5.1 to work by yourself—you will continue to work in the file *Chapter 5 exterior.rvt*.

3.8. When your partner notifies you of an editing request, go to Step 4.32.

## User1:

Start Exercise 4.

## Exercise 4. Create walls

4.1.  Open the PARKING plan view. This view shows the level above as the Underlay, so you see the walls and topography at ground level. Make Site the Active Workset in the Worksets Toolbar. Select Wall from the Basics tab of the Design Bar.

4.2.  Click the Properties icon on the Options Bar. Set the Type value of the new walls to **Foundation – 12" Concrete**. Set the Base Constraint to Level **PARKING**, with a Base Offset of **0' 0"**. Set the Unconnected Height to **11' 0"** (see Figure 5–33). Choose OK.

Revit defines foundation walls from the top down; it's easier in this case to create the walls from the bottom up in a view that contains all the relevant information.

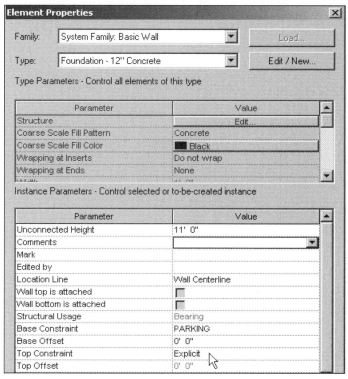

**Figure 5–33**  *Create Foundation walls*

4.3.  Check the Chain Option on the Options Bar and trace over the green model lines to create four walls. Hit ESC twice to terminate the Wall tool.

4.4.  Select the Floor icon from the Basics tab of the Design Bar. The Design Bar will change to Sketch mode. Select Walls will be selected by default on the Design Bar. Uncheck Extend into wall (into core) on the Options Bar.

Select the walls you just created. Select Floor Properties from the Design Bar. Make the Floor Type **Generic – 12"**; make the Level value **PARKING** with a **0' 0"** Height Offset.

4.5. Choose OK in the Element Properties dialog. Choose Finish Sketch on the Design Bar.

## Create a sloped floor using the Slope Arrow

4.6. Zoom in Region to the right side of the new walls. Choose the Floor tool again. The Design Bar will change to Sketch mode. Select Lines on the Sketch tab of the Design Bar. Select the Pick icon (arrow) on the Options Bar.

4.7. Two Model lines have previously been placed in the Central File to indicate edges of a ramp from the driveway on the 1ST FLOOR level down to the PARKING level. Select those two lines to start the outline of the ramp, as shown in Figure 5–34.

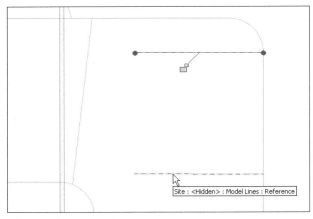

**Figure 5–34** *Select two pre-drawn lines to start the floor sketch*

4.8. Select Draw on the Options Bar. Draw two vertical lines to create left and right edges for the floor outline, as shown in Figure 5–35.

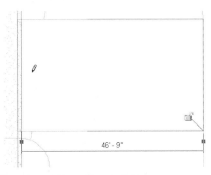

**Figure 5–35** *Draw two lines to continue the sketch*

4.9.    Trim the four new lines to complete a rectangular outline. Choose Slope
        Arrow from the Sketch tab of the Design Bar, as shown in Figure 5–36.
        Select the midpoint of the left side of the outline as the start point. Pull
        the cursor horizontally to the right to define the Slope Arrow, as shown in
        Figure 5–37. Select the midpoint of the right side of the floor outline for the
        end point of the arrow.

**Figure 5–36** *The Slope Arrow tool*

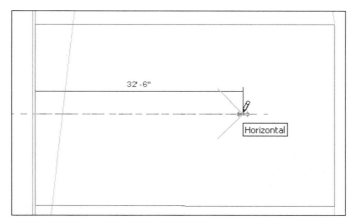

**Figure 5–37** *Draw the slope arrow from the right-side line to the left-side line*

4.10.   Select the new Slope Arrow. Choose the Properties icon from the Options
        Bar. Make the Level at Tail value as **PARKING**, with a Height Offset of
        **0' 0"**. Make the Level at Head value **IST FLOOR** (see Figure 5–38).
        Choose OK. Select Floor Properties. Make the Type value **Concrete
        walkway 4"**. Choose OK. Choose Finish Sketch. The ramp will display
        according to the cut line for the view.

| Parameter | Value |
|---|---|
| Workset | Site |
| Edited by | Chris |
| Specify | Height at Tail |
| Level at Tail | PARKING |
| Height Offset at Tail | 0' 0" |
| Level at Head | 1ST FLOOR |
| Height Offset at Head | 0' 0" |
| Rise/12" | 4" |
| Length | 46' 6" |

**Figure 5–38** *Properties for the head and tail of the ramp Slope Arrow*

4.11.    Select the Split tool from the Tool Bar. Check Delete Inner Segment on the Options Bar. Split the wall that crosses the left end of the ramp at the edges of the ramp, as shown in Figure 5–39. Revit will delete the wall segment at the end of the ramp.

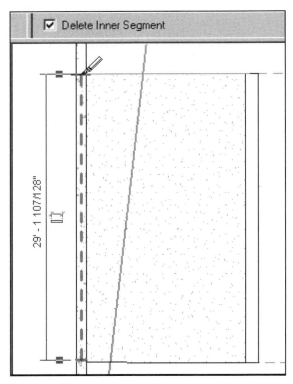

**Figure 5–39** *Splitting the wall to remove a section*

## Create sloped walls and an opening for the ramp

4.12.    Select Wall from the Basics tab of the Design Bar. From the Type Selector drop-down list, select **Basic Wall: Retaining – 12" Concrete**. Use Properties to make the Base Constraint **PARKING**. Make the Base Offset 0. Make the Top Constraint Explicit. Accept the current Unconnected Height setting. Create two horizontal walls 47' long along the sides of the ramp.

4.13.    Select Modify from the Basics tab of the Design Bar. Choose the two new walls. Select Properties. Set the Top Constraint value to **Up to level: IST FLOOR**, as shown in Figure 5–40. Choose OK.

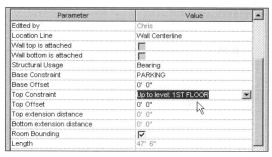

| Parameter | Value |
|---|---|
| Edited by | Chris |
| Location Line | Wall Centerline |
| Wall top is attached | ☐ |
| Wall bottom is attached | ☐ |
| Structural Usage | Bearing |
| Base Constraint | PARKING |
| Base Offset | 0' 0" |
| Top Constraint | Up to level: 1ST FLOOR |
| Top Offset | 0' 0" |
| Top extension distance | 0' 0" |
| Bottom extension distance | 0' 0" |
| Room Bounding | ☑ |
| Length | 47' 6" |

**Figure 5–40** *Adjust the wall height*

4.14.   With the walls still highlighted, select Attach Top/Bottom from the Options Bar, as shown in Figure 5–41.

**Figure 5–41** *Attach the bottom of the new walls*

4.15.   The Options Bar will change appearance. Select the Bottom radio button as the Attach Wall choice. Select the ramp. The walls will change appearance and end at the cut line, consistent with the ramp (see figures 5–42 and 5–43).

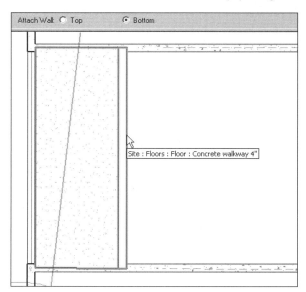

**Figure 5–42** *Select the ramp to attach the wall bottoms*

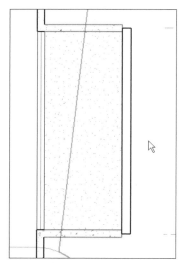

**Figure 5–43** *The walls have been attached to the ramp*

4.16. Zoom to Fit in the PARKING plan view. Open the Elevations South view. Type **VG**. On the Model Categories tab, make sure Topography is unchecked. On the Worksets tab, make sure that Exterior, Shared levels and Grids, and Site are checked. Choose OK.

4.17. Zoom in Region to the right side of the view. Examine the ramp and side walls you have just created (see Figure 5–44).

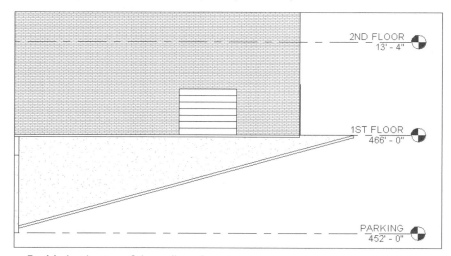

**Figure 5–44** *A side view of the walls and ramp*

4.18. Open the Floor Plan view SITE. Type **VG**. On the Model Categories tab of the Visibility dialog, make sure that Topography is checked. On the Worksets tab, make sure that Exterior, Shared levels and Grids, and Site are checked. Choose OK.

4.19. Zoom to Fit. Use a crossing selection box to select one corner of the parking lot. Filter so that Lines (<Hidden>) and Lines (Lines) are the only selections. Choose OK. You have selected the sketch lines you used to make the Parking Level walls. Delete the lines.

4.20. Make the Site tab on the Design Bar active. Choose Split Surface from the Site tab on the Design Bar.

4.21. Select the L-shaped Toposurface that defines the area between the main drive and the shuttle bus loop/handicapped parking access (see Figure 5–45).

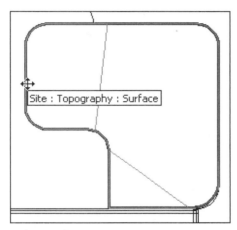

**Figure 5–45** *The topography to edit*

4.22. The Site tab will change to Sketch mode. Type **WF** to enter Wireframe View mode. If a warning about View Worksets appears, choose OK. Select the Rectangle tool on the Options Bar, as shown in Figure 5–46.

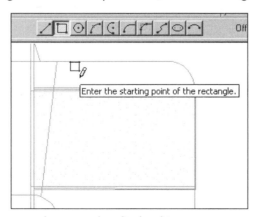

**Figure 5–46** *About to start the rectangle split sketch*

4.23. Select two corners at the edges of the new ramp side walls, as shown in Figure 5–47. Select Finish Sketch.

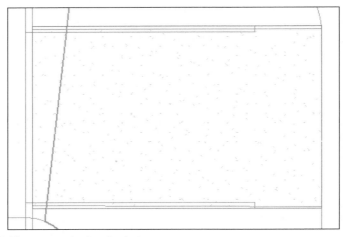

**Figure 5–47** *The rectangle sketched*

4.24. Type **HL** to return to Hidden Line mode. Select the new rectangular Topo-surface you split off from the existing one. Delete it. The ramp and walls will appear, as shown in Figure 5–48.

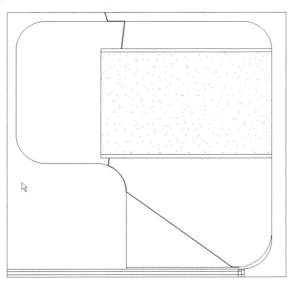

**Figure 5–48** *The ramp and walls appear in hidden line mode*

## Sharing Worksets: Editing requests

4.25. Open the Floor Plan View 1ST FLOOR. Type VP. In the View Properties dialog, make the Underlay value Level Below (see Figure 5–49). If a Warning appears Choose OK. Choose Edit in the Visibility field. On the Model Categories tab of the Visibility dialog, check Topography. On the Worksets Tab, check Site. Choose OK twice.

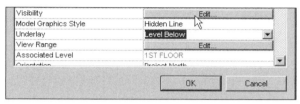

**Figure 5–49** *Change the Underlay in this view*

4.26. Zoom in Region to the lower-right corner of the now-visible parking garage walls. In the Worksets dialog, make Exterior the current Workset, even though it is not editable in this local file (see Figure 5–50).

**Figure 5–50** *Make the Exterior Workset the active one*

4.27. Make the Modeling tab active on the Design Bar. We are going to create an opening in the walkway for a future stair down to the parking garage. Select the Opening tool. In the Options dialog that opens, check Pick a roof, floor, or ceiling and cut vertically, as shown in Figure 5–51. Choose OK.

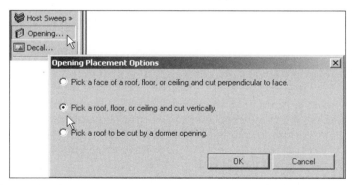

**Figure 5–51** *Use the Opening tool to cut a hole in the walkway, which is a floor*

4.28. The cursor will change shape. Select the exterior walkway. The cursor will change shape again, and the Design Bar will switch to Sketch mode. Lines will be the default option.

4.29. Select the rectangle option from the Options Bar. Sketch a 6' x 10' rectangle as shown in Figure 5–52, starting at the intersection of the walkway edge and the wall below.

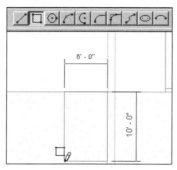

**Figure 5–52** *Sketch the cut in the floor*

4.30. Choose Finish Sketch from the Design Bar. An error message will appear. Choose Make Elements Editable. A question box will appear. Choose Yes to request permission from your partner to make a change in the Exterior Workset in this exercise (see Figure 5–53).

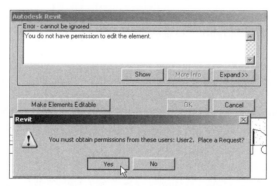

**Figure 5–53** *Request to change a non-editable Workset entity*

4.31. Revit will display a confirmation dialog with options to continue work or to check for a grant to your editing request. If you are working this exercise alone, move to the second session of Revit that has the file *chapter 5 exterior.rvt* open. If you are working with a partner, notify her or him that you have reached this point in your exercise.

## User2:

4.32. You are working in the file *chapter 5 exterior.rvt*. When your partner notifies you (or if working alone, when you have made the editing request in Step 4.31), choose the Editing Requests icon from the Worksets toolbar (see Figure 5–54).

**Figure 5–54** *Open the Editing Request dialog*

4.33.   The Workset Editing Requests dialog will open. You can expand the list to see who is requesting what change in the Worksets you control from your local file. Select the one from your partner (or yourself) and choose Grant (see Figure 5–55).

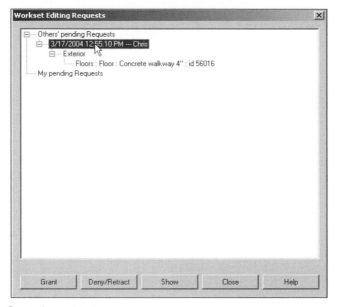

**Figure 5–55** *Grant the request*

4.34.   Choose Close and continue with your work in Exercise 5.

## User1:

4.35.   Once you (or your partner) has completed Steps 4.32 to 4.34, choose Check Now in the Check Editability Grants dialog. Choose OK in the confirmation dialog (see Figure 5–56). Revit will create the opening in the walkway.

**Figure 5–56** *Check the request and finish the sketch*

You could have chosen Continue in the Grant Check dialog and continued with other work in your project file, rather than requesting an immediate grant. In a real-world situation, the demands of work flow, and a possibly far-flung design team, may very well mean that requests accumulate while other work goes on.

4.36.    Open the Worksets dialog. Note that your name now appears as a Borrower(s) of the Exterior Workset (see Figure 5–57). Choose Cancel to exit the dialog.

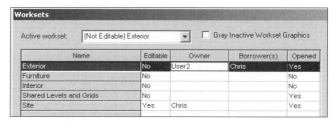

**Figure 5–57**  *You are now a borrower of the Exterior Workset in this local file*

4.37.    Open the default 3D view. Type **VG**. On the Model Categories tab of the Visibility dialog, check **Topography**. On the Worksets Tab, check **Site**. If a warning appears, select Make Editable. Choose OK. From the View menu, choose View>Orient>Northeast.

4.38.    From the View menu, choose View>Shading (see Figure 5–58).

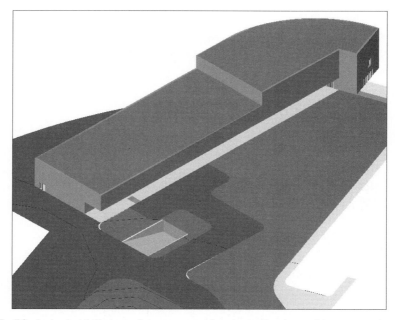

**Figure 5–58**  *A shaded NE view of the ramp and stair cut for the parking garage in place*

4.39.   Choose the Save to Central icon from the Toolbar. In the Save to Central dialog, check all the options to relinquish element editability after the save. Check the option to save the local file. In the Comment field, type **Site plan with underground parking area and ramp started**. Choose OK.

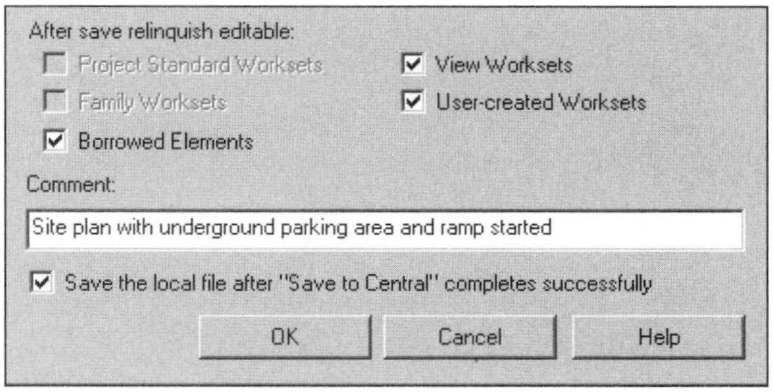

**Figure 5–59** *Save to Central with a comment*

4.40.   Close the File. Move to Step 5.1 unless you have already worked exercise 5.

## User2:

If User1 has Saved to Central before you get to this step, Revit will ask you to reload Worksets before continuing work. Certain views may therefore look different from the illustrations in this chapter.

## Exercise 5. Create and edit a curved curtain wall

You are working in file *chapter 5 exterior.rvt*. Open the Floor Plans 1ST FLOOR view. Zoom in Region to the left side of the model.

5.1.   Choose the Worksets icon on the Toolbar. Make Exterior and Interior Editable. Choose OK. Keep Exterior the Active Workset (see Figure 5–60).

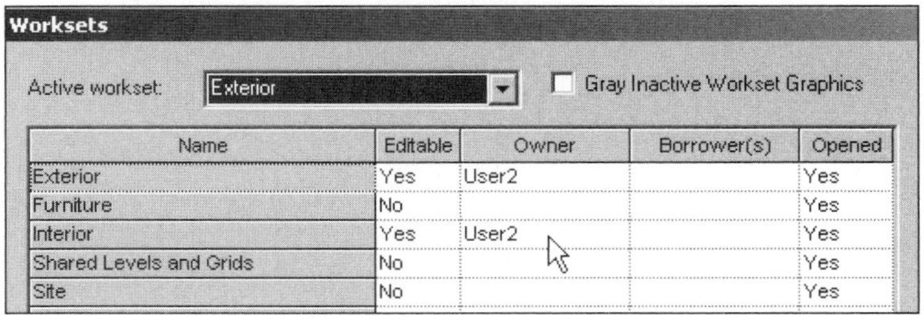

**Figure 5–60** *Make Worksets editable*

5.2.   Activate the Split tool. Split the curved wall on the left side of the building, as shown in Figure 5–61. The exact location is not critical; wall lengths will be adjusted in the next few steps.

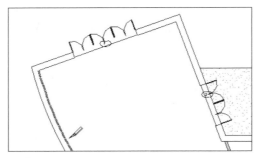

**Figure 5–61** *Split the wall*

5.3. Choose Modify from the Basics tab of the Design Bar. Select the short segment you just created. The wall will highlight red, and blue control dots will appear. Select the lower control dot and drag it up to separate the wall ends. Left-click when the temporary arc dimension reads close to **5°** (see Figure 5–62).

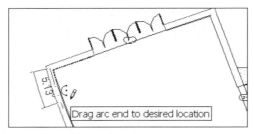

**Figure 5–62** *Adjust the position of the wall end*

5.4. If a warning appears about walls overlapping, choose OK. Drag the end of the longer wall to meet the end of the short one.

5.5. Select the other curved wall segment. Choose Activate Dimensions from the Toolbar. Select the angular dimension text so that the text field opens for input. Type **70** in the field and hit ENTER, as shown in Figure 5–63. The wall length will adjust.

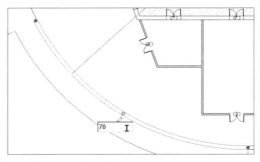

**Figure 5–63** *Adjust the wall arc angle value*

5.6. Select the short wall segment you previously created. Drag its lower end so it aligns with the upper end of the 70° curved wall segment. The wall centerline will highlight with a dashed green line when the alignment is correct (see Figure 5–64).

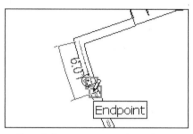

**Figure 5–64** *Pull the short wall end down to meet the other one*

5.7. Open the Default 3D view. From the View menu, choose View>Orient>Southwest. Zoom in Region at the curved wall section

5.8. Select the Split tool. As you move the cursor over the building exterior, the two curved wall segments will highlight in turn. When the larger segment highlights, move the cursor up the wall, as in Figure 5–65. Select a point near but below the roof to split the wall into two segments vertically.

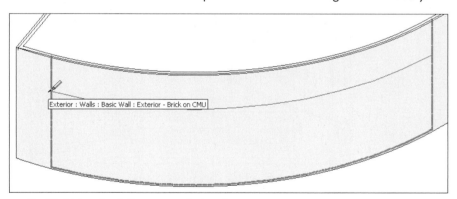

**Figure 5–65** *Using the Split on a vertical wall*

5.9. Select Modify from the Basics tab of the Design Bar. Select the large curved wall at the floor line. Choose the Properties icon. Make the Top Constraint Explicit. Make the Unconnected Height value **40**. Revit will adjust the field value to 40' 0". Choose OK.

5.10. With the wall still selected, use the drop-down list in the Type Selector to select Curtain Wall Curtain Wall 1 (the bottom item on the list). The wall displays as a single straight (not curved) transparent panel. Make the Modeling tab of the Design Bar active. Choose Curtain Grid from the Modeling tab, as shown in Figure 5–66.

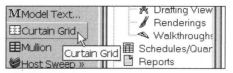

**Figure 5–66** *The Curtain Grid tool*

5.11. Place the cursor to the left end of the curtain wall and start moving it to the right. Angular dimensions will appear. When the angle values read **5.00°/65.00°**, left-click to place a curtain grid.

 **Note:** The temporary dimension precision level is keyed to the Zoom scale. Use the mouse scroll wheel, if so equipped, to move the zoom in or out so that the angular dimensions read in whole degrees. You can do this while using the grid placement tool.

5.12. Continue pulling the cursor to the right and place a curtain grid element every 5° around the arc. Revit will snap weakly to whole degree increments (see Figure 5–67).

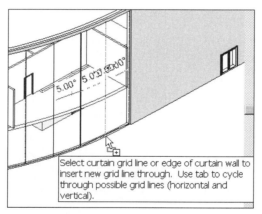

**Figure 5–67** *Adding curtain grids at 5°*

5.13. Repeat the process, placing vertical curtain grids at **2.50°** arc increments between the 5° grid elements. Revit will snap weakly to the midpoints and one-third divisions of the grid panels.

5.14. Move the cursor to the left side of the curtain wall. Move it up the left side and Revit will show vertical dimensions. Place horizontal curtain grids at **8'**, **16'**, **24'**, and **32'** up the wall (8' up from the bottom or previous grid), as shown in Figure 5–68.

 **Tip:** If you locate a grid line in the wrong place, you can select the Undo arrow on the Toolbar to step back one (or more) selections while remaining in the command.

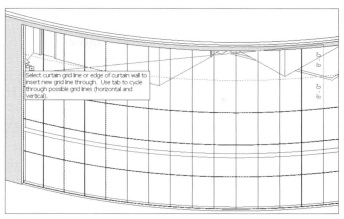

**Figure 5–68** *Adding horizontal curtain grids at 8' increments*

5.15. Open the South Elevation view. Zoom in Region to the right side of the curtain wall.

5.16. Select the first (right-most) vertical curtain grid so that it highlights with a dashed green line. The Options Bar will change its options. Select Add or Remove Segments. Select the lowest segment of the curtain grid. It will highlight with a dashed green line, as shown in Figure 5–69. Hit ESC twice to terminate the command. The grid segment will disappear.

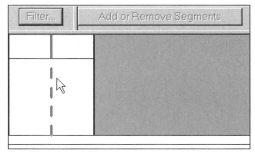

**Figure 5–69** *Removing a curtain grid segment*

5.17. Select Curtain Grid from the Modeling tab of the Design Bar. Choose One Segment from the Options Bar. Place a horizontal curtain grid at **7'** elevation in the open grid you just created, as shown in Figure 5–70.

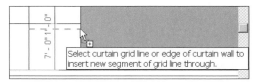

**Figure 5–70** *Adding a curtain grid segment*

5.18. Zoom to Fit. Choose Mullion from the Modeling tab of the Design Bar. From the Options Bar, choose All Empty Segments. Accept the default Mullion type in the Type Selector. Put the cursor over the curtain wall and the whole wall will highlight. Left-click on the wall. Revit will create mullions throughout the curtain wall.

5.19. Expand the Families section of the Project Browser. Expand the Curtain Panels Section under Families. Note that it contains four types of panels: Empty Panel, Empty System Panel, Glazed Panel, and System Panel. We need a door to go in a curtain wall panel.

5.20. From the Modeling tab, select Door. On the Options Bar, choose Load. Navigate to the *Content/Imperial Library/Doors* folder. Select *Curtain Wall-Store Front-Dbl.rfa* (see Figure 5–71). Choose Open.

**Figure 5–71** *The door type in the Imperial Library*

5.21. Revit will attempt to place a door, but not the type you just loaded, and the new door type will not be in the Type Selector list. Press ESC. The Families/Curtain Panels section of the Browser will now show **Curtain Wall-Store Front-Dbl** (see Figure 5–72). The Door family has been loaded into the file, but does not work with the Door tool. There is another way to specify the correct door type in the curtain wall panel.

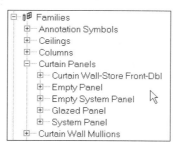

**Figure 5–72** *The new curtain panel type in the Project Browser*

5.22. Zoom in Region around the open panel you created earlier. Place the cursor over a mullion at the edge of the panel. Hit TAB until the four sides of the panel pre-highlight and the tooltip identifies the System Panel (see Figure 5–73). Left-click to select it.

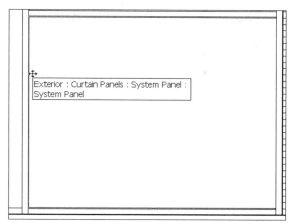

**Figure 5–73** *The System Panel selected*

5.23.   From the Type Selector drop-down list, choose Curtain Wall-Store Front-Double Door. Revit will fit a double door into the panel, as shown in Figure 5–74.

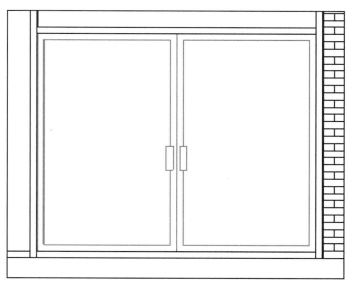

**Figure 5–74** *The panel type set to double door*

5.24.   Open the 3D View. Zoom to Fit. Open the Visibility/Graphics dialog. On the Model Categories tab, check Topography. On the Worksets tab, check Site. If a Workset warning appears, choose Make Editable. From the View menu, select View>Shading.

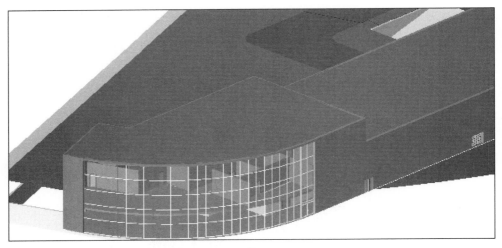

**Figure 5–75** *A shaded view of the curtain wall*

5.25. Choose Save to Central from the Toolbar. Check User-created Worksets and Family Worksets to relinquish the Worksets for editing by others. Check the option to save the local file automatically. In the Comment field, type: **Exterior with curtain wall**, as shown in Figure 5–67. Choose OK. If you have not yet worked Exercise 4, move to Step 4.1.

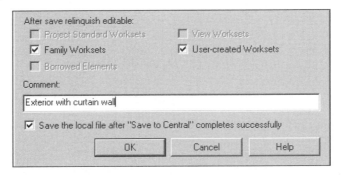

**Figure 5–76** *Save Central with a comment*

This ends the tandem-user section of exercises in this chapter. For the rest of the exercise steps there will be no user names specified. Close all open files. If you have worked the tandem-user exercises by yourself, close one session or Revit. From the Menu Bar, choose Settings>Options to set the Worksets Username to its previous value.

## Exercise 6. Worksets in a linked file

6.1 Open a new Revit project. Select File>Worksets from the File menu to activate Project Sharing. Choose OK in the Project Sharing dialog.

6.2   In the Worksets dialog, choose New. Name the new Workset **Links**. Uncheck Visible by default in all views. Choose OK. In the question box that appears, choose No, as shown in Figure 5–77. Workset1 will be the active Workset.

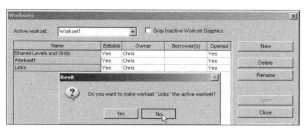

**Figure 5–77** *Create a Links Workset, but do not make it active*

6.3   From the File menu, select File>Import/Link>RVT. Navigate to the folder that holds *Chapter 5 central.rvt*. Select the file so that it highlights and appears in the File Name field. In the Open Worksets field, select All. Accept the Positioning defaults, as shown in Figure 5–78. Choose Open.

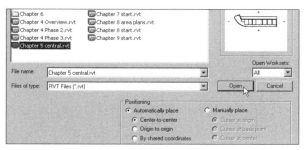

**Figure 5–78** *Open a linked file*

6.4   Zoom to Fit. Type **VG**. In the Visibility/Graphic Overrides dialog, check the Worksets tab. Note that Links is not checked. Leave it for now. On the Linked RVT Categories tab, expand the list under the linked file name. Note that all model categories in the linked file are checked. Choose OK.

 **Note:** The Site Workset in the linked Central File was not set to display in all views by default. At this time there is no way for Revit to override that condition in a linked file, so the topography of the link will not display.

6.5   Select the linked file. Choose Copy from the toolbar. Make a copy 100' down (−90° vertically) on the screen.

6.6   From the File menu, select File>Manage Links. On the RVT tab, highlight the name of the linked file. Choose Unload. Place the cursor on the title bar of the dialog, left-click, and drag the dialog to the right of the drawing window so that you can see what is under it. Both instances of the linked file have disappeared from the screen.

6.7    Choose Reload. Both link instances will reappear. Select Reload From. In the Add Link dialog, make sure that the same file name appears in the File name field and that the Open Worksets value is Specify. Choose Open.

6.8    In the Linking Worksets dialog, select Furniture, Interior, Shared Levels and Grids and Site, as shown in Figure 5–70. Choose Close.

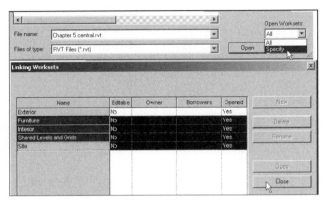

**Figure 5–79** *Select the Worksets to close (not make visible) in the link*

Limiting the visibility of Workset elements in linked files reduces the amount of system memory Revit has to use.

6.9    Choose OK. The linked files will appear without the interior walls and doors visible, as shown in Figure 5–80. Choose OK to exit the Manage Links dialog.

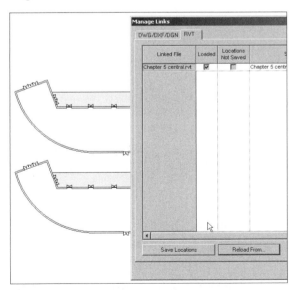

**Figure 5–80** *Two instances of a linked file—the Interior Workset is not open*

6.10   Select the lower of the two instances of the linked file. Select Properties. In the Properties dialog, change the Workset value to Links, as shown in Figure 5–81. Choose OK. The instance will disappear.

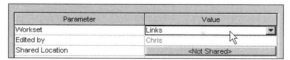

**Figure 5–81** *Place one linked file instance in the Links Workset*

6.11   Open the Level 2 Floor Plan view. In this view the linked file instance in Workset1 is visible. The instance in Links Workset is not. Type **VG**. In the Visibility/Graphic Overrides dialog, open the Worksets tab. Check Links. Choose OK. The second link instance appears.

6.12   Open Level 1 Floor Plan. Only one link instance is visible.

6.13   Close the file without saving.

This is an example of a way to organize linked files in a campus development project file using Worksets. If your file contains models of three or four housing types, for instance, using a Workset specifically for links allows you to control visibility of more than one link at a time, rather than unloading or reloading files in the Manage Links dialog.

Revit gives the user a number of different ways to control the visibility of project elements or entire models, as we have just seen. The Visibility/Graphics Overrides dialog allows you to adjust the visibility of elements in a view by object category in the project file, by Workset, and (within limits) by category within a linked file.

### Exercise 7. Phasing

7.1   Open the file *Chapter 5 central.rvt*. Choose File>Save As from the File menu. Save the file in a location specified by your instructor under the name *Chapter 5 interior.rvt*. In the Save As>Options, make sure that Make this the Central location after save is unchecked, so that you create a local file.

7.2   Open the 2ND FLOOR plan view. Type **VV** to open the Visibility/Graphic Overrides dialog. On the Worksets tab, check Furniture and Interior. Uncheck Shared Levels and Grids. On the Annotation Categories tab, uncheck Door Tags (see Figure 5–82). Choose OK. Select Make Elements Editable in the Warning box to save the view properties changes.

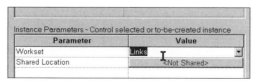

**Figure 5–82** *Editing Visibility—turn off Door Tags*

7.3 Choose the Worksets icon on the Toolbar to open the Worksets dialog. Make Furniture and Interior open and editable. Close Site and Shared levels and Grids. Choose OK. Make Interior the active Workset.

7.4 From the Settings menu, choose Settings>Phases, as shown in Figure 5–83. On the Project Phases tab of the Phasing dialog, two default Phases will display: (1) Existing and (2) New Construction. Select the Name field for Phase 2 and change it to Building Shell. Select After in the Insert section of the dialog to add a new line. Revit will create Phase 1 on line 3.

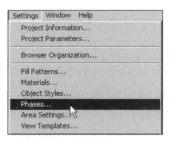

**Figure 5–83** *Phases from the Settings menu*

7.5 Make the Name for Phase 1 **Temporary Classrooms**. In the Description field, type **2nd floor used for classes during Miller Hall renovations**. Select After button again. In the Name field for Phase 1 (on line 4), type **Offices**. Type the following in the Description field: **2nd floor offices and labs** (see Figure 5–84).

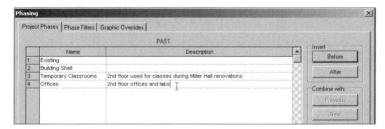

**Figure 5–84** *One Phase renamed, two new ones added and named with descriptions*

7.6 Select the Phase Filters tab. Study the default Phase display settings for a moment.

7.7 Select the Graphic Overrides tab and study it so that you gain an understanding of the display characteristics for the different Phase conditions.

7.8 Choose OK to exit the dialog. Select Make Elements Editable in the Warning box that appears.

7.9 Choose two points on either side of the screen to select everything visible in a crossing pick box. Note that the exterior walls and elevations do not highlight, since the pick box setting with Worksets activated includes an option for Editable Only, which is checked by default.

7.10 Select Filter. Uncheck Floors. Choose OK. The selection set will include only interior walls and doors. Choose Properties. Note that the Phase Created value shows Building Shell, the edited name for Phase 2. Change that value to **Temporary Classrooms**, as shown in Figure 5–85. Choose OK.

**Figure 5–85** *Change the Phase of a view in the View Properties*

7.11 The walls and doors will disappear. Type **VP** to enter the View Properties dialog. Change the Phase Value for the View to Temporary Classrooms. Choose OK.

7.12 Select walls and doors as shown in Figure 5–86.

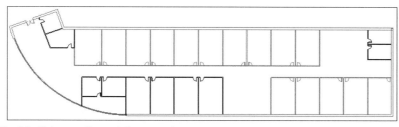

**Figure 5–86** *Select walls and doors to designate as temporary*

7.13 Select Properties. Change the Phase Demolished value to Offices. Choose OK.

7.14 Right-click on 2ND FLOOR in the Project Browser. Select Rename. Name the view **2ND FLOOR CLASSROOMS**. Choose No in the rename level query.

7.15 Right-click on 2ND FLOOR CLASSROOMS in the Project Browser. Choose Duplicate. Rename the new view **2ND FLOOR OFFICES**. Type **VP**. Change the Phase of the View to Offices. Change the Phase Filter to Show Previous + New, as shown in Figure 5–87. Choose OK. The walls that you designated to be removed in the Office construction Phase disappear. The Walls constructed in the previous phase (Temporary Classrooms) that remain in the phase for this view appear halftone.

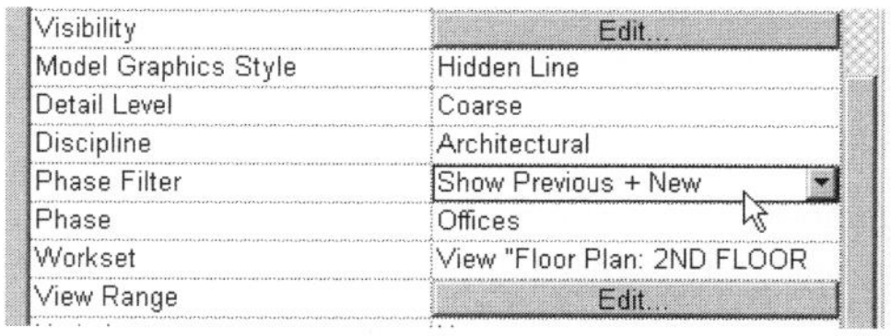

**Figure 5–87** *Set the Phase Filter for the View*

7.16  Create 20 Office spaces each 11' wide, as shown in figures 5–88 and 5–89. Do not draw the dimensions. Make the walls Generic 6" and the doors Single Flush 36" x 84".

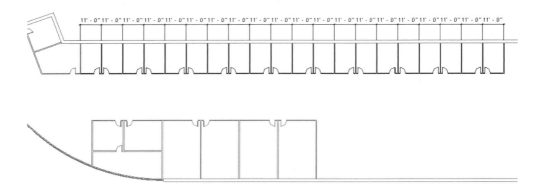

**Figure 5–88** *New offices on the second floor*

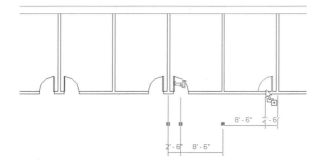

**Figure 5–89** *Typical spacing for doors*

  **Tip:** Try using the array tool to avoid drawing 20 copies of walls and doors. Uncheck Group and Associate and use an 11' interval distance.

## PLACE, EDIT, AND GROUP OFFICE FURNITURE

7.17  Use the Workset tool on the Toolbar to make Furniture the Current Workset.

7.18  Select Component from the Basics tab on the Design Bar. Select Load on the Options Bar. Navigate to the *Content/Imperial Library/Furniture System* folder and select *Work Station Cubicle.rfa* (See Figure 5–90). Choose Open.

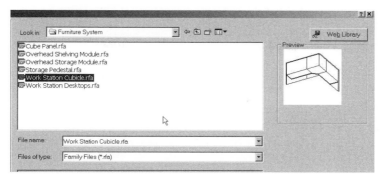

**Figure 5–90** *Find a Work Station Cubicle component to insert*

7.19    Locate an instance of the workstation as shown. Do not draw the dimensions.

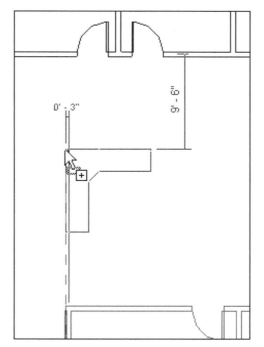

**Figure 5–91** *Locate the first cubicle*

7.20    Choose Modify from the Design Bar to terminate component placement. Select the new workstation. Pick Array from the Toolbar. Make a linear array of the workstation as shown in Figure 5–92 (24 copies x 8'-2" move distance). Uncheck Group and Associate on the Options Bar. Choose a convenient point to start the distance, and pull the cursor to the right until the temporary dimension reads 8'-2", then left-click to accept the distance.

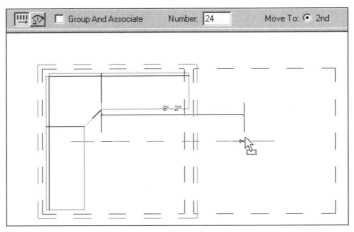

**Figure 5–92** *Getting ready to array the workstation*

7.21 Zoom in Region to the right side of the model. Copy the right most cubicle 25' down (−90°). The exact location is not critical.

7.22 The new component will still be highlighted. Select the Properties icon from the Options Bar. Choose Edit/New. In the Type Properties dialog, choose Duplicate. In the Name box, type **108" x 108"** and choose OK.

7.23 In the Type Properties dialog, change the Panel1 Length and Panel2 Length values to **6**—Revit will translate the value to 6' 0", as shown in Figure 5–93. Choose OK twice to exit the edit dialogs. The cubicle will grow.

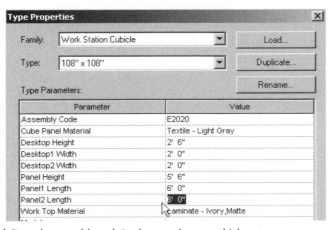

**Figure 5–93** *Adjust the panel length in the new larger cubicle type*

7.24 Make three copies of the cubicle. Rotate the copies 90°, 180°, and 270° in turn. Move the copies (using panel endpoints for snap points) to create a four-sided cubicle arrangement as shown in Figure 5–94.

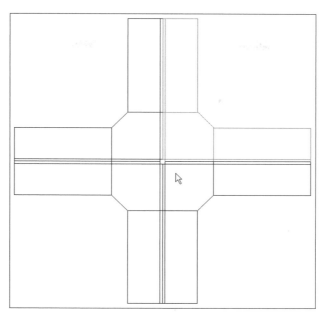

**Figure 5–94** *The large cubicles arranged back to back to back*

 **Instructor's Note:** Symmetrical arrangements can be quickly created using the Mirror tool, but as of press time for this text the workstation cubicle family object was exhibiting inconsistencies of behavior that made Copy/Rotate/Move the safer choice for this exercise.

7.25    Select all four cubicles. Choose Group from the Toolbar. Revit will make a Group of the four cubicles (which are, themselves, Groups). The Type Selector will display Group: Group 1. Select Properties. In the Element Properties dialog, choose Edit/New. In the Type Properties dialog, choose Rename. In the Rename dialog, change the name of the Group to **Quad Workstation 108"**. Choose OK three times to exit the dialogs.

**CAD Manager Note:**    Groups are a very useful tool in Revit. When combined with Work-sets they introduce a level of complexity that needs to be carefully managed. Groups can be created with elements from different Worksets; this can create headaches later, as designs develop and teams interact.

7.26    The new Group will be highlighted. Select Rotate from the Toolbar. Accept the default location of the rotation icon and rotate the Group 45°. Choose Copy. Uncheck Constrain on the Options Bar. Check Multiple. Copy the Group to the left and right at a distance interval of **25'** four times, as shown in Figure 5–95. When the copies have been created, hit ESC twice to terminate the Copy tool.

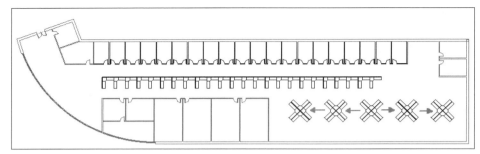

**Figure 5–95** *Workstations in place in the Furniture Plan*

7.27 Zoom to Fit. Save to Central. Relinquish Family Worksets and User-Created Worksets. For the Comment field, type **Interior with two phases on 2nd floor**. Save the local file. Close the file.

## Phases and file linking

7.28 Open a new file. Enable Worksets in the new file. Do not create any new Worksets. Choose OK.

7.29 From the File menu, select File>Import/Link>RVT. Find the *Chapter 5 interior.rvt* file you just created and link it, accepting the defaults. Zoom to Fit.

7.30 From the Settings menu, select Settings>Phases. Note on the Project Phases tab that linking a file does not make the Phases in that file accessible in the host file.

7.31 Check the Project Browser. There are no Floor Plans corresponding to the Phased Floor Plans you created in the previous file.

7.32 Select the Worksets Icon. Note that Worksets in a linked file are not available in the host file.

7.33 Close the file without saving it.

---

## SUMMARY

This chapter's exercises have introduced you to the uses of Worksets and Phasing, two important mechanisms in Revit for organizing a building model project file. Worksets divide a project's contents according to whatever system you decide works best for you. The Save to Central/Save Local function allows numerous users to work on a project simultaneously and coordinate their efforts. Many firms find that using Worksets in projects without multiple users is worthwhile for the control it gives designers over component visibility and organization.

Phases provide a way to show changes in a project over time. Phases, together with their associated Filters and Graphic Overrides, give a complete way to determine view visibility characteristics—new work, old work, temporary work, and future work can all be displayed or hidden according to your preferences.

# REVIEW QUESTIONS

## Multiple Choice

1. A Workset can be

    a) open but not Editable

    b) editable but not Open

    c) active but not Editable

    d) all of the above

2. To select elements in a Workset-enabled file,

    a) at least one Workset must be editable

    b) Editable Only must be unchecked on the Options Bar

    c) either a or b

    d) both a and b

3. Properties of Slope Arrows for floors include

    a) Width and Rotation

    b) Height at Tail, Height at Head, Offsets

    c) Thickness

    d) all of the above

4. When placing grid lines in a curtain wall,

    a) you can set vertical and horizontal lines in any order

    b) you must work right to left for vertical lines

    c) you must work top–down for horizontal lines

    d) Revit disables snaps

5. To place a door in a curtain wall,

    a) create a hole in the curtain wall and place a standard door

    b) create a section of standard wall inside the curtain wall

    c) load the curtain wall door family you want to use, and edit the curtain wall panel to the correct door type

    d) you can't place doors in curtain walls

**True/False**

6. In order to make a change in a model object that is part of a Workset already being edited by another user, you must make an Editing Request of that user.

7. The user needs to create Worksets for annotation types, as Revit does not create those Worksets automatically.

8. When a Workset-enabled file is linked into a host file, Revit displays only the Worksets from the link that were originally set to display in all views.

9. When you Attach the bottom of a wall to a roof or floor below the wall, Revit will split the wall—but at the floor line—and you have to delete the extra section by hand.

10. Revit will let you create as many Phases in a project as you want.

 Answers will be found on the CD.

CHAPTER 6

# Managing Annotation

## INTRODUCTION

So far we have been working on the design tools integrated into Revit. In previous chapters we have massed buildings, and added walls, doors, and windows. We have linked models and sites, and shared models with others. The main purpose of building a model is to communicate to the construction team what the design team wants to build. Traditionally, our communication as designers has been in the form of construction documents: printed collections of plans, sections, elevations, and specifications. This may not be the only way we deliver or communicate design intent in the future, but documents of a certain widely accepted appearance are today's standard and will not soon be completely replaced. If a wonderful model of a wonderful building cannot be understood by those who have to approve, pay for, or build it, the building won't be built. Proper annotation is essential to effective design practice.

Revit is comprehensive building information modeling software for the full spectrum of architectural practice, so we shall now explore its tools for documentation of our designs. This chapter will contain several sections to illustrate different types of standard documentation and the management of each technique. In the following exercises we will demonstrate techniques for adding, editing, and managing annotation. In the interest of keeping the exercises moving, we will focus on one area of our hypothetical project. Please note that many of these exercises are organized to show you techniques for setting up and managing annotations, and therefore a certain similarity will become evident. In a working environment, once annotations have been set up in a project, the settings can be saved in company standard templates and therefore will not have to be repeated.

For the sake of simplicity, while previous chapters included worksets (a one-way process, and a normal step in Revit design team environment), the model file that starts this chapter has been created without worksets.

## OBJECTIVES

- Create, insert, manage, and modify tags—room, door, window, sections, elevation, and grids
- Insert, manage, and modify annotation—notes and dimensions
- Transition entire set from Schematic annotation style to Design Development annotation style

### REVIT COMMANDS AND SKILLS

*Creation and customization of: Room Tags, Section Indicators, Elevation Symbols, Titleblocks, Dimensions, Units, Font Styles, Notes, and Grids*

*Working with Phases*

*View Templates*

*Project Browser as a Tool*

*Views and Sheets*

*View Titles*

*Tag All Not Tagged*

One of the great benefits of using Revit is its coordination of all objects in the model, and coordination when it comes to tagging your object is no exception. Because the tag is merely a view or "looking glass" into one specific piece of information about the object it is annotating, the coordination of that information is handled by the object and the computer. Very little user input is necessary, which eliminates opportunities for user error. There are tags for just about any Revit object you can think of, and customizing them to match company standards is a simple task.

 **Note:** In an effort to demonstrate the management of annotation and its appearance in different phases of design, we will assume that we are in our project's Schematic Design phase during the first portion of the exercises. Our annotation during this phase will be loose in appearance. We will be using the Comic Sans MS font style (standard on Windows PCs) to illustrate this. Midway through the chapter we will switch to a Design Development phase appearance with more formal "hard-line" tags and annotations. We will make this change in a few short steps; it's Revit magic!

In many cases you will want your annotations to convey a different image at different times. You may want to show or not show certain information at different stages of design. The best way to do that is through the customization of the tag that reports that information. In the following exercise we will create some custom tags for a Schematic Design look.

### Exercise 1. Customizing Tags: Room Tags

1.1 Launch Revit. From the File pulldown menu, select File>Open. In the dialog that opens, navigate to the folder that holds the CD files for Chapter 6. Select *Room Tag With Area.rfa* (see Figure 6–1). Select File>Save As from the File menu. Select the Options button. Set the Retain at most <> backups value to 1. Save the file as **Room Tag With Number - Area – SD.rfa** (the SD in this filename stands for Schematic Design).

# Room name

101

## 150 SF

**Figure 6–1** *The Room tag file before changes*

    1.2    Select and delete the lines that form the box around the room number. Select the Room Name label and choose the Properties button in the Options toolbar. Select Edit/New and change the Text Font to Comic Sans MS. Check the Bold, Italic, and Underline checkboxes, as shown in Figure 6–2. Choose OK twice to see the results.

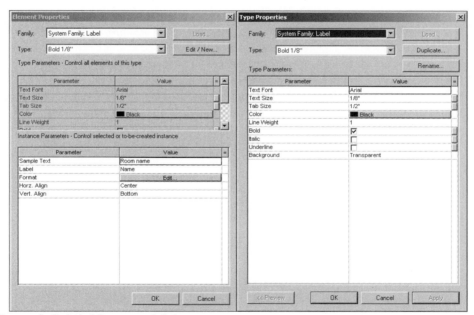

**Figure 6–2** *Label font control dialog*

    1.3    Repeat the steps above with the Room Number label. Change the Text Font in the Room Number label to Comic Sans MS, and select the Italic checkbox. Choose OK twice to see the change.

    **Note:** Both the Room Number and Room Area label text fonts are changed. This is because they are both driven by the same Type Parameters.

    1.4    Type **VG** to open the View Graphics dialog. Select the Annotation Categories tab and check the Reference Planes checkbox. Choose OK.

1.5      The intersection of the reference planes represents the insertion point of the tag. Move the Room Number and Room Area labels closer together so they look more like a single text item than individual lines (see Figure 6–3). Save and close the file.

*Room name*
*101*
*150 SF*

**Figure 6–3** *Room tag with room number and area set up for Schematic Design annotation*

## Room Tags

1.6      Open the file *Chapter 6 start.rvt*. It will open to the 1ST FLOOR – FLOOR PLAN view that shows exterior and interior walls.

1.7      From the File menu, select File>Load From Library>Load Family. Navigate to find the *Room Tag With Number - Area – SD.rfa* you just created and choose Open. This room tag will already be loaded in the *Chapter 6 start* file; choose OK to overwrite it with your newer customized version.

1.8      From the Basics Design Bar, select the Room Tag tool. Make sure Room Tag with Number - Area – SD is the active tag. Place a room tag in all of the rooms that do not already have one, plus the open curved area at the left of the plan.

1.9      Zoom in Region to the far left of the plan. Select the Room Tag in the large curved area, click on the Room label text, and it will become editable. Change the name to **Lobby**. Do the same for the Number label and change it to **101**. Continue from left to right and name the rooms as follows: **102 – Office, 103 – Office, 104 – Storage, 105 – Men, 106 – Women, 151 – Janitor** (as shown in Figure 6–4), and **152 – Elevator**. Actual room names and numbers are not critical.

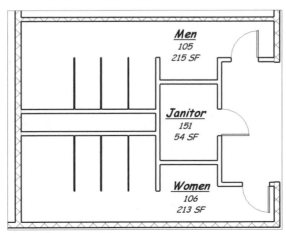

**Figure 6–4** *Room tags with room numbers and area*

## Section Indicators

1.10    From the Basics Design Bar, choose the Section tool. In the Type Selector, select Section Building Section from the list, and add three sections as shown in Figure 6–5.

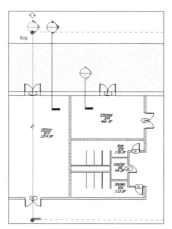

**Figure 6–5** *Section indicators—the blue control allows you to flip the section head, split the line, and cycle through alternate head/tail symbols*

1.11    Select the far right section indicator and select the Properties icon from the Option toolbar. Choose Edit/New in the Element Properties dialog, and then select Duplicate. In the Name field, type **Wall Section** as shown in Figure 6–6, and choose OK. Change the Section Head to Section Head – Open and choose OK twice. The new section will now have an open section head.

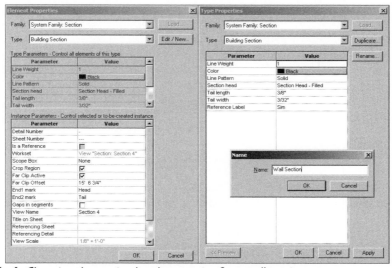

**Figure 6–6** *Changing the section head properties for a wall section*

1.12 From the File menu, select File>Load From Library>Load Family. Navigate to the *Imperial/Annotations* folder and select *Section Head – 1 point Filled.rfa*. Choose Open (see Figure 6–7).

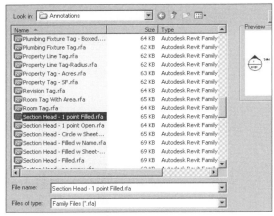

**Figure 6–7** *Select the Section Head with one filled arrow*

1.13 Select the middle section indicator and select the Properties icon from the Option toolbar. Select Edit/New in the Element Properties dialog; then choose Duplicate. In the Name field, type **Temp Section** and choose OK. Change the Section Head and Section Tail fields to Section Head – 1 point Filled (the Family you just loaded into the file), as shown in Figure 6–8. Choose OK twice.

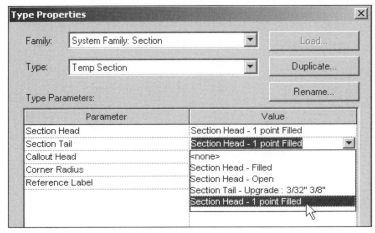

**Figure 6–8** *Changing the section head properties for a temporary section*

1.14 You will now have four section types to choose from when adding a section, plus a Detail View, as shown in Figure 6–9. The Building Section and Filled Arrow section types were previously loaded in the project.

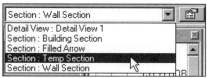

**Figure 6–9** *Section type selection*

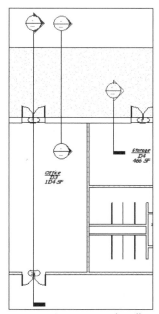

**Figure 6–10** *Building section, temporary section, and wall section indicators*

 **Note:** Many offices differentiate wall sections and building sections with different section heads. Thus the solid fill and no fill section heads. Additionally, many experienced Revit users find that using a section to study a model and a different one for documentation purposes is useful. Some offices develop a temporary section that is graphically different to indicate to other users that it is permissible to move, edit, or even delete it without consequences to the documentation set. In this case we are using a double head indicator with a single arrow.

## Elevation Indicators

1.15    From the View Design Bar tab (right-click over the Design Bar and check View from the list if necessary), select the Elevation tool. Place an elevation indicator on the bottom of the women's room, as shown in Figure 6–10. Hit ESC to terminate the tool.

 **Tip:** Revit will snap the point of the elevation to the nearest wall. You can also use TAB to cycle through pointer locations before placing the elevation indicator.

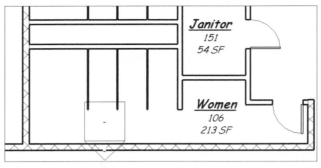

**Figure 6–11** *Interior elevation indicator*

1.16 Select the square portion of the elevation. Select the Properties icon from the Options toolbar. Select Edit/New from the Element Properties dialog. Choose Duplicate from the Type Properties dialog. Type **Interior Elevation – SD** in the Name dialog and choose OK. Change the Width, Shape, and Text Position values as shown in Figure 6–11. Place a check in the Filled value and change the Text Font to Comic Sans. Choose OK twice to close the dialog box.

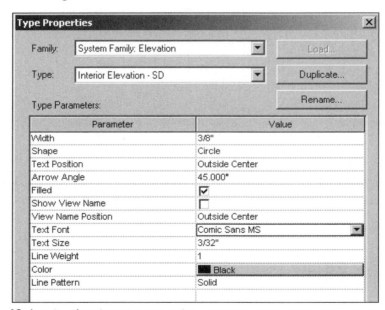

**Figure 6–12** *Interior elevation type properties*

1.17 Select the elevation marker (circle); checkboxes will appear at the quadrants. Check the three boxes that are not checked (see Figure 6–13).

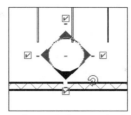

**Figure 6–13** *Checkboxes at the quadrants indicate whether an elevation has been generated or not*

 **Note:** Elevation views will be added to your Project Browser (see Figure 6–14).

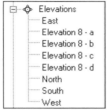

**Figure 6–14** *Elevation Views are automatically added when you check the associated checkboxes*

1.18 Double-click on any one of the arrow portions of the interior elevation indicator. This will open the associated interior elevation. Type **VP** to open the Elevation View Element Properties dialog. Change the View Scale to 3/8" = 1' – 0" and choose OK.

1.19 Choose the View pulldown and select Save As View Template. Type **Interior Elevation – 3/8"** and choose OK. In the View Templates dialog that opens, change the Detail Level value to Fine. Choose OK (see Figure 6–15).

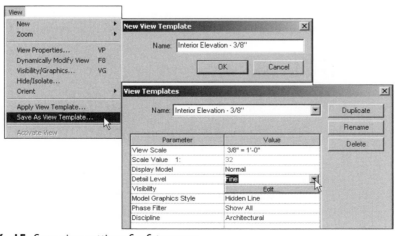

**Figure 6–15** *Save view settings for future use*

1.20 Open a different interior elevation view. Select the View pulldown menu and select Apply View Template. In the Select View Template dialog, select Interior Elevation – 3/8". Choose OK. Repeat for the other two interior elevations (see Figure 6–16). Note that there is an option to apply the template you choose to new views as you create them.

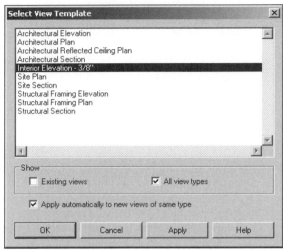

**Figure 6–16** *Apply the new View Template to the other interior elevation views*

 **Note:** In this case we are using the View Template to simply store the scale of the Interior Elevations, but this tool can be used to manage any number of visibility settings including linestyles, lineweights, colors, and detail levels. Just about anything you can change regarding view characteristics or visibility can be stored and managed using view templates. It will be well worth your time to explore these settings on this or any other model.

1.21 Save your project file.

## Exercise 2. Text and Notes

As you recall, we are hypothetically in Schematic Design, and all of our annotation text so far has used the Comic Sans MS font. We will need to create a note style that matches.

2.1 Open the IST FLOOR – FLOOR PLAN view. Set the Zoom to include the washrooms and open area to the right.

2.2 From the Basics tab of the Design Bar, select the Text tool. Select Text: 3/32" Arial from the Type Selector pulldown, and select the Properties icon. Choose Edit/New. Select Rename, and change the name of the Text Type to **3/32" Note**. Choose OK in the Rename dialog. Change the Leader Arrowhead value to Heavy End 1/8". Change the Text Font to Comic Sans MS and check the Italic checkbox, as shown in Figure 6–17. Choose OK twice.

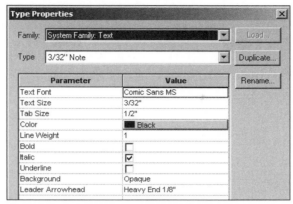

**Figure 6–17** *Modifying text properties*

> 2.3 While still in the text command, choose the Two Segment leader and the Left Justified Text Alignment from the Options Bar, as shown in Figure 6–18.

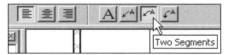

**Figure 6–18** *Leader type and Text Alignment*

> 2.4 To start a text leader, click a spot on the bottom washroom wall, then a point down and to the right of that, and finally select a point directly to the right of the second point. Type **8" MASONRY WALL, FULL HEIGHT. PROVIDE FULL FIRE SEALANT**. Select a point outside of the text to end the text string, and hit ESC twice to get back to the Modify tool.

> 2.5 Select the text you just entered to bring up the blue grips and drag the right-hand grip to the left to adjust the width of the text box.

**Note:** There are many ways to move the text you have created with different results. First select the text and put the cursor over the highlight around the text. Drag the text box to drag the text and leader as a stationary unit. Now select the text and drag the blue four-headed arrow at the top left of the text box. Note that the text moves but the leader arrowhead remains stationary (see Figure 6–19).

**Figure 6–19** *Moving text two ways*

2.6 Continue adding notes as shown in Figure 6–20. Revit automatically snaps the second and endpoints of the leader line to align with the adjacent leaders and notes.

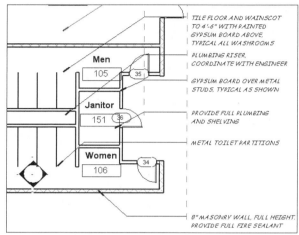

**Figure 6–20** *Schematic Design notes and leaders*

## Dimensions: Linear

 **Note:** For the purpose of linear dimensions, we will use an enlarged plan of the washrooms.

2.7 From the View Tab on the Design Bar, select the Callout tool. Click a point at the upper-left of the washrooms and click again at the bottom right. Move the callout head to the bottom of the callout lines by selecting the callout and dragging the grip at the intersection of the callout head and the leader to the bottom of the callout, as shown in Figure 6–21. Hit ESC once to get back to the Modify command.

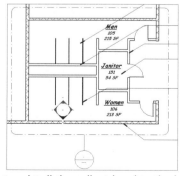

**Figure 6–21** *Create a callout and pull the callout head to the bottom of the callout*

2.8    Open the callout by double-clicking the blue callout head. In the callout view, type **VP** to open the View Properties dialog. Choose Edit/New. Choose Duplicate. Change the new View Type name to **ENLARGED WASHROOM PLAN.** Choose OK. Set the View Name to **ENLARGED WASHROOM PLAN**, the View Scale value to 1/4"=1' - 0", and uncheck the Crop Region Visible checkbox, as shown in Figure 6–22.

2.9    Select Edit from the Visibility Parameter (this is the same as typing **VG** from the view). Select the Annotation Categories tab and uncheck the Elevations checkbox. Choose OK twice to see the results.

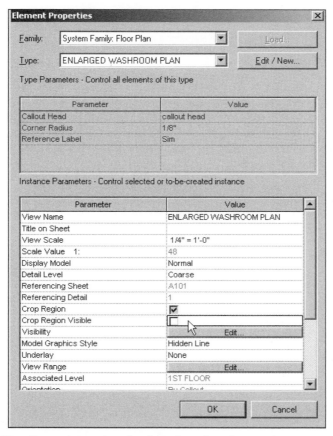

**Figure 6–22** *View Properties settings for the enlarged washroom plan*

2.10    Add room tags to the Men's, Women's, and Janitor's rooms in this view. The Room Tag tool is on the Basics tab of the Design Bar. See Figure 6–25 for appropriate locations. Notice that the room information is filled in for you. You have defined that space as a room; Revit recognizes it as an object and therefore "reports" the information through the room tag.

2.11    From the Basics or Drafting Design Bar, select the Dimension tool. Add dimensions to the plan similar to Figure 6–25. (You will need to switch between Prefer wall centerlines and Prefer wall faces on the Options Bar during the dimension selections.)

2.12    Hit ESC twice to get back to the Modify tool. Select the entire view, use the Filter tool, and uncheck all of the categories except Dimensions, as shown in Figure 6–23.

**Figure 6–23** *Filter out all categories except the Dimensions*

2.13    From the Options toolbar, select the Properties icon. Select Edit/New, then Rename. In the New Name field, name the dimension style **Linear - SD** and choose OK. Set the Parameters per Figure 6–24. Choose OK twice when you are finished to see the results. (You will notice that there is no Italic checkbox for the text.)

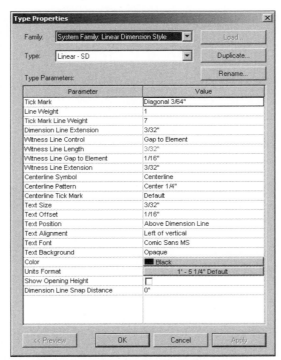

**Figure 6–24** *Schematic Design Linear Dimension Style Parameters*

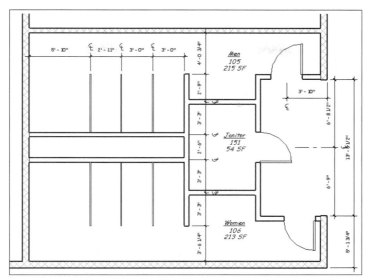

**Figure 6–25** *Schematic Design dimensions on the washroom callout plan*

## Angular Dimensions

2.14 Go back to the 1ST FLOOR – FLOOR PLAN view and Zoom in Region to the curtain wall near the doors. From the Basics Design Bar, select the Dimension tool and choose the Angular icon, as shown in Figure 6–26.

**Figure 6–26** *Angular icon*

2.15 Dimension the curtain wall grids as shown in Figure 6–27. Select Modify from the Basics Design Bar and select the dimension string and modify its Parameters using the technique outlined above. Name the type **Angular – SD**, and use all of the same Parameters you used for the linear dimensions.

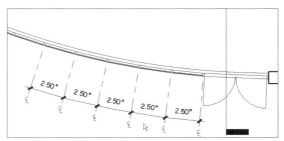

**Figure 6–27** *Angular Schematic Design dimensions at curved curtain wall*

2.16 Now it is time to dimension the radius of the curtain wall. From the Basics toolbar, select the Dimension tool and check the Radial icon. Select the centerline of the curtain wall, it will pre-highlight with a light dashed line and the status bar will read Walls: Curtain Wall: Curtain Wall 1. If it does not automatically pre-highlight, tap the TAB key while the cursor is over the center of the wall. When the wall centerline does highlight, select the wall and then select a location for the dimension line location (see Figure 6–28). Hit ESC twice to exit the command and put you back in the Modify tool.

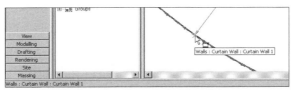

**Figure 6–28** *Status bar feedback and wall centerline highlight*

2.17 Select the radial dimension you just created. Choose the Properties icon from the Options Bar. Choose Edit/New, and then select Rename. Rename the Dimension Type **Radial – SD**. Modify its Parameters to match Figure 6–29. Before closing the Type Properties dialog box, select Units Format, uncheck the Use project settings checkbox, and change the Rounding value to read To the nearest 1", as shown in Figure 6–29. Choose OK three times to exit the dimension dialog boxes.

**Figure 6–29** *Radial Dimension Parameters and Units/Rounding Format*

You now have Schematic Design dimension styles for all three dimension types, to use anywhere in this project or transfer to other projects.

## Grids

Just like some of these other annotation families we have been modifying, Revit's grids also hold the parametric controls necessary to change their appearance with a few simple modifications. In this next exercise we will create a new grid head from scratch.

2.18    Use the File pulldown and select File>New>Family, as shown in Figure 6–30. The New Family browser will open for you to select a template file.

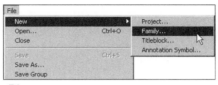

**Figure 6–30** *New Family File*

2.19    Navigate to the folder with the Chapter 6 library files and open *Grid Head.rft*. This will open the template file to create any grid head geometry you want. In this case we will create a Schematic Design head to match our other annotations.

2.20    From the Family tab (the only one) on the Design Bar, choose the Label tool. Make sure that the Center and Middle icons are selected in the Text Alignment section of the Options Bar, as shown in Figure 6–31. Place the label centered on the vertical reference plane and above the horizontal reference plane. The label will snap weakly to the reference planes. Revit will ask you for the label name; there is only one Parameter to choose from on a grid, Name. Choose OK.

**Figure 6–31** *Label Text Alignment*

2.21    Select the Modify tool from the Design Bar and choose the Name label. Select the Properties icon from the Options toolbar. Select Edit/New in the Element Properties dialog, and change the Font to Comic Sans MS. Check the Bold, Underline, and Italic checkboxes, as shown in Figure 6–32. Choose OK twice to see the results.

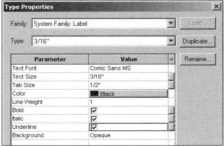

**Figure 6–32** *Label text properties*

2.22 The red text and "dummy" line are there for your use in orienting the grid. They are no longer necessary—delete them. Save the file as *Grid Head – SD.rfa* in a location determined by your instructor, and close the file.

2.23 Back in the project file, from the File menu select File>Load from Library>Load Family, as shown in Figure 6–33. Navigate to the *Grid Head – SD.rfa* file you just created, and select Open. The screen will not change appearance.

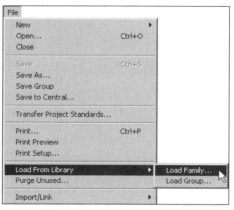

**Figure 6–33** *Load grid head from library*

2.24 From the Basics tab on the Design Bar, select the Grid tool and draw a gridline by selecting a point at the bottom left of the project and another straight above the first. The grid head will appear as a circle with the number I in it. Hit ESC twice to get back to the Modify tool.

2.25 Select the grid, select the Properties icon from the Options toolbar, and choose Edit/New in the Element Properties dialog. Select Duplicate and name the new type **Grid – SD**. In the Grid Head Value, select Grid Head – SD, as shown in Figure 6–34. Choose OK twice to see the changes.

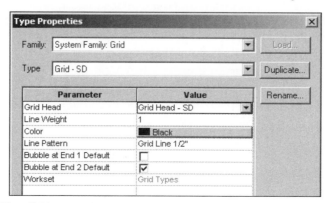

**Figure 6–34** *New Grid properties*

2.26 From the Basics Design Bar, select the Grid tool. In the Type Options selector, select Grid – SD, and continue placing vertical grids from left to right on your project. Specific locations are not critical. Note that the end points will snap to alignment; additionally, when you drag grid endpoints all of the other aligned grid endpoints will maintain their alignment.

2.27 Select any grid line and choose the Copy tool. Choose an arbitrary point in space, drag the cursor horizontally to the right, type 1, and hit ENTER. You have just copied a grid line 1' – 0" to the right of the original. The text at the grid head will be overlapped. Zoom in Region to an area just around the grid heads.

2.28 Select the new grid and drag the second (lower) grip on the grid line to the right. The text will move and there will be an angled extension line on the grid line, as shown in Figure 6–35. Select the Grid label text to edit it.

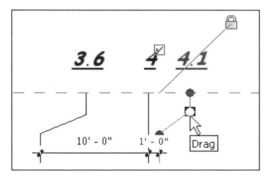

**Figure 6–35** *New Schematic Design grids with offset heads*

2.29 Save your file.

## Exercise 3. Titleblocks

 **Note:** One of the most common requirements in customizing standard annotation stems from the fact that every firm has an individualized titleblock or sheet frame to incorporate into plan sets. We are going to show you how to create a Schematic Design (horizontal) titleblock easily using an existing titleblock and its labels.

3.1 From the File menu, select File>Open. Navigate to the Chapter 6 library folder and open the file *36 x 48 Horizontal - DDCD.rfa*. This is the generic Revit titleblock that we will use to create our custom Schematic Design titleblock.

3.2 From the File pulldown, select File> Save As and save the file as *36 x 48 Horizontal – SD.rvt,* in a location determined by your instructor.

3.3 Zoom in Region around the lower-right corner of the sheet. While holding down the CTRL key, select the text and labels circled in Figure 6–36.

**Figure 6–36** *Text and labels to be reused in the Schematic Design titleblock*

3.4 Use the Move command to move the selection to the middle of the sheet. Repeat this technique to move the Autodesk Revit logo and Address from the top right of the titleblock to the middle of the sheet. Your arrangement will look similar to Figure 6–37.

**Figure 6–37** *Place all of the necessary labels and text in the middle of the sheet to modify them*

3.5 Delete all of the lines and remaining text from the right side of the sheet; leave the outside perimeter line for size reference. We are now going to change the appearance of the remaining text and labels.

3.6 Select the Autodesk Revit logo and choose Properties. This is a bitmap image with a height and width. Change the Width value to **7"** and choose OK.

3.7 Select the Address label and change the value in the Type Selector to Text: 1/4". Notice that the text is now compressed in a narrow box. Select the blue grip on the left of the box and drag it out to the left until the address is back to 5 lines. You will have to do this on most of the labels and text boxes in the next several steps.

3.8 Select the Client Name label and select the Properties icon from the Options toolbar. Select Edit/New in the Element Properties dialog. Select Duplicate in the Type Properties dialog. Name the new type **3/4" Comic** and choose OK. Change the Type Parameter Values as shown in Figure 6–38. Choose OK twice.

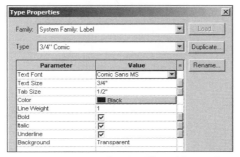

**Figure 6–38** *3/4" Comic label Parameters for the Client Name, Project Name, and Sheet Name labels*

3.9 Drag the right-hand grip well to the right to allow for a long client name.

 **Note:** Remember that the labels represent information that will come from the project and may be longer in some projects than others. If you plan to re-use titleblocks in the future, allow enough space for labels.

3.10 While holding the CTRL key down, select the Project Name and Sheet Name labels. Change the value in the Type Selector to Label: 3/4" Comic.

3.11 Edit the Project Number label (not the text) similarly. Create a **3/8" Comic** Label type, using the Parameters shown in Figure 6–39. Change the type of the Date and Scale labels by selecting them and changing the value in the Type Selector to Label: 3/8" Comic. Stretch the grips on any label that is not long enough to contain its contents in one line.

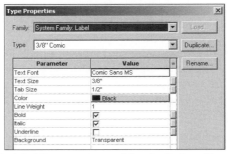

**Figure 6–39** *3/8" Comic label Parameters used for the Project Number, Sheet Date, and Scale labels*

3.12 Repeat the above step on the Sheet Number label. Create a **1" Comic** Label Type using the Parameters from Figure 6–40. Before closing the Element Properties dialog, set the Horz. Align value to Right and the Vert. Align value to Bottom. Choose OK.

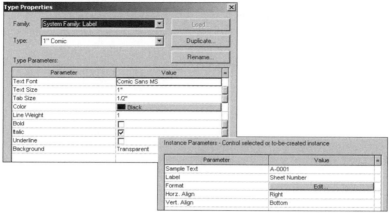

**Figure 6–40** *1" Comic label Parameters used for the sheet number label*

3.13 Edit the vertical Date/Time Stamp label similarly. Create a **1/4" Comic** Label Type using the Parameters shown in Figure 6–41.

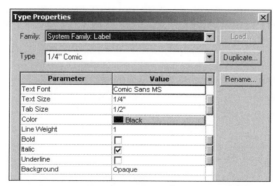

**Figure 6–41** *1/4" Comic label Parameters used for the vertical Date/Time Stamp date label*

3.14 Finally, for the plain text, select the Project Number text and select the Properties icon from the Options toolbar. Select Edit/New, and then select Rename. Rename the type **3/8" Comic** and change the Parameters as shown in Figure 6–42. Choose OK twice. The Date and Scale text will also display the Type changes we just made.

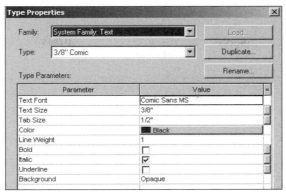

**Figure 6–42** *3/8" Comic text Parameters used for the straight text*

3.15    Stretch the three text objects as necessary to make each fit on one line with the text border of an appropriate size. You now should have a mess in the middle of your sheet—maybe even more congested than what's shown in Figure 6–43. Relax—we will straighten that out right away.

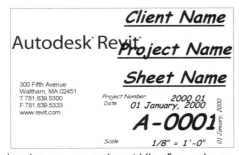

**Figure 6–43** *New label and text types at the middle of your sheet—quite a mess!*

3.16    Zoom to Fit (**ZF**) to see your whole sheet.

3.17    From the Options toolbar, select the Offset tool. Set the type to Numerical, the Offset to **1"** (be sure to add the **"**), and check the Copy checkbox, as shown in Figure 6–44.

**Figure 6–44** *Offset settings*

3.18    Move your cursor over the top of one of the boundary lines. A green dashed line will appear on one side or the other, indicating where Revit is going to place your offset line. Before selecting, tap the TAB key once to link all four boundary lines and select the interior side of the line. We will use this line as a guide and delete it later.

3.19 Offset another, single line, **3 1/2"** up from the bottom (change the value of the Offset on the Options Bar). This will also be a guideline to be deleted later.

3.20 Draw a line from the midpoint of the horizontal boundary, vertically, the full height of the sheet.

3.21 Now move the text and labels into position, as shown in Figure 6–45. Location is not too critical. You may notice that Revit text and labels do not have snap handles per se—eyeballing the text into location is acceptable here. Revit text and labels do, however, align with each other, so once one is in place the others will align.

**Figure 6–45** *Text and label locations for our new titleblock; guidelines are still in place*

Many firms have solid or otherwise wide linework on their titleblocks. The most effective way to re-create this look is by using fills in various shapes and colors.

3.22 From the Design Bar, select the Filled Region tool. From the Options toolbar, select the Rectangles option, as shown in Figure 6–46. If you do not want a black border on the edge of your band, make sure you are sketching with <Invisible lines> selected in the Type Selector drop-down list.

**Figure 6–46** *Rectangle sketch tool*

3.23 We will be creating a long narrow rectangular blue band from the horizontal guideline, the full width of the printable area of the sheet, 3/8" high. Select the end point on the guideline above the Autodesk logo, and select another point on the right-hand vertical guideline above the horizontal guideline. You will be left with a rectangle that is 3' – 10" wide by an ambiguous height. Hit ESC twice to get back to the Modify tool, select the top line of the rectangle, and change the temporary dimension value between the top and bottom of the rectangle to **3/8"**.

3.24 From the Sketch Design Bar, select the Region Properties tool and choose Edit/New. Change the Parameters to the values shown in Figure 6–47. Choose OK twice, and then select Finish Sketch on the Sketch Design Bar.

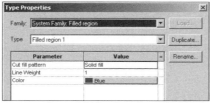

**Figure 6–47** *Solid fill Parameters for solid blue band on titleblock*

3.25   Select Label from the Family Design Bar. Select the Center and Bottom
       alignment icons, as shown in Figure 6–48.

**Figure 6–48** *Text Alignment icons for label alignment*

3.26   Select a point directly below the Client Name label. Revit will show you
       green dashed alignment lines to assist in the placement. In the Select Param-
       eter dialog, select the Project Status Parameter as shown in Figure 6–49,
       and choose OK.

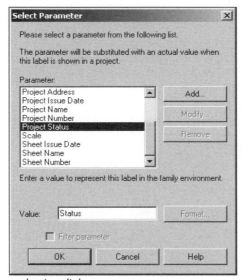

**Figure 6–49** *Parameter selection dialog*

3.27   Hit ESC to terminate the Label tool. Select the label you just created, and
       use the Type Selector to change it to Label: 3/8" Comic. Adjust its position
       if necessary.

3.28   Erase all of the guidelines and save the file.

3.29   Congratulations, you have now created a custom titleblock that can be
       used on any project in the Schematic Design phase (see Figure 6–50). Close
       the file.

**Figure 6–50** *Finished Schematic Design titleblock family*

## Exercise 4. Sheet Layout

4.1 Now we'll use that Schematic Design style Titleblock in our project file. Open or return to the *Chapter 6 Start.rvt* file from the previous exercises. From the File pulldown menu, select File>Load from Library>Load Family. Navigate to the *36 x 48 Horizontal – SD* family you just created, select it, and choose Open.

4.2 Repeat the last step. Navigate to the Chapter 6 library folder and load the generic titleblock family file *36 x 48 Horizontal - DDCD.rfa* into the project. We will use both when making the transition from Schematic Design to Design Development annotations.

4.3 Activate the View tab on the Design Bar and select the Sheet tool. Select the 36 x 48 Horizontal – SD sheet from the list. Revit creates and opens a Sheet view named A101 – Unnamed.

4.4 Click on the IST FLOOR – FLOOR PLAN view from the Project Browser. Drag it onto the sheet and locate it in the center towards the bottom. Drag the ENLARGED WASHROOM PLAN view from the browser to the area in the upper-right side of the sheet.

4.5 Select the Titleblock. Select Properties from the Options toolbar and change the Parameters as shown in Figure 6–51.

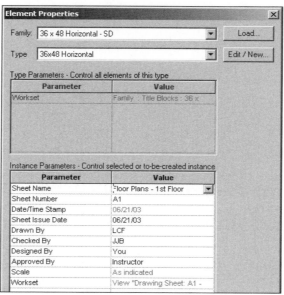

**Figure 6–51** *Titleblock Parameters for sheet A1*

 **Note:** The Drawn By, Checked By, Designed By, and Approved By labels are not used in the Schematic Design titleblock but will appear in the next phase.

You will notice that the labels for the floor plan views on this sheet do not use our Schematic Design look with the Comic Sans font. We have preloaded a Schematic Design view title for your use.

4.6    Choose either floor plan view (which will display a red border), and select Properties from the Options toolbar. Select Edit/New, and change the Parameters to those shown in Figure 6–52. Choose OK twice to exit the dialogs.

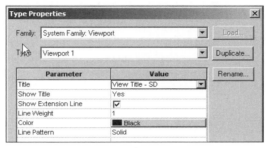

**Figure 6–52** *Change the View Title to the SD style*

4.7    From the Views Design Bar, use the Sheet tool to create a new sheet. Note that Revit numbers the new Sheet A2—it picks up the numbering system from the existing sheet. Delete the new sheet and change the number of the existing sheet to **A1.1**

4.8    From the Views Design Bar, use the Sheet tool to create six new sheets. Modify the sheet names and numbers per Figure 6–53.

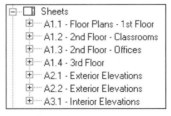

**Figure 6–53** *New sheets names and numbers*

4.9    Once the new sheets have been added, rename the four interior elevation views and create an enlarged floor plan view of the elevator area. Drag the appropriate views to the new sheets and arrange them as you see fit. Figure 6–54 shows the sheet names, numbers, and views on each of the sheets in our completed set.

**Tip:** Before you move views to sheets, right-click on the name of each floor plan view, right-click, and select Properties, which opens up the View Properties dialog. Select the Edit button in the Visibility field, and make sure that Elevations is not checked on the Annotation Categories tab. Do this for each floor plan, and they will take up less space on the sheets. For the Elevation views, uncheck Topography on the Model Categories tab.

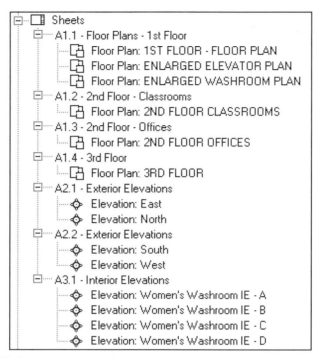

**Figure 6–54** *All sheets, names, numbers, and subsequent views*

You now have what represents a mini-set of Schematic Design drawings. Take a few minutes to look at each sheet; print them out if possible. This set could be 10 sheets or it could be 100 sheets, but they all look the same and are coordinated with one another. Now it is time for that "Revit Magic" mentioned at the beginning of the chapter. As you recall, this exercise to date has been focused on the Schematic Design look and feel. It is now time to jump to the Design Development phase of our hypothetical project. With that jump comes a completely different set of graphics, tags, notes, and sheets—everything needs to look "tighter" (more condensed and composed on the page) and "hard-lined" (not hand-drawn or sketchy in appearance).

It is important to note that we will be using two different techniques to make this change. In exercises in the first part of this chapter, when in the Element Properties dialog you selected Edit/New and then did one of two things. One option was selecting Duplicate (to create a new type) and then modifying Parameters for the new type. The other option was just to modify the Parameters of the existing object. The reason this is important now is that we will be using two different, corresponding techniques to change the appearance of annotation objects. For the items we duplicated (created new annotation sets), we will "swap out" the Schematic Design annotation for the Design Development annotation. For the annotation objects we edited individually, we will re-edit them to change their appearance accordingly.

## Exercise 5. "Swap-out" annotation types

5.1    Open the view IST FLOOR - FLOOR PLAN. From the Window menu,
select Window>Close Hidden Windows. In the Project Browser, scroll
down towards the bottom and select the + sign next to Families. The
family list will expand, showing all of the family categories. Select the + sign
next to Annotation Symbols to expand this category. Expand the Room
Tag with the Number - Area – SD category, and right-click on the Room
Tag with Area – SD type, as shown below. Choose Select All Instances, as
shown in Figure 6–55.

It is important to note that this is a global selection and will select room tags
from all views whether they are visible or not. It is a very powerful tool.

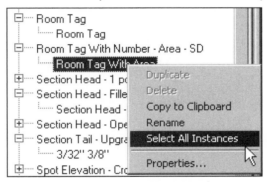

**Figure 6–55** *Select All Instances of the Room Tag with Number – Area – SD*

5.2    You will notice that all of the room tags in the adjacent view are now
selected and highlighted in red. From the Type Selector pulldown, change
the type to Room Tag. Notice that the fonts in the room tags are now all
Arial, there is a rectangular box around the room number, and the area field
is no longer shown.

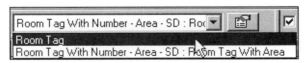

**Figure 6–56** *Room Tag Type*

5.3    Repeat the selection step as in Step 5.I with the Titleblocks, at the top
of the list under Families>Annotation Symbols in the Project Browser.
Select 36 x 48 Horizontal – SD, right-click to Select All Instances, and use
the Type Selector drop-down list to select 36 x 48 Horizontal – DDCD
(DDCD – Design Development/Construction Documents), as shown in
Figure 6–57. Revit will apply this titleblock to all sheets.

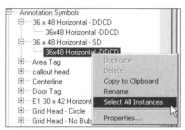

**Figure 6–57** *Select All Instances of 36x48 Horizontal – SD titleblock*

5.4 In the Project Browser, select Annotation Symbols>Grid Head – SD> Grid Head – SD. Right-click and choose Select All Instances, as shown in Figure 6–58.

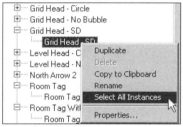

**Figure 6–58** *Select All Instances of Grid Head – SD*

5.5 Select the Properties icon on the Options toolbar. In the Element Properties dialog, select Edit/New. Make the Grid Head value Grid Head – Circle, as shown in Figure 6–59. Choose OK twice.

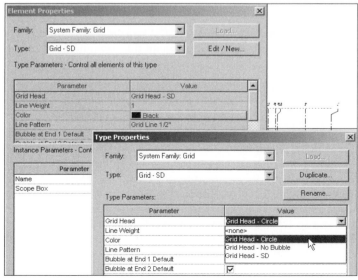

**Figure 6–59** *Change the Grid Head to the original Grid Head – Circle type*

## Changing Existing Parameters

The second technique we will use to manage the annotation graphics is simply changing the Parameters of the annotation families.

5.6 Open the 1ST FLOOR – FLOOR PLAN view and select one of the notes we created at the washrooms. Select the Properties icon from the Options toolbar. Select Edit New, and change the Text Font to Arial, uncheck the Italic checkbox, and change the Leader Arrowhead to Arrow Filled 30 Degree, as shown in Figure 6–60. Choose OK twice.

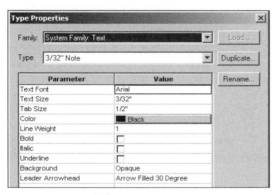

**Figure 6–60** *New Parameters for 3/32" Note*

5.7 Repeat the previous step on the interior elevation symbol, changing the Text Font to Arial, as shown in Figure 6–61. Note that you must select the center of the symbol, as the arrows are separate entities and have properties of their own. Choose OK twice to exit the dialog.

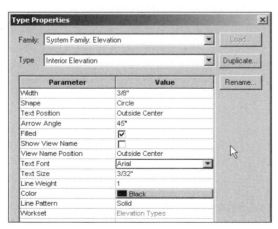

**Figure 6–61** *Interior elevation Parameters*

5.8    Open any sheet view and select any viewport. Repeat the previous steps on the viewport, changing the Title value to View Title, as shown in Figure 6–62. Choose OK twice to exit the Type Properties and Element Properties dialogs.

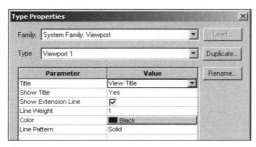

**Figure 6–62** *Viewport Parameters*

5.9    Open the view 1ST FLOOR - FLOOR PLAN. Repeat the previous process on all three types of dimensions (linear, angular, and radial). In the Type Properties dialog for each, change the Text Font value to Arial.

5.10   Save the file.

In a working environment, well-designed project templates will carry annotation style or appearance information to eliminate even the little bit of item-by-item editing that we just went through.

The replacement of our SD titleblock that has sheet information along its bottom with the generic version that has its project data fields along the right side will make the views on sheets appear a little unbalanced. There is still a little house cleaning to be done to make these sheets completely ready for client presentation, but you have dramatically changed the appearance of an entire set of drawings in a few quick and easy steps.

Keep in mind, these techniques can be used on any annotation family, not just the few we used today. Imagine that these 7 sheets were a 250-sheet set (not an unusual size for a building of the size of our hypothetical project), and you changed everything about its appearance in 20 minutes. Now, that is Revit Magic!

## SUMMARY

We have just walked through a small portion of Revit's annotation capabilities. While there is not enough room in this book to cover all of the possibilities and techniques with each annotation tool, what you have just completed will set a solid foundation for creating, editing, and managing the other annotation types.

# REVIEW QUESTIONS

## Multiple Choice

1. Changing Type Parameters in a Room Tag family lets you control

    a) the appearance of all instances of that Tag

    b) the size of rooms

    c) the room numbers

    d) the appearance of Interior Elevations for that room

2. Adding a Callout to a Floor Plan

    a) creates a new sheet in the project

    b) deletes any dimensions that the callout touches

    c) creates a separate plan view showing the area enclosed by the callout border

    d) means the end of civilization as we know it

3. Titleblock Text and Labels

    a) can only be erased from a titleblock family file, not added

    b) can only appear in one titleblock family at a time in any single project

    c) can only be Left Justified

    d) none of the above

4. Titleblock Labels that are going to read-in values from Project Parameters

    a) have to be created before any text

    b) should be of appropriate size for long names

    c) can be formatted Left, Right, or Center

    d) b and c, but not a

5. View Templates

    a) are only for elevation views, not plans or sections

    b) save and apply common view settings

    c) can't be applied after they are changed

    d) can only be used once per file

## True/False

6. You can only adjust a grid line header offset when you first create the grid line.

7. Elevation indicators can be square or circular.

8. Dimension Styles for Linear, Angular, and Radial dimensions are edited separately.

9. Revit will align text leaders to model objects, but not to other leaders.

10. You create a room when you place a room tag into a space whose boundary consists of either three or more room-bounding walls or three or more room separation lines.

 Answers found on the accompanying CD.

# Schedules

## INTRODUCTION

It has been said that one of the most tedious and unrewarding tasks in an AEC (Architecture/Engineering/Construction) firm is compiling, counting, and organizing schedules. Whether a schedule lists doors, windows, vents, parking stalls, sheets, or anything else, time spent devising and populating the schedule is much better spent elsewhere. Does the thought that "today I am going to count, categorize, and organize all of the windows on this project" sound depressingly familiar? Computers are unambiguously excellent at these tasks. The developers of Revit understood this when they set out to develop a software package for the AEC industry. One of the founding principles of Revit is to let the computer handle the tedious, trivial tasks, and let the designer design. The Scheduling module is the perfect example of this concept. Because Revit is a central database of building information, scheduling is quick, accurate, and simple. The fact that you can create an accurate custom schedule in a matter of minutes means you can spend more time designing and less worrying about the count and sizes of your windows. For that matter, you can create several custom schedules and provide more specific information for better communication, resulting in less confusion in the field. Virtually every object in Revit can be scheduled.

The following exercises will take you through a variety of schedule types and illustrate several techniques in creating, editing, and managing your schedules.

## OBJECTIVES

- Understanding the Schedule Properties Dialogs
- Create, modify, and manage a single category schedule
- Create, modify, and manage a multiple category schedule

## REVIT COMMANDS AND SKILLS

*Create a Schedule*

*Add and remove fields*

*Create new fields*

*Work with shared and project parameters*

*Sort Schedule Rows*

*Create and modify headers and footers*

*Group Schedule rows and columns*

*Format Schedule font, alignments, and orientations*

*Manage Schedule appearance*

*Create and modify a multi-category Schedule*

Scheduling is one of the greatest features of Revit. You can add, modify, remove, and change any or all Revit components in the model and Revit tracks them, no matter what. Revit's scheduling abilities can track location, size, number, or any other parameter associated with virtually any building component.

## THE SCHEDULE DIALOG BOX

The power of Revit scheduling is controlled by the Schedule Properties dialog box with its five critical tabs. Because all schedules are controlled by these tabs, it is imperative that you understand what each tab contains and the functions each controls. We will start out by outlining these tabs, and follow with the exercises.

You can examine the Schedule Properties tabs we are about to discuss in a Revit project file by selecting View>New>Schedule/Quantities from the Menu Bar, or selecting Schedule/Quantities from the View tab of the Design Bar. Choose OK in the New Schedule dialog to open the Schedule Properties dialog, as shown in the following illustrations.

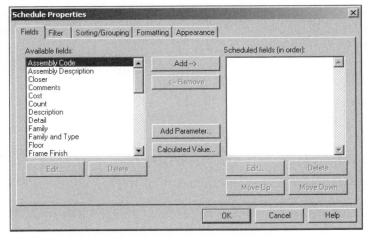

**Figure 7–1** *The Fields Tab in a Multi-Category Schedule*

## FIELDS TAB

The Fields tab controls the fields that will be in your schedule (see Figure 7–1).

The Available fields area contains a list of the default Family Parameters that can be included in your schedule. This list will be different for each schedule type. The schedule type shown in Figure 7–1 is multi-category. The Schedule Type is selected in the New Schedule dialog. The Scheduled fields area contains the list of parameters (in order) that will be included in your schedule. The top of the list will be the far-left column, with the next on the list being the next column to the right, and so on.

The Add button adds the highlighted parameter in the Available fields list to the Scheduled fields area, and the Remove button removes the parameter from the Scheduled fields area.

The Add Parameter button allows you to create a custom field, whether it is a project parameter or shared parameter. We will look at these two parameter types later.

The Calculated Value button creates a field whose value is calculated from a formula based on other fields in the schedule. For instance, a Width parameter value could be calculated as 1/2 the Length parameter.

The Edit Field buttons—note that there are two—allow you to edit user-created parameters, and the Delete buttons—also two—allow you to delete a user-created parameter.

The Move Up and Move Down buttons move the highlighted parameter up or down the list and therefore left and right in the schedule.

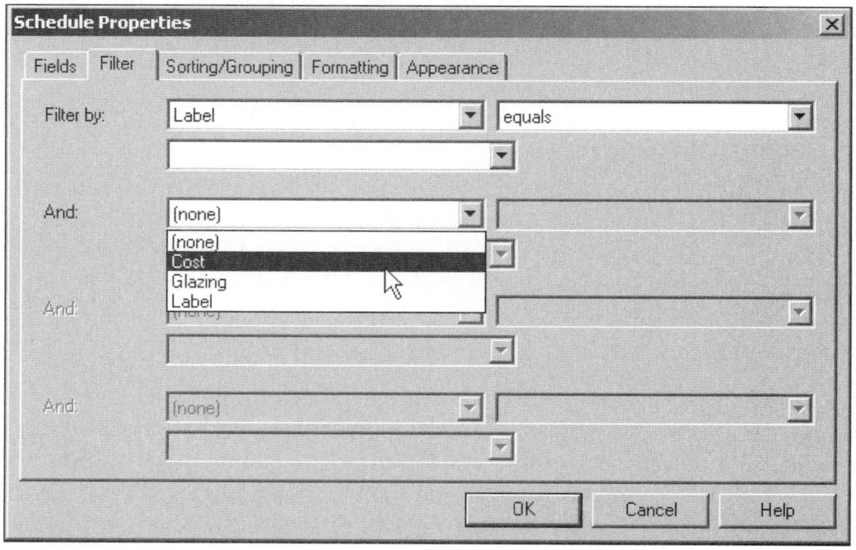

**Figure 7–2** *The Filter Tab*

## THE FILTER TAB

The Filter tab shown in Figure 7–2 allows you to restrict what elements display in simple or multi-category schedules (and also to view lists, drawing lists, and note blocks). You can set up to four filters on this tab, and all the filters must be satisfied for elements to display. You can use displayed or hidden schedule fields as filters. Certain fields—mostly Type parameters—can't be used as filters. An example of a filtered schedule would be a door schedule filtered by floor.

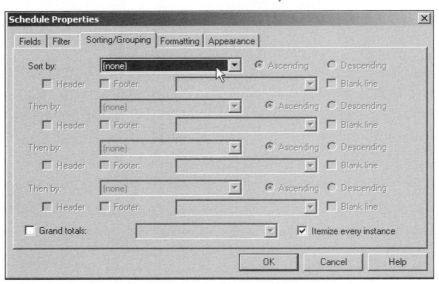

**Figure 7–3** *The Sorting/Grouping Tab*

## THE SORTING/GROUPING TAB

The Sorting/Grouping tab shown in Figure 7–3 is a very powerful tab that works well with Revit's underlying ease of use. There are some concepts that you will have to understand before jumping in. This tab, simply put, controls the order in which the information will be displayed—sorting. An example of this in a door schedule might be sorting by door number. Revit will display all of the doors in numerical order by door number (which is most often keyed to the room number).

Grouping allows you to group items together. In the door schedule example, you might want to group doors by level or by a particular zone in a building. By virtue of this you could tell Revit to group the doors by zone and then sort them by door number. To activate grouping you will need to check the Header or Footer checkbox. The Grand totals checkbox will provide a grand total at the bottom of your schedule, where it applies. The Itemize every instance checkbox allows you to toggle between a list of every item (doors, in our example) and a list that might just show the type and a count of each.

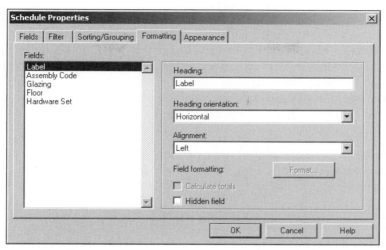

**Figure 7–4** *The Formatting Tab*

## THE FORMATTING TAB

The Formatting tab shown in Figure 7–4 controls the formatting of the individual columns. There is a field to change characteristics of the displayed heading—Heading, Heading orientation, and Alignment. In addition you can format the units shown for numerical fields using the Format button, and have Revit calculate the totals with a selection of the Calculate totals checkbox. The Hidden field checkbox allows you to hide a column. This is useful if you want to group items by field (level, for instance), but don't want to repeat the information in the schedule.

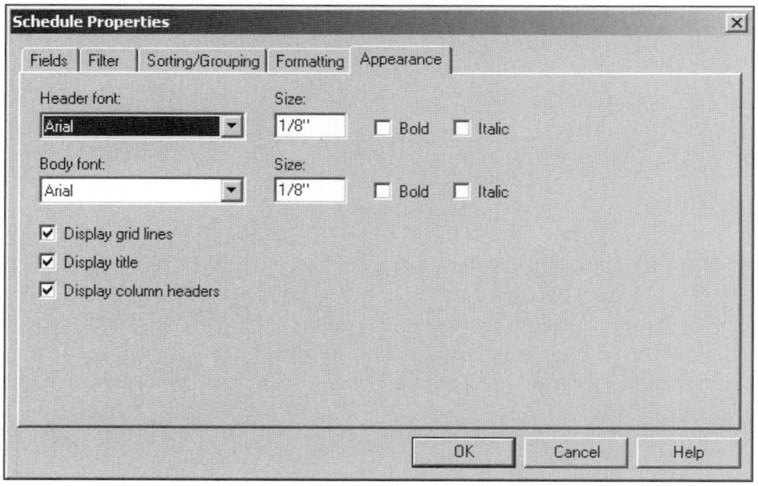

**Figure 7–5** *The Appearance Tab*

## THE APPEARANCE TAB

The Appearance tab shown in Figure 7–5 controls the overall graphic appearance of the schedule. The Header font controls manage the font type, its size, and whether it is bold, italic, or both. This will affect all headers in the schedule. The Body font controls manage the appearance of the remaining fonts. You have the ability to display grid lines, a title, and column headers or not, by using the checkboxes at the bottom of the dialog box. We will use a majority of these settings in the exercises to follow.

As stated earlier, Revit enables you to schedule virtually any object or combination of objects in your project. For the purposes of this next exercise, we will demonstrate the schedule features using a door schedule. We will begin with the basics and add complexity as we go.

### Exercise 1. Create a Schedule for Doors

1.1 Launch Revit and open *Chapter 7 Start.rvt*. This file is the continuation of the exercises in Chapter 6 with a few modifications to door tags and numbers.

1.2 From the View tab on the Design Bar, select the Schedules/Quantities tool as shown in Figure 7–6.

**Figure 7–6** *Schedule/Quantities Tool*

1.3 In the New Schedule dialog, select Doors from the Category list. Leave the Schedule Name as Door Schedule, select the Schedule building components radio button, and set the Phase to Offices, as shown in Figure 7–7. Choose OK.

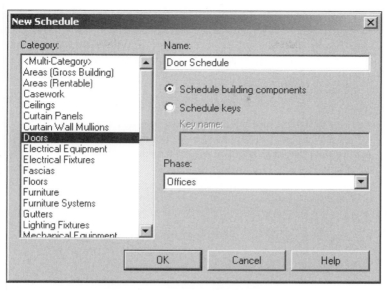

**Figure 7–7** *Starting a new Door Schedule for the Offices Phase*

1.4 In the Schedule Properties dialog, while holding down the CTRL key, select Height, Mark, Thickness, and Width, and choose the Add button.

1.5 In the Selected fields list, highlight Mark and select Move Up until Mark is at the top of the list. Using the Move Up and Move Down buttons, arrange the parameters to match what you see in Figure 7–8. Choose OK.

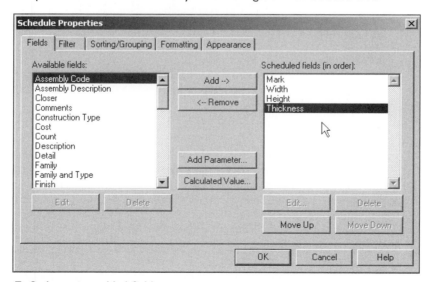

**Figure 7–8** *Arranging added fields*

A Schedule view will open. You will now have a list of all of the doors in this building project. Please notice that it is in random order, and that there are some doors labeled using a room number and a sequential letter, and others that are just numbers. This is intentional, to demonstrate the fact that Revit schedules recognize phases.

1.6 In the Schedule view, right-click and select View Properties to bring up the Element Properties dialog. In the Phase Filter pulldown, select Show Previous + New. The Phase parameter was already set to Offices, as shown in Figure 7–9. Choose OK to see the results. Your schedule should now include only doors that are numbered by room number, plus a sequential letter designation, but the numbering is still in random order.

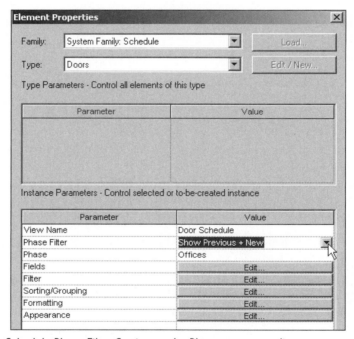

**Figure 7–9** *Schedule Phase Filter Settings—the Phase was set earlier*

1.7 In the Schedule view, right-click and select View Properties again. In the Element Properties dialog, select the Edit button for the Sorting/Grouping value. The Schedule Properties dialog opens to the Sorting/Grouping tab. In the Sort by pulldown, select Mark and choose OK twice. You now have a basic door schedule that represents all of the doors in the office phase, in ascending numeric order from top to bottom. Now you can really move.

1.8 Right-click in the view again, select View Properties as before, and set the properties for Fields, Sorting/Grouping, Formatting, and Appearance, as shown in figures 7–10 through 7–13. Choose OK twice when the adjustments are complete.

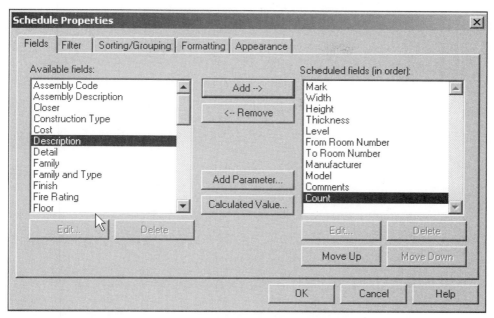

**Figure 7–10** *Fields Settings—Add and Remove the additional fields*

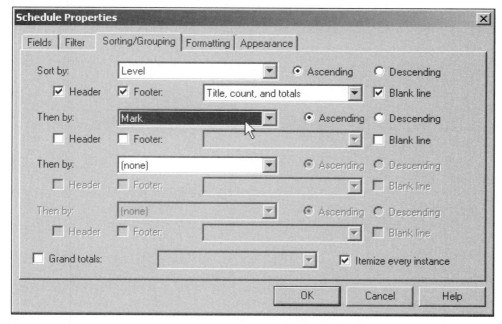

**Figure 7–11** *Sorting/Grouping Settings—Sort by Level, check Header and Footer, then Sort by Mark*

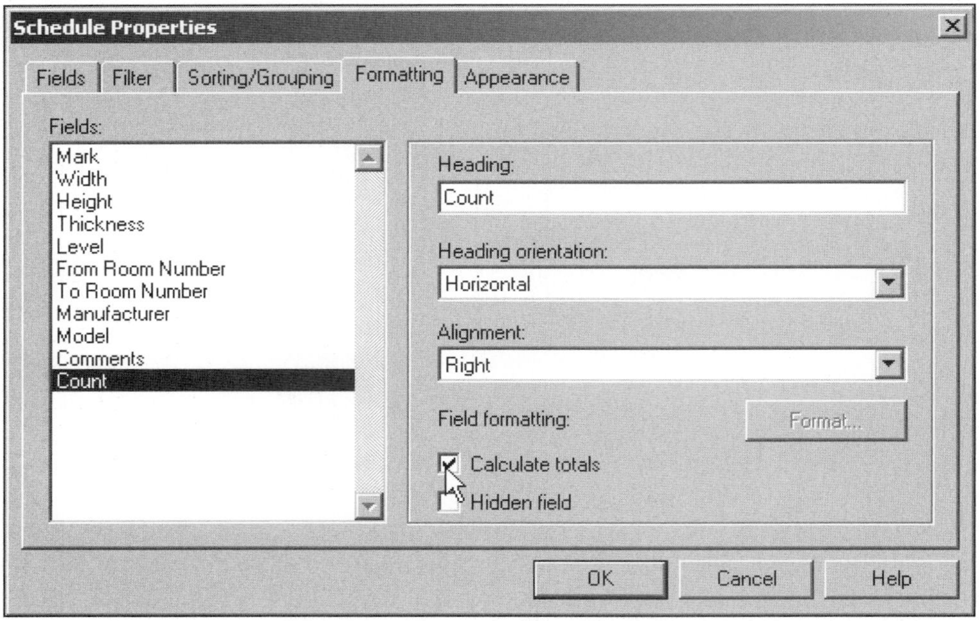

**Figure 7–12** *Formatting Settings—set Count to Align at Right, and select Calculate totals*

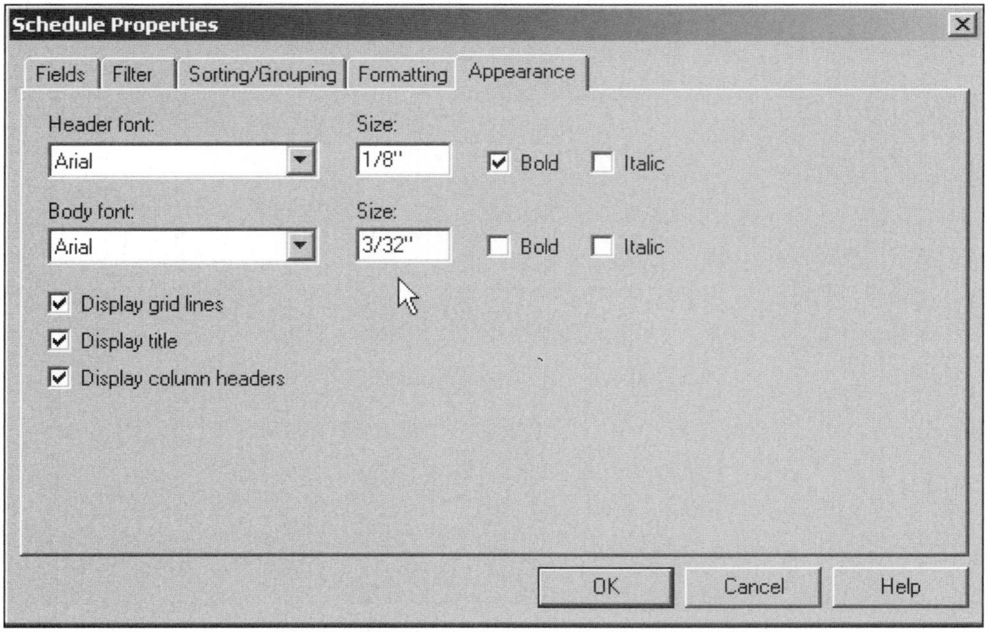

**Figure 7–13** *Appearance Settings—set Header font to Bold and Body font to 3/32"*

You will notice that the Appearance tab does not affect the font appearance in the view. Revit only displays this when the schedule is placed on a sheet. Additionally, you will see that the level information is redundant, as it is already in the header and in each row.

1.9 Right-click and select View Properties to get back to the Element Properties dialog. Select Edit in the Formatting parameter. In the Formatting tab, select Level and check the Hidden field checkbox; then choose OK twice.

**Tip:** You may also right-click over the top of any of the cells in the level column and select Hide Column(s).

1.10 From the View tab on the Design Bar, select Sheet, select the 36x48 Horizontal – DDCD titleblock, and choose OK. Revit will create a new sheet view and open it. Drag the Door Schedule view from the Project Browser onto the new sheet. Notice that the columns are all of equal width. Select the schedule to bring up the display controls. You will have triangles at the top and a squiggle line at the mid-point of the far right-hand column, as shown in Figure 7–14.

| Mark | Width | Height | Thickness | From Room Number | To Room Number | Manufacturer | Model | Comments | Count |
|------|-------|--------|-----------|------------------|----------------|--------------|-------|----------|-------|
| 1STFLOOR | | | | | | | | | |
| 6-16 | 6'-2" | 9'-6 1/2" | | | 6-1 | | | | 1 |
| 6-1B | 6'-6" | 7'-6" | 6'-2" | 6-1 | | | | | 1 |
| 6-1C | 6'-6" | 7'-6" | 6'-2" | 6-1 | | | | | 1 |
| 6-1D | 6'-6" | 7'-6" | 6'-2" | 6-1 | | | | | 1 |
| 6-1E | 6'-6" | 7'-6" | 6'-2" | 6-1 | | | | | 1 |
| 6-1F | 6'-6" | 7'-6" | 6'-2" | 6-1 | | | | | 1 |
| 6-1G | 6'-6" | 7'-6" | 6'-2" | 6-2 | 6-1 | | | | 1 |
| 6-1H | 6'-6" | 7'-6" | 6'-2" | 6-1 | 6-1 | | | | 1 |
| 6-2C | 6'-6" | 7'-6" | 6'-2" | 6-2 | | | | | 1 |

**Figure 7–14** *Schedule display controls—triangles control column width and a squiggle symbol splits the schedule*

1.11 Drag the triangular grips left and right to set the column widths.

**Tip:** It is best to set widths starting with the columns at the left and move right.

There will be occasions when you will want to display the same schedule but separate the grouped areas in different locations on the sheet. We will separate level 1 from levels 2 and 3.

1.12 Select the squiggle symbol at the right of the schedule; this will split the schedule into two individual views. Use the Move grip (double-headed arrow) in the middle of the bottom schedule to drag the lower view to the right of the upper portion. Drag the grip at the bottom center of the right-hand view up, until it is set at the gap between levels 1 and 2. The tops of the schedules will snap to alignment.

**Note:** These adjustments require you to be mindful of doors as they are added and deleted. As the level 1 schedule grows in length it will begin to populate on the level 2 side, and you will need to stretch the grip to adjust it.

One of the advantages to having a schedule is that information can be viewed and edited in a format conducive to annotative information on a broad brush scale.

1.13 Return to the Door Schedule view. Select the cell under Manufacturer in the door 101B row, and type in **Judson Door and Window Co.** This change will be applied to all doors of this type—a notice will appear. Choose OK.

1.14 Repeat the above step for doors 105A – **Fox and Sons**, 125B – **Delmar Door Corp.**, and 204A – **Lucas Woodworks**. Use the same technique to fill out model information. Fill in individual comments as you see fit, as shown in Figure 7–15.

1.15 The column widths may not display all of the information you have input. Move the cursor over the top of one of the vertical column lines to get a double-headed arrow (see Figure 7–15) and drag the column edge to the appropriate width. Note that this does not affect the schedule display on the sheet.

| 2ND FLOOR | | | | | | | | | |
|-----------|--------|--------|--------|-----|-----|------------------------------|-------------|-----------------|
| 201A | 6' - 0" | 7' - 0" | 0' - 2" | | 201 | Judson Door and Window Co. | 7284-2-ABC | Verify Clearance |
| 202A | 3' - 0" | 6' - 8" | 0' - 2" | 202 | 201 | Western Door and Window | ✛ 1236-KJ-RW | Under cut 1" |
| 203A | 3' - 0" | 6' - 8" | 0' - 2" | 203 | 201 | Western Door and Window | 1236-KJ-RW | |
| 204A | 3' - 0" | 7' - 0" | 0' - 2" | 211 | 204 | Lucas Woodworks | CDW-3672-PSS | Paint Red |

**Figure 7–15** *Schedule—manufacturer, model, and comments*

1.16 Go back to the sheet view that you placed the schedule on and make sure that the new information is showing correctly—a single line of text per schedule row. Use the triangulated grips to adjust the column widths if necessary.

 **Note:** It is now time to add key card information to the schedule, but it is not on the list of parameters for this schedule.

1.17 Right-click on the schedule in the sheet view and select Edit Schedule. The Door Schedule view will open. Right-click and select View Properties.

## Add a Parameter, Change the Format

1.18 Select the Edit button for Fields and select Add Parameter. In the Name field of the Parameter Properties dialog, type **Key Card**, set the Type to Yes/No, and select the Instance radio button (as shown in Figure 7–16), as doors of the same type will have different Key Card values (in this case a simple Yes/No). Choose OK.

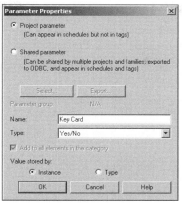

**Figure 7–16** *New Field Parameters—Key Card*

1.19 Move Key Card to just above the Count parameter in the order of scheduled fields.

1.20 Go to the Formatting tab and set the Key Card Heading orientation to Vertical and the Alignment to Right. Set the Count Heading orientation to Vertical as well. Choose OK twice. Revit creates a column for Key Card, with grayed out check boxes. Click on a check box to activate it, and click again to toggle the check off (turn the Yes to No in the Door Schedule sheet view).

1.21 Open the sheet with the Door Schedule on it. Adjust the column widths for Key Card and Count to take advantage of the vertical header alignment, as shown in Figure 7–17.

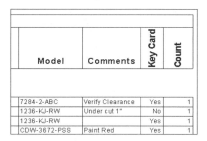

**Figure 7–17** *Vertical heading orientation*

**Note:** Traditionally, architects provided a door schedule once, in one format, and it was up to the contractor and door manufacturers to sort through this information to get the information they needed for their particular use. Because Revit is a centralized database, information is always up-to-date; the architect using Revit has the ability to group, sort, and provide it in a concise manner. These next few steps fall under the "Revit Magic" category, allowing you to manipulate the same data and provide it in a completely different manner for different uses.

## Create a Second Door Schedule—More Formatting Changes

1.22 From the Project Browser, right-click on the Door Schedule and select Duplicate. This will make a view called Copy of Door Schedule the active view.

1.23 In the new view, right-click, and select View Properties. Change the settings as shown in figures 7–18 through 7–21.

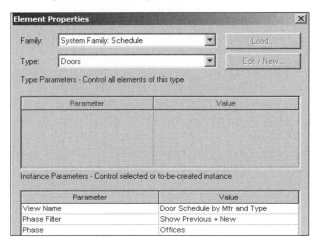

**Figure 7–18** *View Properties—name the new schedule*

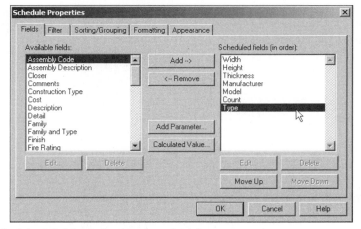

**Figure 7–19** *Scheduled fields for the new schedule*

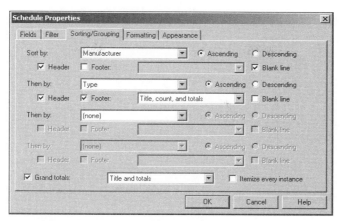

**Figure 7–20** *Sorting/Grouping in the new schedule*

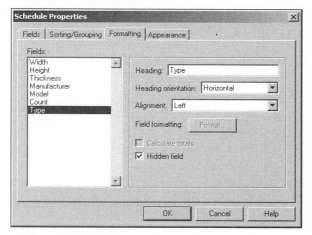

**Figure 7–21** *Formatting—hide the Type*

1.24 From the Formatting tab, select Manufacturer and check the Hidden field checkbox. From the Appearance tab uncheck the Display grid lines checkbox. Choose OK twice.

1.25 Open the sheet view with the first door schedule on it and drag the new Door Schedule by Manufacturer and Type onto the schedule sheet. Note that it appears markedly different from the first schedule.

1.26 From the Project Browser, open the 2ND FLOOR OFFICES – FLOOR PLAN. Select door number 212A and select properties from the Options bar. Note that all of the type and instance parameter information we input on the schedule is displayed, as well as the Key Card Parameter we defined in the schedule (see Figure 7–22). Uncheck the Key Card value, and choose OK.

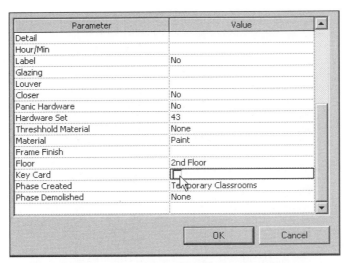

| Parameter | Value |
|---|---|
| Detail | |
| Hour/Min | |
| Label | No |
| Glazing | |
| Louver | |
| Closer | No |
| Panic Hardware | No |
| Hardware Set | 43 |
| Threshhold Material | None |
| Material | Paint |
| Frame Finish | |
| Floor | 2nd Floor |
| Key Card | |
| Phase Created | Temporary Classrooms |
| Phase Demolished | None |

**Figure 7–22** *Door Parameters—note that the Key Card now appears*

 **Note:** On many occasions your schedule will have several columns that are related to each other. Many firms like to call attention to this by putting a header above the related columns. Revit has the capability to achieve this by grouping columns.

1.27 From the Project Browser, open the Door Schedule view. Click and drag your cursor over the Width, Height, and Thickness column heads, as shown in Figure 7–23. Select the Group button from the Options bar.

**Figure 7–23** *Grouping column headers*

1.28 In the cell that is created, type **Door**. Repeat for To Room Number and From Room Number, using **Room** for the group header.

1.29 Highlight the words To Room Number and type **To** in the Heading box, and do the same for From Room Number, typing in **From** instead. Choose OK.

 **Note:** This can also be accomplished in the Schedule (View Properties) dialog, as shown in Figure 7–24.

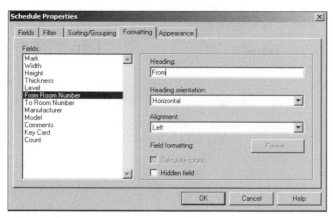

**Figure 7–24** *Changing the display name on a parameter*

1.30 Open the schedule sheet to see the results.

1.31 From the Project Browser, open the Door Schedule by Mfr and Type that we just created and modify the width on the 36" x 84" doors supplied by Lucas Woodworks to 3' - 6". All of the doors of that type change in the schedule.

1.32 From the Project Browser, open view 2ND FLOOR OFFICES – FLOOR PLAN, and note that the adjusted door widths of the doors for rooms 212–234 are now represented graphically in the plan as well.

1.33 Save the file.

## MULTI-CATEGORY SCHEDULES

Most schedules you produce will be similar to the schedules you have just completed: door, window, wall, curtain panel, furniture, and so on. These types of schedules report information on a single category type. On occasion you will need to schedule components from different categories that might have the same parameter. In addition to single category schedules, Revit allows you to schedule components from different categories, known as Multi-Category Schedules. One example that comes to mind is the electrical requirements for particular components or building parts. Any number of building components have electrical requirements and you may be asked or, better yet, *paid*, to assimilate this information. The following exercise is designed to expose you to the process of creating a multi-category schedule.

To enable the multi-category schedule functionality, the categories of the components must have a common parameter. Revit calls these parameters "shared parameters." Shared parameters are parameters that you can share with multiple Autodesk Revit families and projects. You define them once and save them to a shared parameter file. You can use shared parameters in multi- and single-category schedules, and to tag elements.

There are a few key concepts to understand before using multi-category schedules. Multi-category schedules consider *all* components within the model and return the desired field information. Of course, a schedule on all of the components in a model is of little use. When you set up a multi-category schedule, you will make use of the filter tab in Schedule Properties. This tab allows you to filter out all of the components that do not contain the filter parameter(s), so that you schedule only the components that do.

## Exercise 2. Creating a Shared Parameter File

2.1   Open or continue working with the file from the previous exercise. From the File pulldown menu, select Shared Parameters.

2.2   In the Edit Shared Parameters dialog, select Create and type **Chapter 7 – Shared Parameters** in the Name field. Save it in a location as indicated by your instructor.

2.3   From the Groups area, select New and type **Electrical** in the Name field, as shown in Figure 7–25.

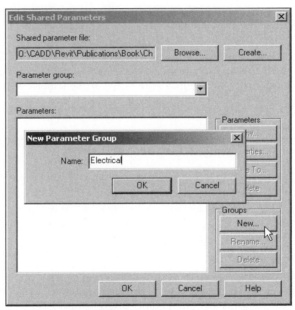

**Figure 7–25** *New Shared Parameter Group—Electrical*

2.4   From the Parameters area of the dialog, select New and type **Amps** in the Name field. Use the pulldown list to tell Revit that this parameter will be a Number, and choose OK (see Figure 7–26).

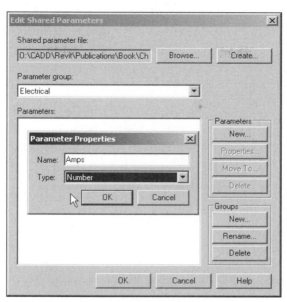

**Figure 7–26** *Parameter Properties—it's important to specify the type here, as it is not editable afterwards*

2.5 Repeat the last step adding **Watts** (also a Number) to the Electrical group. Choose OK.

 **Note:** You have now created two new parameters that will be available within this project, as well as other projects you create. This is a great way to create standard custom parameters for a company that you know will be maintained throughout your projects.

2.6 You will now need to indicate which categories have the parameters. From the Settings menu, select Project Parameters, select Add, select the Shared Parameter radio button, choose Select, highlight Amps, and choose OK.

2.7 Back in the Parameter Properties dialog, select the Type radio button next to Value stored by, and check the following categories: Furniture Systems, Lighting Fixtures, Mechanical Equipment, and Specialty Equipment (see Figure 7–27). Choose OK.

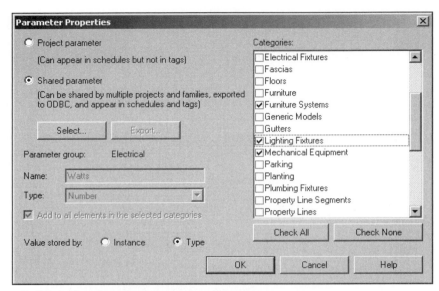

**Figure 7–27** *Adding shared parameters to the project.*

You have now added a parameter to the categories checked above that will be included in all components in those categories.

 **Note:** Selecting Instance or Type values is only available when you add the parameter to the project.

2.8 Repeat the above steps and add Watts to the project parameters. Remember to select the Type radio button. Choose OK twice to exit the dialogs.

2.9 From the Project Browser, open the 2ND FLOOR – FLOOR PLAN. Select one of the Workstation Cubicles and select the Properties icon from the Options bar. Select Edit/New, and enter a numerical value for Amps and Watts. The actual numbers don't matter in this case. Choose OK twice.

2.10 Repeat the above step for the Computer Monitor in the Workstation and the Copier out in the Work Room area.

2.11 From the Project Browser, open the Roof view, and repeat the steps above on the Air Conditioner and Boiler.

2.12 From the Menu Bar, select View>New>Schedule/Quantities. From the New Schedule dialog box, choose <Multi-Category> and choose OK, as shown in Figure 7–28. There is no option to set the Phase for the schedule at this point.

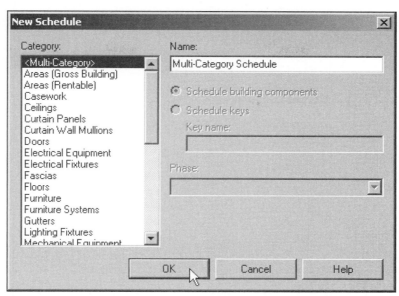

**Figure 7–28** *Create a new Multi-Category Schedule*

2.13 Use the Fields tab to add and order fields for the new schedule, as shown in Figure 7–29: Family, Type, Room Number, Count, Amps, and Watts.

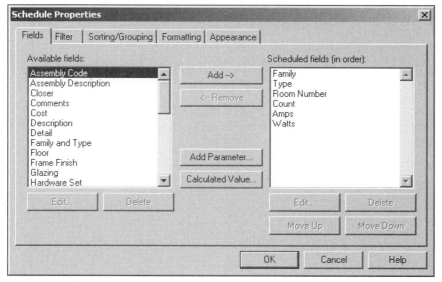

**Figure 7–29** *Multi-Category Fields*

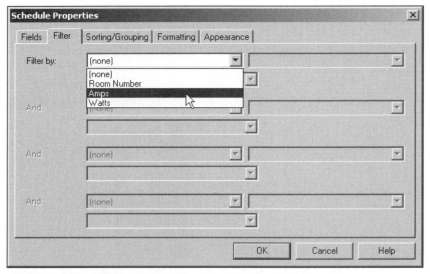

**Figure 7–30** *The Filter Tab*

2.14 On the Filter tab, select Amps from the drop-down list. Once a parameter is selected, the right-hand drop-down list becomes active, so that you can further filter against the parameter value equals/does not equal, greater than/less than, and so forth. Don't make any changes to that option.

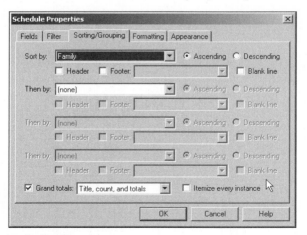

**Figure 7–31** *Sorting by Family*

2.15 On the Sorting/Grouping tab, choose Sort by Family, as shown in Figure 7–31. Uncheck Itemize every instance.

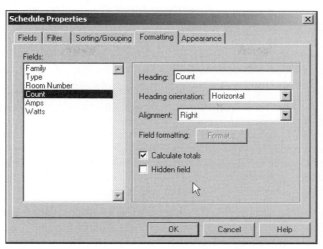

**Figure 7–32** *Formatting*

2.16   On the Format tab, set the Count, Amps, and Watts with Right Alignment, and check the Calculate totals checkbox (see Figure 7–32). Choose OK.

2.17   There you have it—a schedule showing all of the Amperage and Watts for specified equipment in the entire building. If you do not see the Copier, Monitor, and Workstation in your schedule, right-click and select View Properties, then set the Phase for the schedule to Offices. The view that was active (on the screen) controls the initial Phase setting for any Schedule (which is a View) you create.

 **Note:** Keep in mind that this technique can be applied to just about any categories. You can customize the Filter by creating a new Shared Parameter just for this use. While we have covered considerable ground regarding schedules, we have barely scratched the surface of Revit's potential. Continue to explore new uses and combinations; you will never fail to amaze yourself with the power of schedules.

## SUMMARY

Congratulations, you now have the base knowledge to create virtually any schedule type imaginable. These exercises have exposed you to the tools you need to create any schedule type and modify it to suit any customized requirements.

Keep in mind, a schedule is just another view into your project data and there is no need to create schedules with "blinders on"— in other words, like the schedules you have been creating in the past. You have a new tool, so use it. Remember that different people need the same information in different ways. You might have an overall master door schedule, and a separate schedule in which you might group and total all the doors by hardware set for the hardware installer. The same data can be shown in different ways for different uses (see *Sheet A9.91* in *Chapter 7 – Start.rvt*).

Additionally, you can display graphical information in coordination with your schedule. For instance, you could create three different room schedules (by Department, Floor Finish, and Occupancy Type) and put them on the same sheet with color-fill diagrams for analysis (see *Sheet A9.92* in *Chapter 7 – Start.rvt*).

We have included several different schedules and their uses for your review. Open the A9.9 series of sheets and check out some of the schedules that Revit provides.

# REVIEW QUESTIONS

## Multiple Choice

1. The Schedule Properties dialog controls

    a) Fields and Filters

    b) Sorting, Grouping, Formatting, and Appearance

    c) both a and b

    d) b only

2. To Hide a schedule column

    a) use the Hidden Field checkbox on the Schedule Properties Formatting tab

    b) right-click in a field cell in the Schedule View and select Hide Column

    c) delete the Schedule and start over

    d) a or b

3. Which property of the on-screen view, when you create a schedule, affects the schedule output?

    a) zoom

    b) detail level

    c) phase

    d) shading

4. Multi-Category Schedules make use of

    a) Shared Parameters

    b) Schedule Filters

    c) Cost Estimates

    d) a and b, but not c

5. Project or Shared Parameters can include fields of the following types:

    a) Text, Number, and Length

    b) Area, Volume, and URL

    c) Material, Angle, and Yes/No

    d) all of the above

## True/False

6. A Schedule Filter can be set to equal or not equal a certain value

7. Project Parameters can appear in Schedules but not in Tags

8. Shared Parameters can be shared by multiple projects and appear in Schedules and Tags

9. Calculated Parameters are text-only values

10. If you Group Schedule Column Headers, you have to delete the Schedule from any Sheets it appears on and replace it for the change to appear

 Answers found on the accompanying CD.

# CHAPTER 8

# Beyond the Design — Area and Room Plans

## INTRODUCTION

Revit's intelligent building modeler provides tools that fill a couple of basic needs for the people who pay designers to do their work. Design clients, usually the owners of the building being designed, almost always need to know the relative areas reserved for different spaces and functions within the design. Sometimes this information will be used for taxation purposes, most often for rental or other occupancy classifications. Throughout the lifespan of a commercial building in particular, accurate occupancy information can be critical. Revit models can supply this information—as anticipated uses or building configurations change during and after construction—in ways that static 2D drawings or previous 3D models never could. This may have long-range implications for design firms as suppliers of "life-cycle" management (or facilities management) services for buildings long after architectural functions are complete.

One of the easiest ways to collect and display basic area information about a Revit model is the color fill diagram, which can be either an area analysis plan or more sophisticated room plan with associated schedules. These plans are displayed in Views of their own that, like schedules, can be located on sheets.

## OBJECTIVES

- Create and edit a room schedule
- Work with room tags
- Define room styles
- Apply room styles to rooms
- Create and edit a color fill room diagram
- Create multiple color fill diagrams from one set of room tags
- Place color fills and schedules on a sheet
- Create a gross building area plan

- Create a rentable area plan
- Create a custom area plan
- Create and edit a color fill area diagram
- Export Revit information via ODBC

## REVIT COMMANDS AND SKILLS

*Room Schedules*          *Area Settings*

*Schedule Keys*          *Area Plans*

*Room Styles*          *Export ODBC*

*Color Fills*

### Exercise 1. Create a Room Schedule

1.1 Open the file *Chapter 8 start.rvt*. This model file has been created without Worksets to make the exercises in this chapter simpler than would otherwise be the case.

1.2 Floor Plan: 1ST FLOOR should be the active view. Save the file as *Chapter 8 area plans.rvt* in a location specified by your instructor.

1.3 Make Floor Plans: 2ND FLOOR OFFICES the active view. Make Drafting the active tab on the Design Bar.

1.4 From the View menu, choose View>New>Schedule/Quantities. In the New Schedule dialog, select Rooms under Category. Accept the defaults, as shown in Figure 8–1. Choose OK.

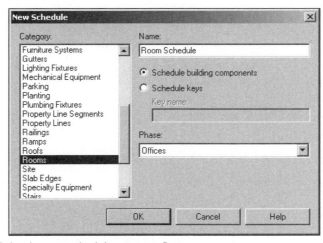

**Figure 8–1** *Make the new schedule category Rooms*

1.5 On the Fields tab of the Schedule Properties dialog, select Area from the Available fields list. Choose Add-> to move it to the Scheduled fields (in order) list.

1.6 Select Department, Name, and Number, and add them to the Scheduled fields, as shown in Figure 8–2.

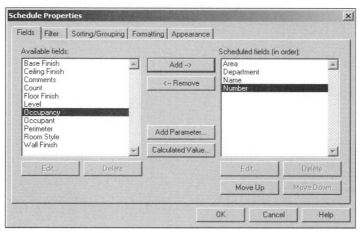

**Figure 8–2** *Move fields from Available to Scheduled*

1.7 Select Name in the Scheduled fields. Choose Move Up twice to put Name at the top of the list. Place Number in the second position (see Figure 8–3).

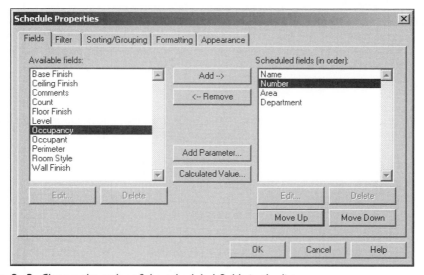

**Figure 8–3** *Change the order of the scheduled fields in the list*

1.8 On the Appearance tab, check Bold and Italic for the Header font, as shown in Figure 8–4. This change will appear when the schedule view is placed on a Sheet. Choose OK.

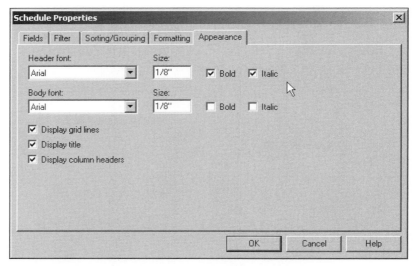

**Figure 8–4** *You can adjust the appearance of the schedule on this tab*

Revit creates and opens a view named Room Schedule under Schedules/ Quantities in the Project Browser and makes it the active view. Eighty rooms are listed.

1.9 Select New on the Options Bar in the Schedule view. Revit will place a room with Name **Room**, Number **81**, and Area Value **Not Tagged** in the last row of the schedule. This blank Room definition has not yet had an area tag placed inside a boundary or room.

1.10 Select New again to create a second un-tagged room in the schedule.

1.11 Open the view 2ND FLOOR OFFICES. On the Drafting tab of the Design Bar, select Room Separation. On the Options Bar, check Draw (the pencil icon) and Chain (see Figure 8–5).

**Figure 8–5** *The Room Separation tool and drawing options*

1.12 Sketch two lines to the left of the corridor door as shown in Figure 8–6. The exact dimensions are not critical.

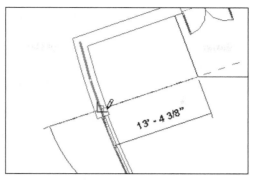

**Figure 8–6** *Sketch a room where there are no walls*

1.13   Sketch two more lines as shown in Figure 8–7. The exact dimensions are not critical. Select Modify. Move the Circulation area tag to the right of the new lines.

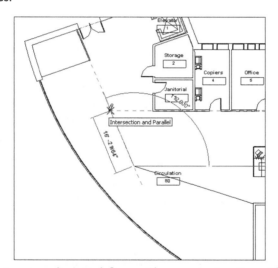

**Figure 8–7** *Sketch two more lines to define another room separation without walls*

1.14   Select Room Tag from the Drafting tab of the Design Bar. On the Options Bar in the Room field, choose the drop-down arrow to expose the list. The two new room definitions are available to use. Select 81 Room, as shown in Figure 8–8. Place the tag in the small rectangular area.

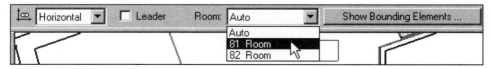

**Figure 8–8** *Choose the room tag values from the Options Bar list*

1.15 Select 82 Room from the Room list and place the tag in the larger boundary area.

1.16 Select Modify. Select the Room Tag for room 81. Select the room name field a second time to make it editable. Change the Name Value to **Reception**.

1.17 Select Room Tag 82. Change the Name Value to **Waiting Area**. Select away from the Tag to exit edit mode.

1.18 Select the following Room Tags and change their Department Values according to the following table:

| Room Tag Number | Department Values |
| --- | --- |
| 4–10 | Human Resources |
| 11–22 | Finances |
| 23, 24 | Student Affairs |
| 30–40 | Student Affairs |
| 41–55 | Finances |
| 56–63 | Human Resources |
| 64–72 | Fundraising |
| 73–79 | Dean's Office |

 **Tip:** Use Crossing or Inside selection windows with Filters as necessary. Once a new Value is typed into a field, Revit remembers the value and makes it available when that field is selected for editing later in another tag.

1.19 Open the Room Schedule view to see the new Department Values. Parameter values that are not driven by the model can be edited in the Schedule View. Select the Name cell for room number 82 and change it to **Lounge**, as shown in Figure 8–9. Your area value for room 82 may be different than shown.

| Conference | 76 | 563 SF | Dean's Office |
| Office | 77 | 333 SF | Dean's Office |
| Office | 78 | 254 SF | Dean's Office |
| Office | 79 | 470 SF | Dean's Office |
| Circulation | 80 | 8979 SF | |
| Reception | 81 | 115 SF | |
| Lounge | 82 | 1161 SF | |

**Figure 8–9** *Edit a Name value in the Schedule view*

## Exercise 2. Create a Schedule Key

2.1 Make the View tab of the Design Bar active. Select Schedule/Quantities.

2.2 In the New Schedule dialog, select Rooms in the Category field. Select the Schedule keys radio button. Accept the Schedule name and Key name default values as shown in Figure 8–10. Select OK.

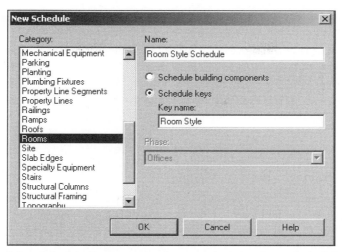

**Figure 8–10** *Select Schedule Keys to make the new schedule a Room Style Schedule*

2.3    In the Schedule Properties dialog, select Available fields Base Finish, Floor Finish, and Wall Finish, and add them in turn to the Scheduled fields. Choose OK.

2.4    Revit will create a Room Style Schedule view and make it active. Select New three times from the Options Bar to create three rows within the schedule.

2.5    Fill in the row cells as shown in Figure 8–11.

| Room Style Schedule | | | |
|---|---|---|---|
| Key Name | Base Finish | Floor Finish | Wall Finish |
| | | | |
| Services | Wood | Tile | Vinyl WC |
| Std Office | Vinyl | Carpet 1 | Paint |
| VP Office | Vinyl | Carpet 2 | Paint |

**Figure 8–11** *Values for the Room Style Schedule*

2.6    Open the view 2ND FLOOR OFFICES. Select the Room Tags in large Offices 77 and 78, the three Conference Rooms, and the Library. Select the Properties icon from the Options Bar. There will be a Room Style Parameter now available. Select VP Office from the drop-down list, as shown in Figure 8–12. Choose OK.

| Perimeter | |
|---|---|
| Room Style | VP Office |
| Comments | |
| Occupancy | |
| Department | Dean's Office |

**Figure 8–12** *Assigning a Room Style to multiple rooms using Properties*

2.7 Select Office 79 and all the Offices along the upper wall. Select Properties. Make the Room Style Parameter Std. Office. Choose OK.

2.8 Select the room tags for the following rooms: Vending 25, Toilets 26 and 28, Services 27, Storage 2, Janitor 3, and Copiers 4. Choose the Properties icon. Make the Room Style Parameter value Services. Choose OK.

2.9 Zoom to Fit. Make the Drafting tab of the Design Bar active. Select Color Fill. When you move the cursor over the Drawing Window, a Color Fill legend will be ghosted under the arrow.

2.10 Place the Color Fill legend to the left of the model. Revit will fill in room boundaries with a color fill pattern. Choose OK in the notice box that appears about visibility of Floors.

2.11 Select the Color Fill legend. Choose the Properties icon on the Options Bar. Choose Edit/New. Select Rename. Change the name from Scheme 1 to **Color by Department**. Choose OK three times.

2.12 Choose Edit Color Scheme from the Options Bar. In the Edit Color Scheme dialog, select the Color field for Finances. Select the Yellow swatch under Custom colors, as shown in Figure 8–13. Choose OK.

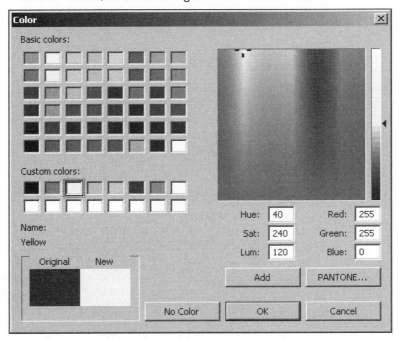

**Figure 8–13** *Selecting the Color for the scheme*

2.13   Select the Color field for Human Resources. Select the Red swatch under Custom colors. Choose OK. (See Figure 8–14 —your colors may vary.)

**Edit Color Scheme**

Color by:   | Department ▼ |

☑ Assign colors automatically

| Value | Color | Fill pattern | Preview |
|-------|-------|--------------|---------|
| Back of House | RGB 128-128-1 | Solid fill | |
| BOH | RGB 064-192-0 | Solid fill | |
| Circulation | RGB 064-192-1 | Solid fill | |
| Dean's Office | RGB 160-096-2 | Solid fill | |
| Finances | Yellow | Solid fill | |
| Fundraising | RGB 032-224-2 | Solid fill | |
| Human Resources | RGB 192-064-0 | Solid fill | |
| Office | RGB 192-064-1 | Solid fill | |
| Public Space | RGB 064-064-1 | Solid fill | |
| Retail | RGB 192-192-0 | Solid fill | |
| Student Affairs | RGB 224-224-2 | Solid fill | |

| OK | Cancel |

**Figure 8–14** *The Color shows in the Edit Color Scheme panel*

2.14   Choose OK to return to the View window. Hit ESC to terminate the edits (see Figure 8–15). Save the file.

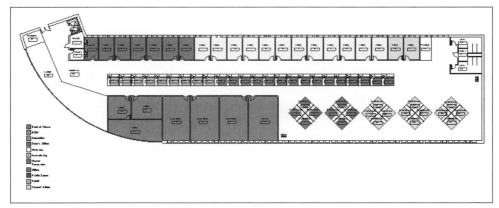

**Figure 8–15** *The finished Departmental Color Fill diagram*

## Exercise 3. Work with Color Fills

### Create a new Schedule Field

3.1 Select the Room Schedule view in the Project Browser. Right-click and select Properties.

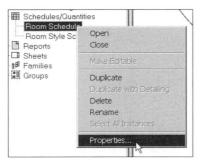

**Figure 8–16** *Edit the Room Schedule Properties*

3.2 In the Element Properties dialog, select Edit in the Fields value. Choose New Field (see Figure 8–17).

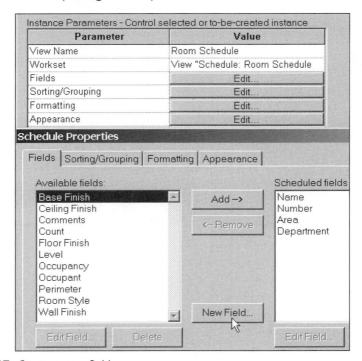

**Figure 8–17** *Create a new field*

3.3 In the Parameter Properties dialog, type **VLAN** in the Name field. Accept the defaults. Choose OK.

VLAN stands for Virtual Local Area Network. The new field will identify the computer network hookup nodes for offices and workstations.

3.4 Choose OK three times to exit the dialogs. Open the Room Schedule view.

3.5 Give the rooms the following VLAN values:

| Room Tag Number | VLAN Value |
| --- | --- |
| 4–15 | 1 |
| 16–24 | 2 |
| 29–49 | 3 |
| 50–63 | 4 |
| 64–73 | 1 |
| 74–77 | 4 |
| 78–82 | 1 |

Figure 8–18 shows a selection of the Schedule view with the VLAN values being applied.

| Name | Number | Area | Department | VLAN |
| --- | --- | --- | --- | --- |
| Janitorial | 3 | 80 SF | | |
| Copiers | 4 | 145 SF | Human Resources | 1 |
| Office | 5 | 165 SF | Human Resources | 1 |
| Office | 6 | 169 SF | Human Resources | 1 |
| Office | 7 | 169 SF | Human Resources | 1 |
| Office | 8 | 169 SF | Human Resources | 1 |
| Office | 9 | 169 SF | Human Resources | 1 |
| Office | 10 | 169 SF | Human Resources | 1 |
| Office | 11 | 169 SF | Finances | 1 |
| Office | 12 | 169 SF | Finances | 1 |
| Office | 13 | 169 SF | Finances | 1 |
| Office | 14 | 169 SF | Finances | 1 |
| Office | 15 | 169 SF | Finances | 1 |
| Office | 16 | 169 SF | Finances | 2 |
| Office | 17 | 169 SF | Finances | 2 |
| Office | 18 | 169 SF | Finances | 2 |

Room Schedule

**Figure 8–18** *Assign VLAN values to rooms*

## Create new Color Fills from the same information

3.6 Select the 2ND FLOOR OFFICE view in the Project Browser. Right-click and select Duplicate, as shown in Figure 8–19.

**Figure 8–19** *Duplicate the view*

 **Note:** When you duplicate a view, the annotations do not duplicate by default. Use Duplicate with Detailing to keep notes, etc. visible in the new view. In this case we don't need to see the Room Tags.

3.7 Revit will open the new view, named Copy of 2ND FLOOR OFFICES. Choose the new view name in the Project Browser, right-click, and select Properties, as shown in Figure 8–20. Change the View Name to **2ND FLOOR OFFICES BY DEPARTMENT**. Change the View Scale value to 1" = 20'-0" (see Figure 8–21). Choose OK.

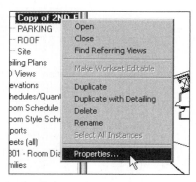

**Figure 8–20** *Edit the View properties*

| Parameter | Value |
| --- | --- |
| View Name | 2ND FLOOR OFFICES BY DEP |
| Title on Sheet | |
| Referencing Sheet | |
| Referencing Detail | |
| View Scale | 1" = 20'-0" |
| Scale Value   1: | 3/16" = 1'-0" |
| Crop Region | 1/8" = 1'-0" |
| Crop Region Visible | 3/32" = 1'-0" |
| Scope Box | 1/16" = 1'-0" |
| | 1" = 10'-0" |
| Display Model | 1" = 20'-0" |
| Visibility | Edit... |

Instance Parameters - Control selected or to-be-created instance

**Figure 8–21** *Change the name and scale*

3.8 Select the 2ND FLOOR OFFICES BY DEPARTMENT view in the Project Browser. Right-click and select Duplicate.

3.9 Revit will open the new view. Select the new view name in the Project Browser, right-click, and choose Properties. Change the view name to **2ND FLOOR OFFICES BY STYLE**. Choose OK.

3.10 Select the Color Fill Legend in the new view. Right-click and select Properties. Choose Edit/New, and then select Duplicate. Make the new Name value by **Color by Style**. Choose OK. Select Edit in the Color Scheme Value field. In the Edit Color Scheme dialog, change the Color by value to **Room Style**. Change the (none) Color to Basic Color White, as shown in Figure 8–22. Choose OK three times to exit the dialogs.

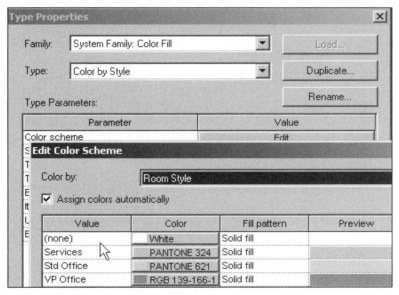

**Figure 8–22** *The colors for the Room Style color fill*

3.11 Select the 2ND FLOOR OFFICE BY STYLE view in the Project Browser. Right-click and select Duplicate.

3.12 Revit will open the new view. Select the new view name in the Project Browser, right-click, and Choose Rename. Change the view Name to **2ND FLOOR VLAN**.

3.13 Select the Color Fill Legend. Right-click and choose Properties. Choose Edit/New, and then choose Duplicate. Make the new Name value **VLAN**. Choose OK. Choose Edit in the Color Scheme Value field. In the Edit Color Scheme dialog, change the Color by value to **VLAN**. Choose OK three times to exit the dialogs.

## Put Color Fills and Schedules on a sheet

3.14 Expand Sheets in the Project Browser. Open view A801.

3.15 In the Project Browser, select view 2ND FLOOR OFFICES BY DEPART-MENT. Drag the view onto the sheet and place it in the upper-left corner.

3.16 Repeat for the views 2ND FLOOR OFFICES BY STYLE and 2ND Floor VLAN (see Figure 8–23).

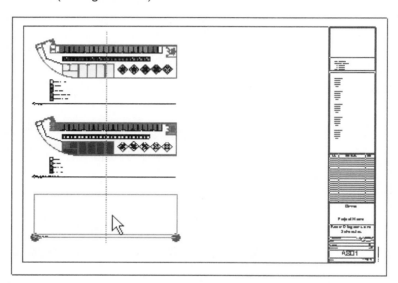

**Figure 8–23** *Drag the views onto the page—Revit will supply alignment planes*

3.17 Select and Drag the Room Schedule view onto the right half of the sheet. The schedule is a little too long to fit. Zoom in Region to the top rows of the schedule. Click the blue arrow-shaped grip that controls the width of the Department field and drag it to the right so that the row text fits on one line per row (see Figure 8–24).

 **Note:** The appearance edits you made to the Room Schedule header earlier appear now that it has been placed on a sheet.

| Room Schedule | | | | Drag to new position |
|---|---|---|---|---|
| Name | Number | Area | Department | |
| Circulation | 81 | 8954 SF | | |
| Elevator | 2 | 51 SF | | |
| Storage | 3 | 100 SF | | |
| Janitor | 4 | 80 SF | | |
| Copier | 5 | 143 SF | | 1 |
| Office | 6 | 187 SF | Human Resources | 1 |

**Figure 8–24** *Adjust the column width for better fit*

**Tip:** Revit provides a control point on schedule views to allow you to split a long schedule to fit on a sheet, as shown in Figure 8–25.

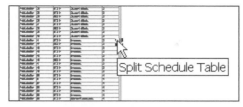

**Figure 8–25** *You can split a long schedule*

3.18 Drag the Room Style Schedule onto the sheet. Place it below the Room Schedule, as shown in Figure 8–26.

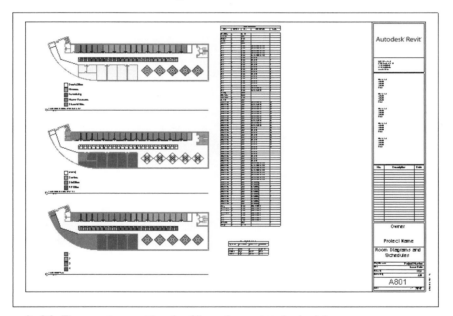

**Figure 8–26** *The new page with color fills and associated schedules*

3.19 Save the file.

You have just quickly created and displayed five views of the same data. Using the floor plan, room tags, and color fills, you have provided graphic information that will be valuable to firms working on the building—interior finish contractors and systems electricians, for example. This same information will also be useful to the facilities manager during the life of the building. The room and room style schedules provide other matrix views of the same information shown in the color fills.

## AREA PLANS AND COLOR FILLS

Revit's Area Plans are views that will appear in the Project Browser once created. Area plan views show spatial relationships of floor-plan levels based on Area Schemes. Area Schemes are definable. Revit creates two types of Area Schemes by default: Gross Building, which cannot be deleted or edited, and Rentable, which can be copied and edited to create additional Area Schemes. Area Boundaries and Area Tags are the building blocks of area plans.

### Exercise 4. Create a Gross Building Area Plan

4.1 Open or continue working with *Chapter 8 area plans.rvt*. Make the IST FLOOR view active.

4.2 Place the cursor over the Design Bar and right-click to bring up the list of available tabs. Select Area Analysis to make that tab available and active, as shown in Figure 8–27.

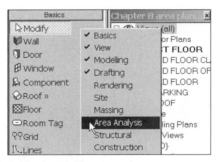

**Figure 8–27** *Open the Area Analysis tab*

4.3 From the Area Analysis tab of the Design Bar, select Area Settings, as shown in Figure 8–28. In the Area Calculations Setting dialog, choose New.

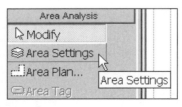

**Figure 8–28** *The Area Settings tool*

4.4 Select the Name Field for the new Area Scheme, and change the Value to **Classrooms and Labs**. Change the Description Value to **Measurements – Temporary Conditions**. We will use this new Area Scheme definition later (see Figure 8–29). Choose OK.

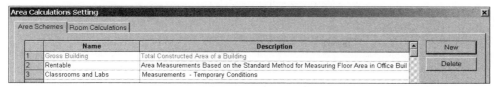

**Figure 8–29** *Name and Description for the new Area Scheme*

4.5 Select Area Plan from the Area Analysis tab of the Design Bar, as shown in Figure 8–30. In the New Area Plan dialog, select Gross Building for the Type and 1ST FLOOR for the level, as shown in Figure 8–31. Choose OK. Choose Yes to create area boundary lines from all external walls, as shown in Figure 8–32.

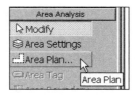

**Figure 8–30** *The Area Plan tool*

**Figure 8–31** *Settings for the first area plan*

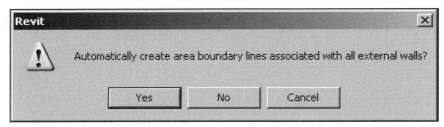

**Figure 8–32** *Revit will read and use the external walls for you*

4.6 Revit will create a new view named Area Plan (Gross Building): 1ST FLOOR, as shown in Figure 8–33, and make it active. Type **VG** to adjust the Visibility settings of the new view. On the Model Categories tab of the Visibility/Graphic Overrides dialog, uncheck Topography. Choose OK. Zoom to Fit.

**Figure 8–33** *The new area plan appears in the Project Browser*

4.7 From the Area Analysis tab of the Design Bar, choose Area Tag (see Figure 8–34). A tag will appear under the cursor. Zoom in Region to the left side of the model so you can read the tag text.

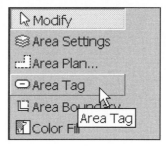

**Figure 8–34** *The Area Tag tool*

4.8 Hold the cursor outside the model. The second line of the tag will read Ambiguous—it is not within an area boundary. Move the cursor inside the exterior walls of the model. The tag will read an area value. This value will be the same whether the cursor is within a room or not, as shown in Figure 8–35.

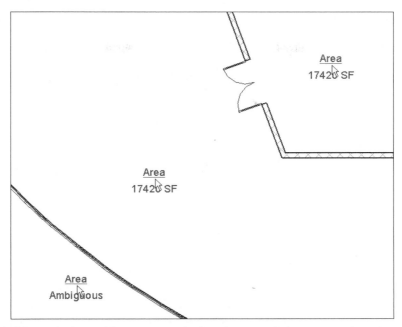

**Figure 8–35** *Inside the model, an area tag displays the area of whatever area boundary it is within*

4.9    Place the cursor inside the model near the curved curtain wall and left-click to place it. Choose Modify from the Area Analysis tab of the Design Bar. Select the new tag.

4.10    Select the Properties icon. Change the Name Value to **Gross Area 1st Floor**, as shown in Figure 8–36. Choose OK. Zoom to Fit.

 **Tip:** You can select the tag, and select the Name text a second time to edit it, as with a dimension. Revit will make fields that are not controlled by the model available for editing in tags and their associated schedules.

**Figure 8–36** *The edited tag name*

## Create a Rentable Area plan

The Gross Building Area scheme contains two simple Area Types: Gross Building Area and Exterior Area. The Rentable Area scheme contains the Exterior area type, plus five others, defined according to the following table:

| | |
|---|---|
| Office Area | Contains tenant personnel, furniture, or both |
| Store Area | Suitable for retail occupancy and use |
| Floor Area | Areas available primarily for use of tenants of the given floor: washrooms, janitorial spaces, utility or mechanical rooms, elevator lobbies, public corridors |
| Building Common Area | Lobbies, conference rooms, atriums, lounges, vending areas, food service facilities, security desks, concierge areas, mail rooms, health/daycare/locker/shower facilities |
| Major Vertical Penetration | Vertical ducts, pipe or heating shafts, elevators, stairs, and their enclosing walls |

4.11   Select Area Plan from the Area Analysis tab of the Design Bar. In the New Area Plan dialog, choose Rentable for the Type and 1ST FLOOR for the level, as shown in Figure 8–37. Choose OK. Answer Yes in the question box that appears, to create area boundary lines from all external walls.

**Figure 8–37** *Settings for a Rentable area plan*

4.12 Revit will create a new view named Area Plan (Rentable): 1ST FLOOR (as shown in Figure 8–38) and make it active. Type **VP** to adjust the Visibility settings of the new view. Change the Underlay to None. Choose Edit in the Visibility field. On the Model Categories tab of the Visibility/Graphic Overrides dialog, deselect Topography. On the Annotations Categories tab, deselect Elevations. Choose OK twice. Zoom to Fit.

**Figure 8–38** *The new area plan appears in the Project Browser*

4.13 From the Area Analysis tab of the Design Bar, select Area Tag. A tag will appear under the cursor. Place it in the same location as on the previous plan.

4.14 Select Area Boundary from the Area Analysis tab of the Design Bar (see Figure 8–39). On the Options Bar, leave the Pick option selected and Apply Area Rules checked. Zoom in Region to the left side of the model.

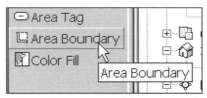

**Figure 8–39** *The Area Boundary tool, for subdividing gross areas*

4.15 Select the two walls of the elevator enclosure facing the interior of the building, as shown in Figure 8–40. Note that once you choose the second wall, the area value of the existing area tag changes.

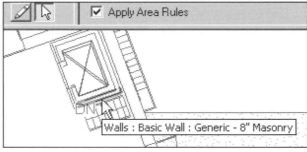

**Figure 8–40** *Select only two walls to define the elevator area*

4.16 Continue selecting all the room-defining interior walls (see Figure 8–41). Do not select walls inside the washroom areas.

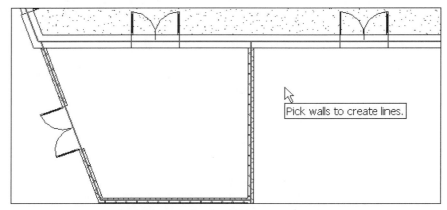

**Figure 8–41** *Select a wall and Revit will create the corresponding boundary line*

4.17 Select the Draw option on the Options Bar. Draw two lines, as shown in Figure 8–42 below, to create an area boundary not defined by walls.

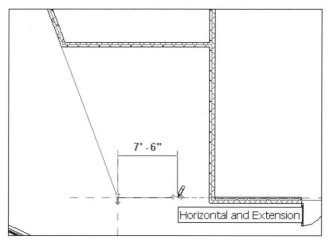

**Figure 8–42** *Switch to the Draw option to create boundaries where there are no walls*

4.18 Draw two sides of a boundary, as shown in the right side of Figure 8–43. Draw a line across the entrance to the washroom area.

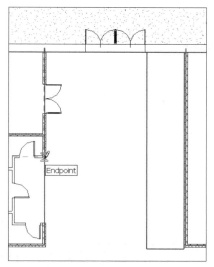

**Figure 8–43** *Boundaries in a wide hallway*

4.19 Draw two sides of a boundary, as shown in Figure 8–44. The actual dimensions are not important.

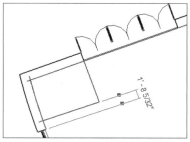

**Figure 8–44** *Another boundary near the double doors*

4.20 Draw two sides of a boundary at the exterior walkway, as shown in Figure 8–45.

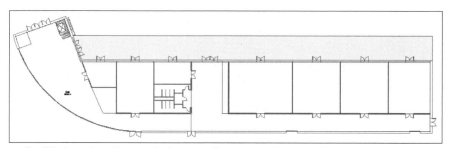

**Figure 8–45** *Boundary lines include the walkway outside the walls*

4.21 Select Area Tag from the Area Analysis tab of the Design Bar. Place area tags inside each room and boundary area you have created. Include the outside walkway (see Figure 8–46).

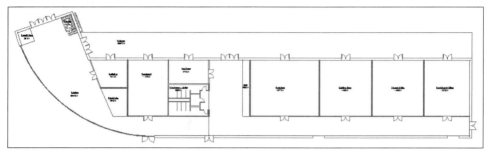

**Figure 8–46** *Area tags in place*

4.22 Edit each area tag Name Value in turn, based on the information contained in figures 8–47 through 8–49 below and the table shown in Step 4.24.

| Parameter | Value |
|---|---|
| Leader Line | ☐ |
| Orientation | Horizontal |
| Name | Elevator |
| Level | 1ST FLOOR |
| Area | 72.60 SF |
| Perimeter | 33' 3 193/256" |
| Comments | |
| Area Type | Major Vertical Penetration |

**Figure 8–47** *Using Properties to edit the area tag in the elevator*

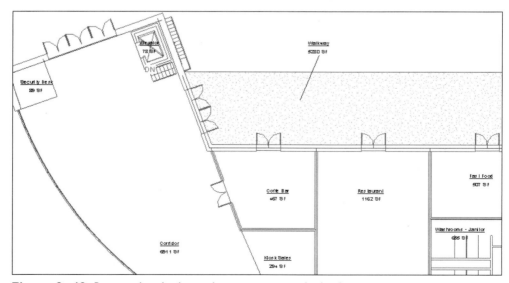

**Figure 8–48** *Renamed and relocated area tag—note the leader*

4.23 For the area tag in the Vending-Seating area, select Vertical orientation from the Options Bar (see Figure 8–49). For the tag inside the walkway, select the Leader option from the Options Bar, then drag the tag outside the walkway boundary (as shown in Figure 8–48), so that it will be sure to show when the color fill is applied in the next steps.

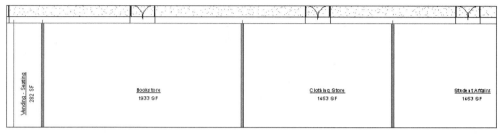

**Figure 8–49** *More area tags—note the vertical orientation at Vending*

4.24 Edit the Area Type values for the area tags according to the following table:

| | |
|---|---|
| Security Desk | Building Common Area |
| Elevator | Major Vertical Penetration |
| Corridor | Floor Area |
| Walkway | Exterior Area |
| Kiosk Sales | Store Area |
| Coffee Shop | Store Area |
| Restaurant | Store Area |
| Fast Food | Store Area |
| Washroom—Janitor | Floor Area |
| Vending—Seating | Building Common Area |
| Bookstore | Store Area |
| Clothing Store | Store Area |
| Student Affairs | Office Area |
| Bank Branch Office | Store Area |

4.25 Choose Color Fill from the Area Analysis tab of the Design Bar. A legend will appear under the cursor. Left-click to locate it to the left of the model, as shown in Figure 8–50.

322

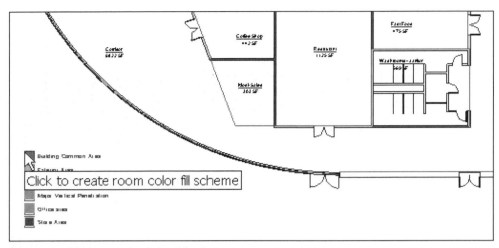

**Figure 8–50** *Locate the Color Fill scheme*

4.26 Revit will supply color fill to the tagged areas, as shown in figures 8–51 and 8–52. The colors on your system may vary. Choose OK in the notification box about floor visibility. Zoom to Fit.

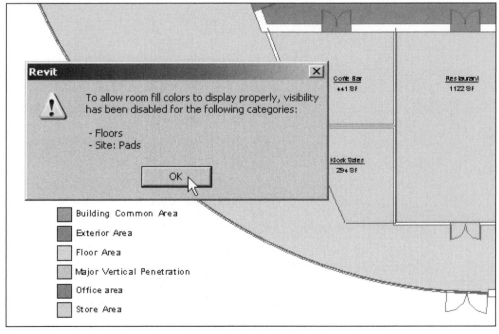

**Figure 8–51** *Revit lets you know it has adjusted visibility settings*

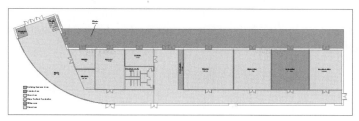

**Figure 8–52** *The Color Fill in place*

4.27   The Color scheme shown in Figure 8–53 is editable. We have edited a color scheme in a previous exercise.

**Figure 8–53** *A close view of the legend (Color Scheme) that controls the ColorFill*

## Exercise 5. Create a Custom Area Plan

5.1   Select Area Plan from the Area Analysis tab of the Design Bar. In the New Area Plan dialog, choose Classrooms and Labs as the Type, and 2ND FLOOR as the Area Plan view's level (see Figure 8–54). Choose OK.

**Figure 8–54** *Settings for the custom Area Plan*

5.2   Choose Yes in the query box about creating boundary lines from the external walls. Choose Unjoin Elements in the Error box that appears, as shown in Figure 8–55.

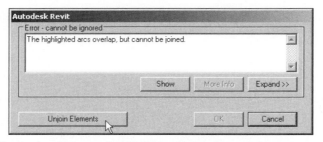

**Figure 8–55** *Unjoin the elements that Revit warns you about in this instance*

5.3   Revit will create a new view named Area Plans (Classrooms and Labs): 2ND FLOOR, as shown in Figure 8–56, and make it active.

**Figure 8–56** *The new Area Plan appears in the Project Browser*

5.4   Type **VP** to open the View Properties dialog. Set the Phase value to Temporary Classrooms. Set the Underlay to None. Choose OK.

5.5   On the Area Analysis tab of the Design Bar, select Area Boundary. Select walls as shown in figures 8–57 and 8–58. Do not select the walls of the washroom in the upper-right corner of the model or the elevator shaft walls. Do not select partition walls between similarly sized spaces.

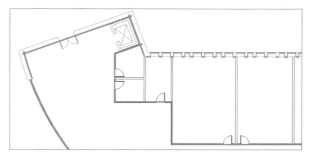

**Figure 8–57** *The first wall selections do not include the elevator shaft walls or walls inside comparable spaces*

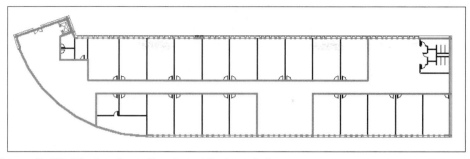

**Figure 8–58** *The interior walls selected for boundaries*

5.6   On the Area Analysis tab of the Design Bar, select Area Tag. Place a tag inside each major area whose walls you previously selected (four in total). See Figure 8–59 for one example. Select Modify to terminate tab placement.

5.7   Zoom in Region to the left side of the model. Choose the area tag shown in Figure 8–59. Choose the Properties icon. Make the Name Value **Permanent Utility Area**. Leave the Area Type Value as **Building Common Area**, as shown in Figure 8–60. Choose OK.

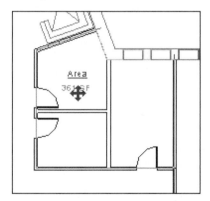

**Figure 8–59** *Edit the tag*

| Parameter | Value |
|---|---|
| Leader Line | ☐ |
| Orientation | Horizontal |
| Name | Permanent Utility Area |
| Level | 2ND FLOOR |
| Area | 352.31 SF |
| Perimeter | 75' 7 13/16" |
| Comments | |
| Area Type | Building Common Area |
| | |

**Figure 8–60** *Name and Area Type values for the area tag*

5.8　Select the area tag shown in the Figure 8–61 and edit its properties. Make the Name Value **Temporary Classroom**. Make the Area Type Value **Floor Area**. (See Figure 8–62.) Choose OK.

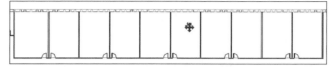

**Figure 8–61** *Edit the tag in the room areas on the north wall*

| Parameter | Value |
|---|---|
| Leader Line | ☐ |
| Orientation | Horizontal |
| Name | Temporary Classrooms |
| Level | 2ND FLOOR |
| Area | 6540.83 SF |
| Perimeter | 466' 9 255/256" |
| Comments | |
| Area Type | Floor Area |

**Figure 8–62** *Name and Area Type values for the Classroom tag*

5.9　Select the area tag shown in Figure 8–63 and edit its properties. Make the Name Value **Temporary Labs**. Make the Area Type Value **Store Area**; these spaces will be considered rentable during their existence (see Figure 8–64). Choose OK.

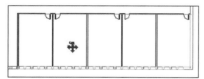

**Figure 8–63** *Edit this tag in the rooms at the southeast corner*

| Parameter | Value |
|---|---|
| Leader Line | ☐ |
| Orientation | Horizontal |
| Name | Temporary Labs |
| Level | 2ND FLOOR |
| Area | 3347.58 SF |
| Perimeter | 260' 1 1/2" |
| Comments | |
| Area Type | Store Area |

**Figure 8–64** *Name and Area Type values for the Labs*

5.10　Select the remaining area tag and edit its properties. Make the Name Value **Permanent Offices**. Make the Area Type Value **Office Area**. Choose OK.

5.11　Zoom to Fit. Choose Color Fill from the Area Analysis tab of the Design Bar. Locate the Color Fill Legend to the left of the model. Choose OK in the visibility notice.

5.12 Zoom in Region around the Color Fill Legend. Select it. Choose Edit Color Scheme from the Options Bar. In the Edit Color Scheme Dialog, change the Color by value to **Name**, as shown in Figure 8–65. Your colors may vary. Choose OK.

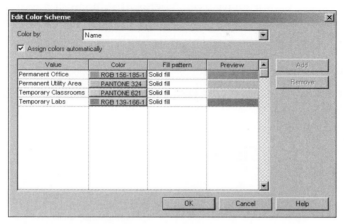

**Figure 8–65** *Color the new scheme by Name*

5.13 The Color Fill Legend will change to show the names you just assigned to the area types in this area plan.

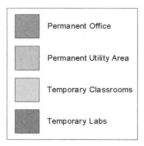

**Figure 8–66** *The legend now shows the assigned Name values*

5.14 Zoom to Fit (see Figure 8–67). Save the file.

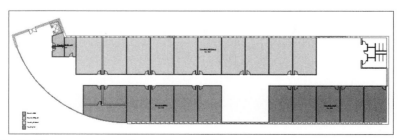

**Figure 8–67** *The custom Area Plan with color fill*

## Exercise 6. Exporting schedule information

Schedule information is valuable inside Revit, but even more useful to many firms when it's taken out to other applications. Revit will export data to any ODBC-compatible (Open Database Connectivity—the Microsoft standard for accessing data from a variety of sources) database management application. This requires setting up the ODBC interface by specifying drivers to use. For this exercise we will use the driver supplied by Revit to export to Microsoft Access. The process would be the same for any ODBC-compliant application.

6.1 From the File menu, choose File>Export>ODBC Database, as shown in Figure 8–68.

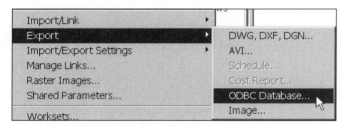

**Figure 8–68** *Export Database information to an external application*

6.2 The Select Data Source dialog opens. It contains two tabs and a file browser window.

ODBC requires Data Sources with Data Source Names (DSN) for connection. Data Sources can be located on the current workstation (Machine Data) or anywhere on the network (File Data). Machine Data Sources cannot be shared, and can be limited to individual users on any given machine. File Data Sources can be shared. Revit supplies a DSN file that appears on the File Data Source tab of the dialog (see Figure 8–69).

6.3 Select the file *revitdsn.dsn* that appears in the dialog window. This will place the file name in the DSN Name field. Choose OK.

**Note:** The ODBC export procedure for versions of Revit older than 6.1 is different than what is shown here.

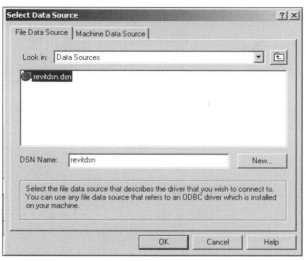

**Figure 8–69** *Create a new Data Source*

6.4   The ODBC Microsoft Access Setup dialog will appear, as shown in Figure 8–70. Leave the System Database option on None. Select Create.

6.5   In the New Database dialog, give the new database the name *chapter 8.mbd*, in a location as directed by your instructor. Accept the defaults in the dialog (see Figure 8–71). Choose OK.

The Locale list provides a list of available languages. The Format selection refers to the Jet engine version inside Access (to allow you to convert to older formats if necessary). System Database and Encryption options allow you to define the database as a System file, which then requires system Administrator status to manage, or to password-protect the file.

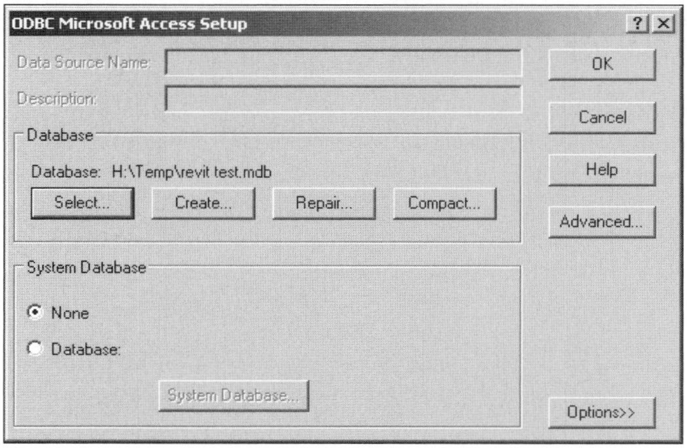

**Figure 8–70** *Next, create a database file to load*

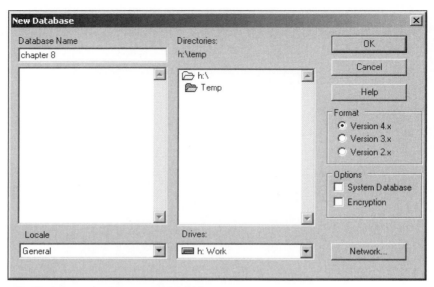

**Figure 8–71** *Locate the file in a useful folder*

    6.6    Revit will display a confirmation box, as shown in Figure 8–72. Choose OK twice to exit the setup.

**Figure 8–72** *Choose OK at the notification*

    6.7    If Microsoft Access is installed on your computer, launch that application and open the *chapter 8.mdb* file you just created (or use your File Manager/Computer Browser to find the new filename and double-click on it to open Access).

**Figure 8–73** *The list of tables in the new Access file*

6.8    Access will open to a window listing the tables created by Revit's export, as shown in Figure 8–73.

6.9    Select Rooms in the Table list and right-click. Choose Open. Revit's table assigns an ID to each room and shows the contents of each cell value as we saw in the schedule, except for Level and Room Style, which are identified by a numerical ID rather than a name (see Figure 8–74).

| | Comments | Perimeter | Level | Occupancy | Department | BaseFinish | CeilingFinish | WallFinish | FloorFinish | Area | Number | Name | Occupant |
|---|---|---|---|---|---|---|---|---|---|---|---|---|---|
| + | | 39.7451391714 | 9946 | | | Wood | | Vinyl WC | Tile | 101.445569247 | 2 | Storage | |
| + | | 36.0839980282 | 9946 | | | Wood | | Vinyl WC | Tile | 80.4199901411 | 3 | Janitorial | |
| + | | 49.5921661111 | 9946 | | Human Resources | Wood | | Vinyl WC | Tile | 144.922728271 | 4 | Copiers | |
| + | | 52.6842173936 | 9946 | | Human Resources | Vinyl | | Paint | Carpet 1 | 164.913659366 | 5 | Office | |
| + | | 53.173800727 | 9946 | | Human Resources | Vinyl | | Paint | Carpet 1 | 168.854148431 | 6 | Office | |
| + | | 53.173800727 | 9946 | | Human Resources | Vinyl | | Paint | Carpet 1 | 168.854148431 | 7 | Office | |
| + | | 53.173800727 | 9946 | | Human Resources | Vinyl | | Paint | Carpet 1 | 168.854148431 | 8 | Office | |
| + | | 53.173800727 | 9946 | | Human Resources | Vinyl | | Paint | Carpet 1 | 168.854148431 | 9 | Office | |

**Figure 8–74** *Revit's room schedule information in the Access database*

6.10    Revit exports the Room Style Schedule we created as a separate table that uses the Room Style key ID shown in the Rooms table (see Figure 8–75).

| | Id | KeyName | BaseFinish | FloorFinish | WallFinish |
|---|---|---|---|---|---|
| + | 75522 | Services | Wood | Tile | Vinyl WC |
| + | 75523 | VP Office | Vinyl | Carpet 2 | Paint |
| + | 75524 | Std Office | Vinyl | Carpet 1 | Paint |

**Figure 8–75** *The Room Style Schedule is translated as well*

6.11 Exit Access.

6.12 Revit does not maintain a live link to this external file, so design changes need to be re-exported. To update a previously created database with new design information, select File>Export>ODBC Database from the File menu, as before.

6.13 In the Select Data Source dialog, select the *revitdsn.dsn* file as before. Choose OK.

6.14 In the ODBC Microsoft Access Setup dialog, choose Select. In the Select Database dialog, select the name of the file you previously created. Choose OK twice to update that file (see Figure 8–76). At this point you could also select another database file to load the Revit information into, or create a new one.

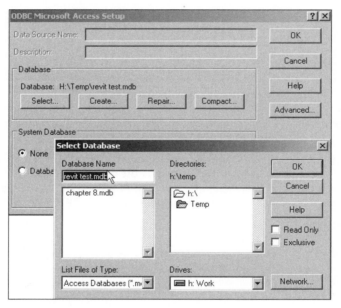

**Figure 8–76** *Choose OK to update the listed file, select another, or create a new one*

6.15 Save the Revit file, or close without saving as directed by your instructor.

## SUMMARY

In this chapter we used Revit's tools for cataloguing areas and rooms. Area Plans and Room Schedules provide quick graphic representations in color-fill diagrams, which are customizable and update automatically (as always in Revit) as the model changes. Revit will export component and schedule information to database applications.

## REVIEW QUESTIONS

### Multiple Choice

1. Room definitions can be created

    a) before any Room Schedules are created, not after

    b) after all Room Schedules are created, not before

    c) only while a Room Schedule is the active view

    d) any time

2. Schedules and Color Fills are created from Plan views, therefore you need to be aware of

    a) the Underlay of the original view

    b) the Detail level of the original view

    c) the Phase of the original view

    d) the Area of the original view

3. Area Plans are based on Area Schemes. Revit supplies these Area Schemes by default:

    a) Gross Building (not editable) and Rentable (editable)

    b) Custom (editable) and Standard (not editable)

    c) Exterior (not editable) and Interior (editable)

    d) all of the above

4. The Rentable Area Scheme contains the following Area Types:

    a) Exterior, Office

    b) Store, Floor

    c) Building Common, Major Vertical Penetration

    d) all of the above

5. Color Fill color schemes can be keyed by

    a) default schedule field parameters only

    b) text schedule field parameters only

    c) numeric schedule field parameters only

    d) any schedule field parameter

## True/False

6. You can select walls to make Area Boundary lines, but not Room Separation lines.

7. Area Plans, unlike Room Plans, can't be placed on Sheets.

8. Color Fills are not editable—once you place a fill, you have to delete it to change a color.

9. Revit will create Room definitions in Schedules without having Room Tags placed in Floor Plan views.

10. Revit can export schedule information into external database files in all recognized formats except Microsoft Access.

 Answers found on the accompanying CD.

# CHAPTER 9

# Illustrating the Design — Graphic Output from Revit

## INTRODUCTION

By now you have worked with a number of building models using Revit and have built pages for construction document sets. Revit's ability to produce informative views of building models is hardly restricted to the plans, elevations, and sections used for standard construction documents. Since all Revit models are built in three dimensions, with information about materials and surfaces as integral parts of the model objects, 3D-perspective views can be quickly and easily obtained. Revit contains a rendering module so that colored, lit (day or night, interior or exterior) pictures can be captured showing any part of the model at any point during design development. Revit incorporates a walkthrough tool that allows the user to create an animation based on a camera moving along a path. Users familiar with rendering software applications can export files—images, models, or animations—that other products can receive.

## OBJECTIVES

- Create and change perspective views, interior and exterior
- Apply and edit material textures to building objects
- Work with render settings—environment, lighting, location
- Create an exterior daylight rendering
- Capture a rendering image
- Export Revit views to external files
- Use the View Section Box to restrict the scope of a view and rendering
- Work with the radiosity tool for interior lighting
- Create an interior nighttime rendering
- Create and edit a walkthrough path/camera combination
- Export a walkthrough animation file

335

- Plot Revit views using the Revit PDF writer
- Prepare Revit views for plotting using standard printers
- Export the Revit model for use in an external renderer

## REVIT COMMANDS AND SKILLS

*Camera properties*

*New camera*

*View grip editing*

*View Properties Eye/Target editing*

*Export Image*

*Dynamic View Walkthrough controls*

*Region Raytrace*

*Raytrace exterior settings: environment, sun, lighting*

*Site components*

*Materials: create and assign*

*Section Box*

*Entourage components*

*Raytrace interior settings: limit model geometry*

*Light Groups, Daylights*

*Radiosity*

*Walkthrough: key frames, camera edit, path edit*

*Export walkthrough*

*Print/Plot*

*Export PDF*

*Export DWG for rendering*

## VIEWING THE MODEL

### Exercise 1. The Camera Tool

1.1 Launch Revit. Open the file *Chapter 9 start.rvt*. The file will open to show the 3D View: {3D} at the default southeast orientation.

1.2 Open Floor Plan: Level 1. Open the Visibility/Graphic Overrides dialog for the view. Uncheck Topography on the Model Categories tab, as shown in Figure 9–1. Choose OK.

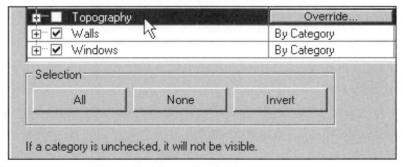

**Figure 9–1** *Turn off topography in this view for clarity*

1.3  Choose 3D View 1 in the Project Browser tree. Right-click and select Show Camera, as shown in Figure 9–2.

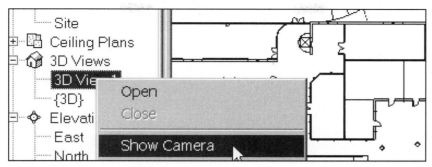

**Figure 9–2** *Show the view camera*

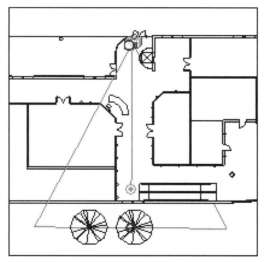

**Figure 9–3** *The camera displayed*

1.4  Place the cursor over the camera/ field of view symbol (see Figure 9–3). Right-click and select Properties. Study the properties of this view tool for a moment (see Figure 9–4). Choose Cancel to exit the dialog without making any changes.

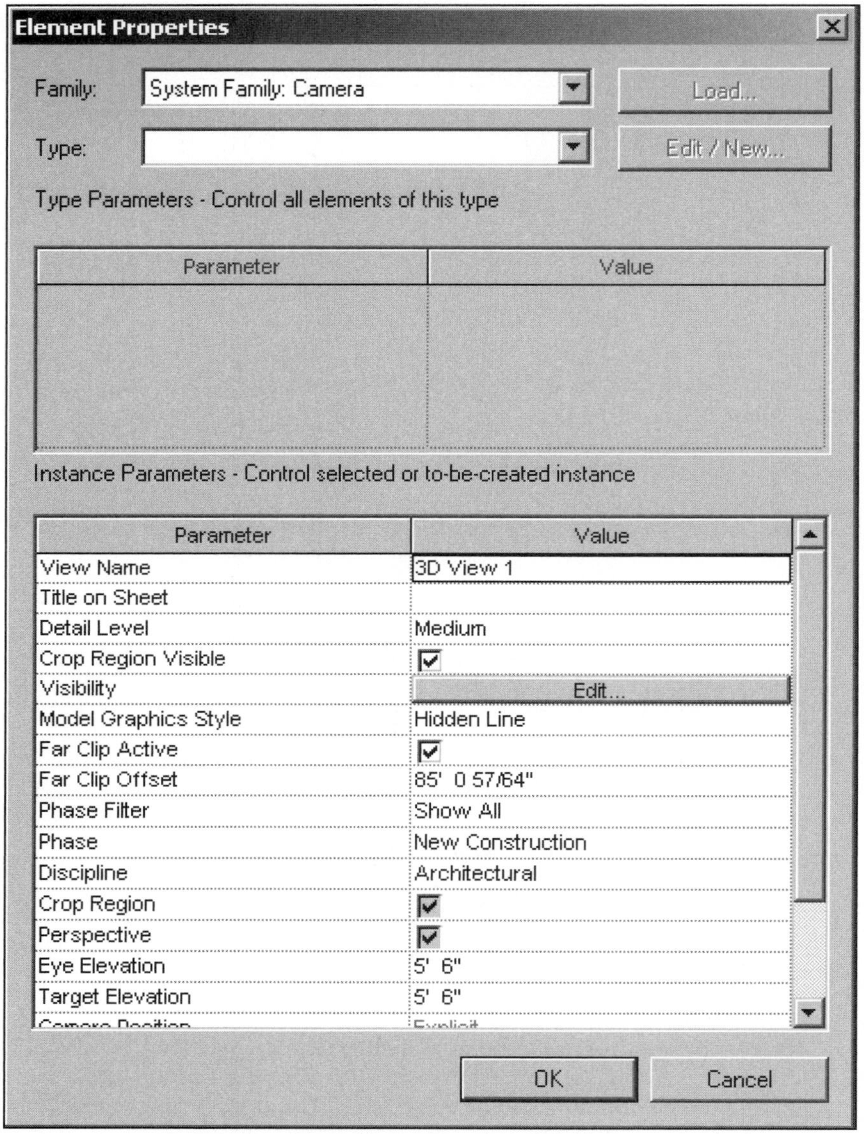

| Element Properties | | × |
|---|---|---|

Family: System Family: Camera ▼  Load...

Type: ▼  Edit / New...

Type Parameters - Control all elements of this type

| Parameter | Value |
|---|---|
| | |

Instance Parameters - Control selected or to-be-created instance

| Parameter | Value |
|---|---|
| View Name | 3D View 1 |
| Title on Sheet | |
| Detail Level | Medium |
| Crop Region Visible | ☑ |
| Visibility | Edit... |
| Model Graphics Style | Hidden Line |
| Far Clip Active | ☑ |
| Far Clip Offset | 85' 0 57/64" |
| Phase Filter | Show All |
| Phase | New Construction |
| Discipline | Architectural |
| Crop Region | ☑ |
| Perspective | ☑ |
| Eye Elevation | 5' 6" |
| Target Elevation | 5' 6" |
| Camera Position | Explicit |

OK    Cancel

**Figure 9–4** *Properties of the camera that controls 3D View 1*

1.5   3D View 1 in the Project Browser tree is pre-selected. Right-click and
       select Open. The view will open with its border highlighted red and blue
       control dots (grips) displayed, as shown in Figure 9–5. Selecting any of the
       grips will allow you to change the field of view.

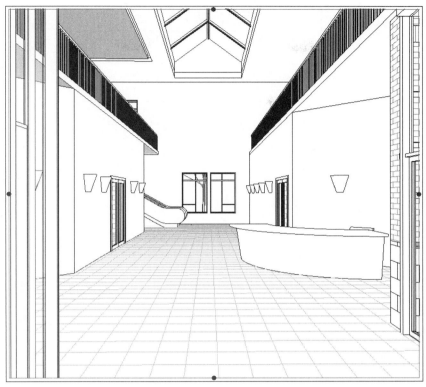

**Figure 9–5** *3D View 1—note the grips for adjusting the field of view*

1.6   Open the Level 1 Floor Plan. From the View menu, choose
      View>New>Camera, as shown in Figure 9–6. Left-click to place the camera
      eye position at the lower-right side of the model. Left-click inside the model
      near the curved desk in the main entry (see Figures 9–7 and 9–8).

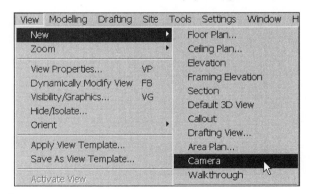

**Figure 9–6** *Create a new camera*

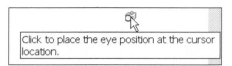

**Figure 9–7** *The first click locates the camera (eye position of the viewer)*

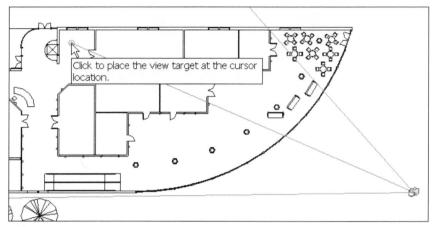

**Figure 9–8** *The second click aims the camera*

1.7   Revit will create and open a new view named 3D View 2, which will need some adjustments (see Figure 9–9).

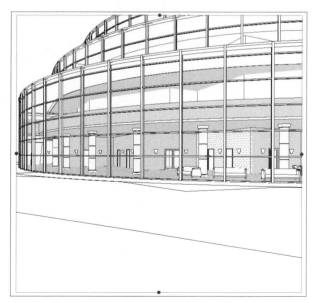

**Figure 9–9** *The new 3D view in its rough state*

1.8 Use the grips on the view to display the model completely, as shown in Figures 9–10 and 9–11. As you change the shape of the view, you may find it useful to Zoom Out (2X) to give yourself room to work, then Zoom to Fit when the adjustments are complete. The appearance of your view may be slightly different from the illustrations shown here, since your camera will not be in exactly the same place.

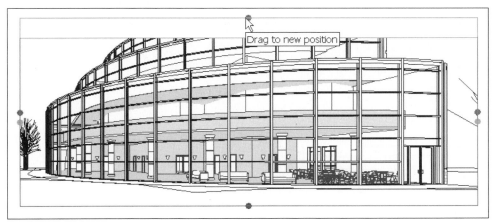

**Figure 9–10** *Drag the grips to alter the field of view*

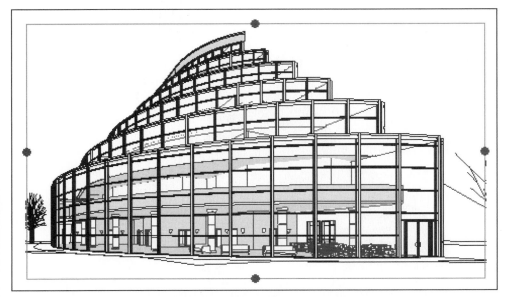

**Figure 9–11** *The view now shows the entire model from this angle*

1.9 With the View grips still active, open the Level 1 Floor Plan. Note the changed appearance of the camera field of view indicator, as shown in Figure 9–12.

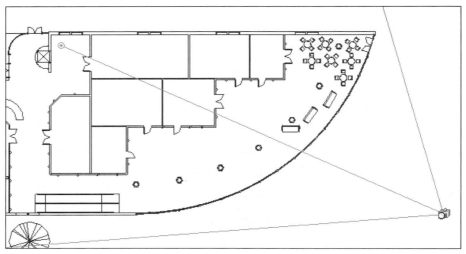

**Figure 9–12** *Note how the field of view has altered*

1.10 Return to 3D View 2. From the View menu, select View>Shading with Edges. Revit will take a second to process the display, shown in Figure 9–13.

 **Tip:** The two-stroke typed shortcut for the Shading with Edges display is SD. HL sets the display to Hidden Line.

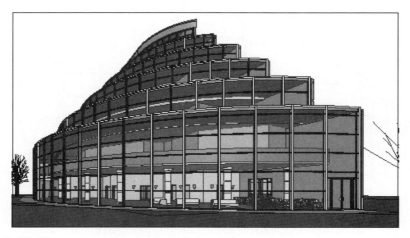

**Figure 9–13** *The perspective view, shaded*

1.11 From the File menu, select File>Export>Image, as shown in Figure 9–14.

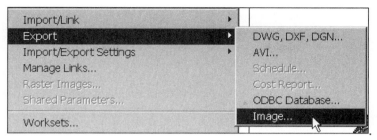

**Figure 9–14** *Choose Image from the Export menu*

1.12 In the Export Image dialog, give the file the name ***Chapter 9 east exterior*** and place it in a folder that you can find again easily. Select Visible portion of current window in the Export Range section. Leave the other settings. Note the output choices for file format, shown in Figure 9–15. Choose OK to create the file.

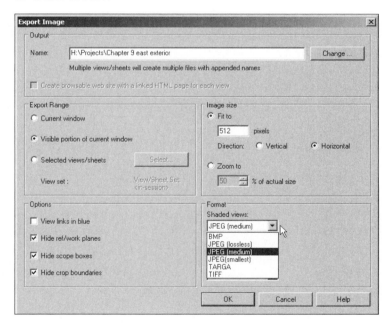

**Figure 9–15** *The Export Image dialog selections*

1.13 Return to Hidden Line mode in the 3D View. Right-click in the view window and select View Properties. In the Element Properties dialog, change the Target Elevation value to **25' 6"**, as shown in Figure 9–16. Choose OK. The angle of the view will change, as shown in Figure 9–17.

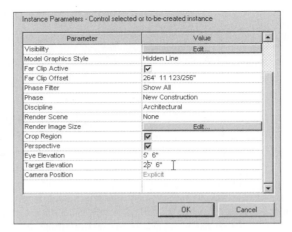

**Figure 9–16** *Change the camera target elevation*

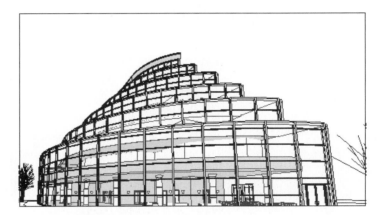

**Figure 9–17** *The results of lifting the target*

1.14   Open the View Properties dialog. Change the Target Elevation value back
to **5' 6"** and change the Eye Elevation value to **55' 6"** (see Figure 9–18).
Choose OK. The angle of the view will change as shown in Figure 9–19.

| Model Graphics Style | Hidden Line |
| Detail Level | Medium |
| Discipline | Architectural |
| Phase Filter | Show All |
| Phase | New Construction |
| Render Scene | None |
| Render Image Size | Edit... |
| Perspective | ☑ |
| Eye Elevation | 55' 6" |
| Target Elevation | 5' 6" |

**Figure 9–18** *Now change the height of the camera*

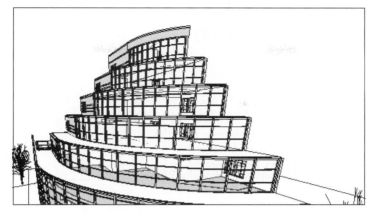

**Figure 9–19** *The results of lifting the camera and looking at the same target*

1.15   Select the Undo tool on the Toolbar to undo the last view edit. Choose the Dynamically Modify View tool on the Toolbar, as shown in Figure 9–20, to open the Dynamic View controls.

**Figure 9–20** *The Dynamic View tool*

1.16   On the Dynamic View dialog, select the Walkthrough tab. Make the Dolly control active, as shown in Figure 9–21. Note the instructions for the controls.

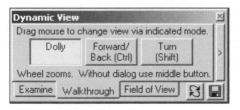

**Figure 9–21** *The Dolly control on the Walkthrough tab*

1.17   Hold down the left mouse button and Dolly (move laterally) the camera up/down and left/right to adjust the view. Moving the cursor up has the effect of moving the camera position down, and vice versa (see Figures 9–22 and 9–23).

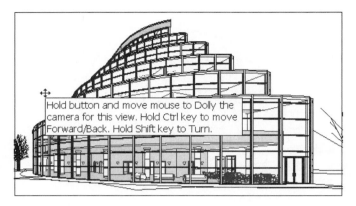

**Figure 9–22** *Preparing to Dolly the camera*

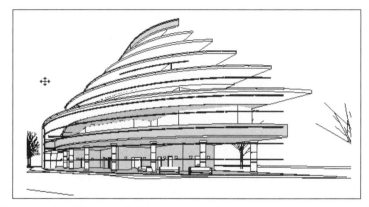

**Figure 9–23** *Revit simplifies the display during motion*

1.18 Hold down the CTRL key to make the Forward/Back control active and repeat—alter the camera position in and out (see Figures 9–24 and 9–25).

**Figure 9–24** *The Forward/Back cursor has a different appearance from the Dolly cursor*

**Figure 9–25** *Results of moving the camera Forward*

1.19 Make the Turn control (SHIFT) active and repeat (see Figures 9–26 and 9–27). Note that the camera holds position and turns right as the cursor moves left.

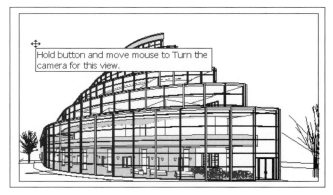

**Figure 9–26** *Preparing to Turn the camera*

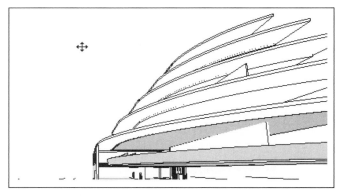

**Figure 9–27** *Turning the camera left*

1.20 Hold the cursor over the blue title bar of the Dynamic View control. Right-click and select X Close ALT F4 to close the dialog. If necessary, use View Properties to return the Target and Eye Elevation values to **5' 6"**, and readjust the field of view using the grips to view the entire model. Your camera may not be in its original position because of your left/right and forward/back adjustments.

## RENDERING

Now that you have experimented with camera placement and have changed a perspective view to show the model from the angle or angles that suit your purposes, you can work on making the model come alive with color rendering.

### Exercise 2. Exterior settings

2.1 Make the Rendering tab of the Design Bar active. Choose the Region Raytrace tool, as shown in Figure 9–28.

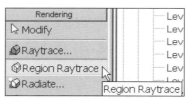

**Figure 9–28** *The Region Raytrace tool*

2.2 Select two points around the double door, as shown in Figures 9–29 and 9–30.

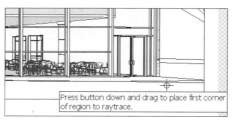

**Figure 9–29** *The first selection*

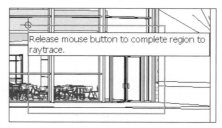

**Figure 9–30** *The second selection for the raytrace region*

2.3 Revit will display the Scene Selection dialog. Select Exterior, as shown in Figure 9–31. Choose OK.

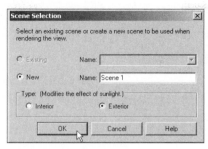

**Figure 9–31** *Select Exterior Scene, then choose OK*

2.4 In the query box that appears, choose Yes to turn off lights inside the building for the render process, as shown in Figure 9–32.

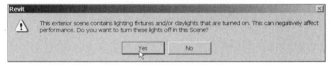

**Figure 9–32** *Turning off lights inside the building*

Revit will produce a rendered region bounded by your previous selections (see Figure 9–33).

**Figure 9–33** *The first rendered test area*

2.5 Choose Settings from the Rendering tab of the Design Bar, as shown in Figure 9–34. In the Render Scene Settings dialog, select Rename. In the Rename dialog, enter **Exterior – Gallery Side**, as shown in Figure 9–35. Choose OK.

**Figure 9–34** *The render Settings tool*

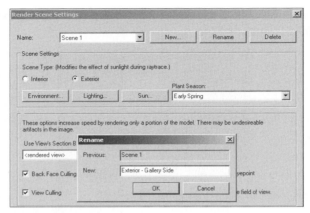

**Figure 9–35** *Rename Scene 1 here*

2.6   Select Environment. In the Environment dialog, accept the default Automatic Sky as the Background Color value. Select Clouds in the Advanced section, as shown in Figure 9–36.

**Figure 9–36** *Turn on Clouds*

2.7   Revit activates the Clouds tab, as shown in Figure 9–37. Choose OK to accept the defaults.

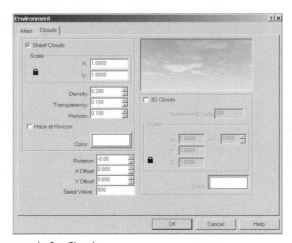

**Figure 9–37** *The controls for Clouds*

2.8 Choose Sun. Accept the defaults on the Date and Time tab. On the Place tab, use the Cities drop-down list to find Nashville, TN, USA, and select it, as shown in Figure 9–38.

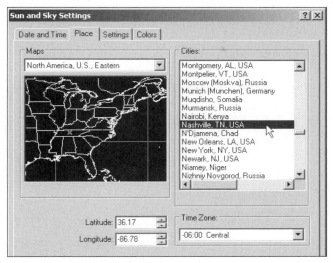

**Figure 9–38** *Assign a location on the Place tab*

2.9 Select the Settings tab and Colors tab in turn to view the controls, as shown in Figures 9–39 and 9–40. Accept the defaults. Choose OK.

 **Note:** The Settings tab allows you to save settings that govern cloudiness, sun and sky intensity, and the direction of solar north, or load in previously saved setting files.

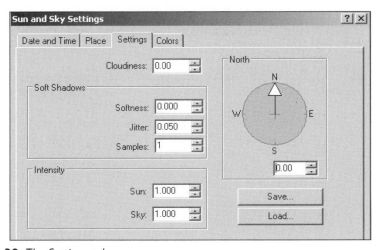

**Figure 9–39** *The Settings tab*

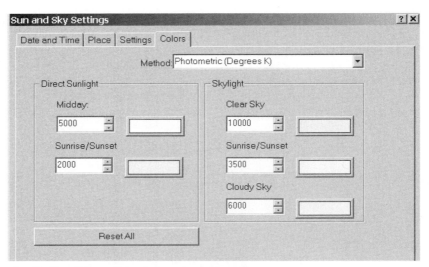

**Figure 9–40** *The Colors of the sky can be adjusted here*

    2.10    Change the Plant Season value on the main Render Scene Settings dialog page to Summer, to match the sun settings (see Figure 9–41). Choose OK to accept the remaining settings.

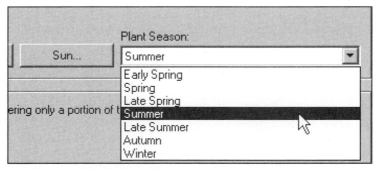

**Figure 9–41** *Match the foliage to the sun angle*

    2.11    Select Image Size on the Rendering tab of the Design Bar. Enter **300** as the Resolution dpi value, shown in Figure 9–42. Hit TAB to view the resulting image size. Choose Cancel.

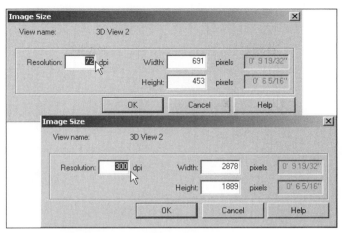

**Figure 9–42** *The Image Size can be controlled here*

2.12 Choose Adjust Image on the Rendering tab of the Design Bar. Note the available Image Controls: Brightness, Contrast, Indirect (ambient light intensity not coming from the sun or other point light sources), and a toggle for dynamic range (see Figure 9–43). Choose Cancel.

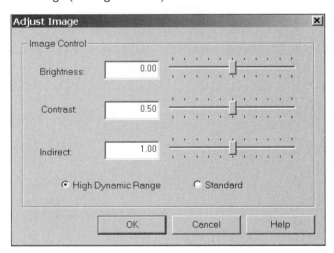

**Figure 9–43** *You can tweak overall picture settings here*

## Materials and plantings

2.13 From the Settings menu on the Menu Bar, choose Settings>Materials, as shown in Figure 9–44. Use the Name drop-down list to find Finishes – Exterior – Curtain Wall Mullions. There is no Texture assigned in the AccuRender area, as shown in Figure 9–45. Choose the select arrow to the right of the Texture field.

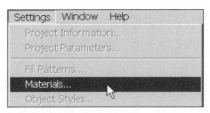

**Figure 9–44** *Choose Materials from the Settings menu*

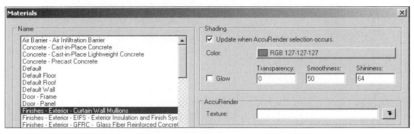

**Figure 9–45** *Select a Texture for the Mullion finish*

2.14 Once you're inside the Material Library dialog, navigate to the
_ACCURENDER\Metals\Aluminum, Anodized\ folder in the tree pane. Select
Bronze, Medium in the Name pane, as shown in Figure 9–46. Choose OK.

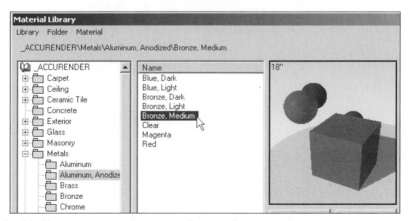

**Figure 9–46** *Navigate among the many selections for materials*

2.15 Use the Name drop-down list to find Glazing – Curtain Wall Glazing. Choose
the select arrow next to the Texture field to change the default value.

2.16 Back inside the extensive Material Library dialog, navigate to the
_ACCURENDER\Glass\Tinted\ folder in the tree pane. Select Bronze,
Medium, Smooth in the Name pane, as shown in Figure 9–47. Choose OK.
Choose OK again to exit the Settings dialog.

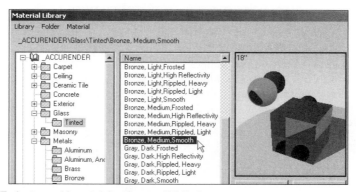

**Figure 9–47** *Assign a Material for Curtain Wall glazing*

2.17 On the Rendering tab of the Design Bar, select Display Model to remove the partial rendering. Select one of the curtain wall mullions, as shown in Figure 9–48. (You may have to use the TAB key to cycle through the possible selections—the Type Selector should read Rectangular Mullion: 2.5" x 5" rectangular.) Choose the Properties icon. Choose Edit/New. Change the Material parameter Value to Finishes – Exterior- Curtain Wall Mullions, as shown in Figure 9–49. Choose OK three times to exit the dialogs.

**Figure 9–48** *Select a curtain wall mullion*

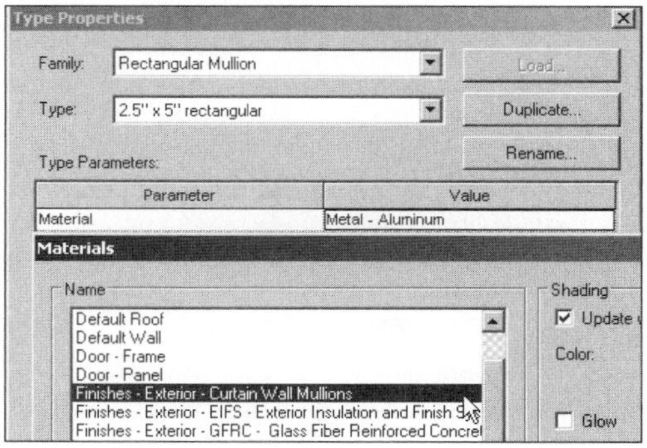

**Figure 9–49** *Assign the material via Properties*

2.18 Carefully choose one of the vertical 5" x 10" mullions. Repeat the Materials assignment to Finishes – Exterior – Curtain Wall Mullions. Choose OK three times to exit the Element Properties dialog.

2.19 Select the double door component. You may have to use TAB to cycle among the possible selections. Choose the Properties icon. Choose Edit/New. Change the Materials parameter Value to Finishes – Exterior – Curtain Wall Mullions, as shown in Figure 9–50. Choose OK three times to exit the Element Properties dialog.

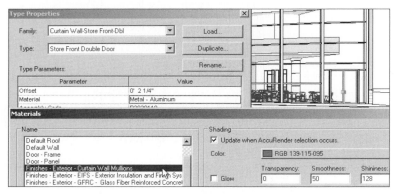

**Figure 9–50** *Assign the same material to the Double Door frame*

2.20 Carefully select a curtain wall panel above the door. You may have to TAB through the possible selections. The Type Selector will read System Panel when your selection is correct. Choose the Properties icon. Choose Edit/New. Change the Materials parameter Value to Glazing – Curtain Wall Glazing, as shown in Figure 9–51. Choose OK three times to exit the Element Properties dialog.

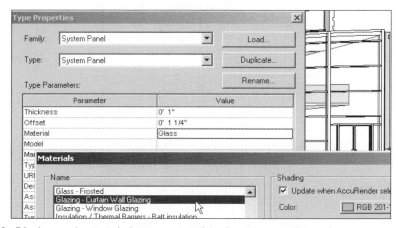

**Figure 9–51** *Assign the tinted glazing material to the curtain wall panels*

2.21   Choose Region Raytrace from the Rendering tab. Select two points around the double door as before. Note the difference in tints now that bronze finishes and glazing have been selected, as shown in Figure 9–52.

**Figure 9–52** *The Region Raytrace shows different colors after materials are changed*

2.22   Choose Display Model, shown in Figure 9–53, to remove the rendered image from the view.

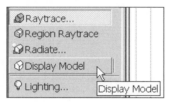

**Figure 9–53** *The Display Model tool*

2.23   Open the Floor Plan Level 1 view. Type **VG** to open the Visibility/Graphics Overrides dialog. Check Topography. Choose OK.

2.24   Make the Site tab active in the Design Bar. Choose Site Component, as shown in Figure 9–54. Select Shrub: Rhodo 24" from the Type Selector on the Options Bar, as shown in Figure 9–55. Select the Properties icon.

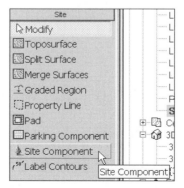

**Figure 9–54** *The Site Component tool on the Site tab*

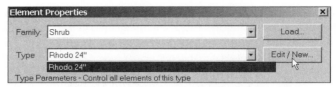

**Figure 9–55** *Make an adjustment to the currently loaded Shrubbery type*

2.25 Choose Edit/New. Choose Rename. Change the Type name to **Rhodo 16"**. Change the Plant Height Value to **16"** (Revit will convert it to 1' 4", as shown in Figure 9–56). Choose OK. Choose OK to exit the Element Properties dialog.

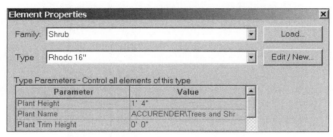

**Figure 9–56** *Values for the new Type*

2.26 Place a number of plants beside the concrete walkway, as shown in Figure 9–57. The exact number and placement are not important. Choose Modify to terminate the component placement.

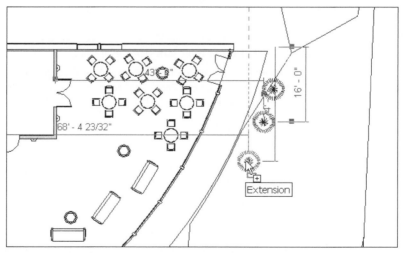

**Figure 9–57** *Place instances near the exterior walkway on the east side*

2.27 Open 3D View 2. Make the Rendering tab active in the Design Bar. Select Raytrace on the Rendering tab, then choose GO on the Options Bar.

Revit will generate a rendering using the sunlight, environment, and materials settings you have previously defined. This may take a few minutes, so be patient. See Figure 9–58 for an example.

**Figure 9–58** *The rendered image*

2.28   When the rendering is complete, select Capture Rendering on the Rendering tab of the Design Bar. Revit will add a view named 3D View 2 under Renderings in the Project Browser.

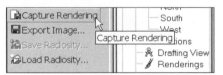

**Figure 9–59** *The Capture Image tool*

2.29   Select Export Image on the Rendering tab of the Design Bar. Place a copy of the image in a location you will be able to find later. Note that you can specify the name and file type of the image using this export routine (see Figure 9–60).

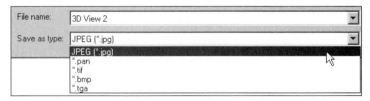

**Figure 9–60** *Export the new image to an external file*

   **Caution:** Image file export bugs have been reported when exporting *.tif* and *.tga* files as this text went to press. When saving to a file format other than *.jpg*, type the full name of the file, including the extension, in the File name field.

2.30   Save the file as *Chapter 9 render.rvt*.

## Exercise 3. Interior settings—section box crop

3.1　Open or continue working with file *Chapter 9 render.rvt*. Open the 3D View {3D}. From the View menu, select View>Shading.

We are going to be working with a view crop tool known as a Section Box, which requires the Shading mode to be visible.

3.2　Right-click and select View Properties, as shown in Figure 9–61.

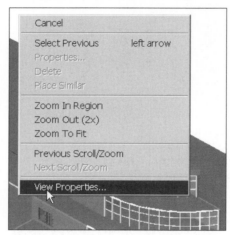

**Figure 9–61** *View Properties for the {3D} view*

3.3　In the View Properties dialog, check Section Box. Choose OK. The Section Box will appear around the model.

| Crop Region Visible | ☑ |
|---|---|
| Visibility | Edit... |
| Model Graphics Style | Shading |
| Far Clip Active | ☐ |
| Phase Filter | Show All |
| Phase | New Construction |
| Discipline | Architectural |
| Render Scene | None |
| Render Image Size | Edit... |
| Section Box | ☑ |
| Perspective | ☐ |
| Eye Elevation | 220' 2 51/64" |
| Target Elevation | 51' 6" |
| Camera Position | Adjusting |

**Figure 9–62** *Activate the Section Box for this view*

3.4   Choose the Section Box. It will highlight red, and blue control dots (grips) will appear. Drag the grips and adjust the boundaries of the Section Box to crop the model. Isolate the west (left) half of the top floor of the tower, as shown in Figures 9–63 and 9–64.

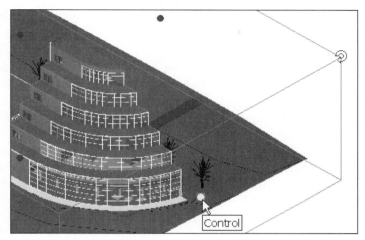

**Figure 9–63** *Choose the grip to drag it*

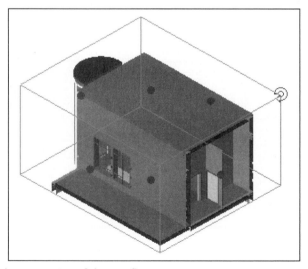

**Figure 9–64** *Isolate a portion of the top floor*

   **Caution:** This is an important step. If you do not correctly set and use the section box, the Rendering and Radiosity calculations in future steps will run extremely slowly, or not at all.

## Components – images of people

Architectural models can look sweeping and grand, or intimate and cozy, when viewed from the proper angle, but even the best spaces tend to appear sterile without human beings or their usual "stuff" in them. Many third-party developers have created quickly-rendered models of people, vehicles, and accessory items to place in rendered scenes to make them come alive for the viewer, and to provide a sense of scale.

3.5 Open the Floor Plan View Level 7. Zoom In Region around the left side of the model, a small lobby or waiting room. From the Rendering tab of the Design Bar, select Component. Choose Load from Library on the Options Bar.

3.6 Navigate to the *Imperial Library\Entourage* folder. Select *RPC Female.rfa*. Hold down the CTRL key and select *RPC Male.rfa*. Choose Open. This will load both families into the current project file.

3.7 Use the Type Selector on the Options Bar to make RPC Male: LaRon the current component. Check Rotate after placement on the Options Bar. Place the instance in the lower-left corner of the room, in front of the elevator doors. Rotate it 135° from the default position, which will point the front of the model to the upper-right of the plan.

RPC person components display in the plan as circles with a radius line representing the front. You can rotate the component after placement if necessary.

3.8 Use the Type Selector to place an instance of RPC Female: Cathy to the right of the desk as shown. Rotate the component −45° so it faces the other RPC person (see Figure 9–65).

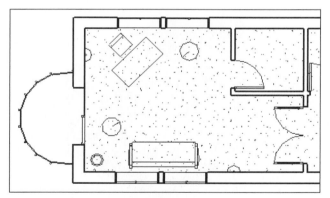

**Figure 9–65** *Placing the new components, facing each other*

3.9 From the View menu, select View>New>Camera. Place the camera to the right of the room, in front of the double doors. Place the target behind the desk, as shown in Figure 9–66. Relocate the Cathy component if necessary, so it will be in the camera view.

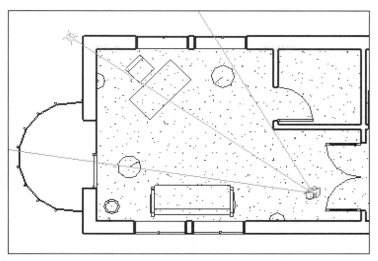

**Figure 9–66** *Place and point a camera in the room*

3.10   Revit will open the new camera view. There is a size control on the Options Bar. Choose the button, which will show the current image size (6" x 6"). Change the width value to **9"** and the height value to **5"**. Leave the Field of view radio button selected, as shown in Figure 9–67. Choose Apply and note the results. Choose OK. Use the field of view grips to adjust the view further if necessary, so that it does not display unnecessary amounts of floor or ceiling. See Figure 9–68 for an example.

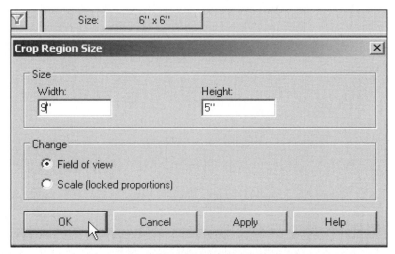

**Figure 9–67** *The view size control on the Options Bar and the Crop Region Size dialog*

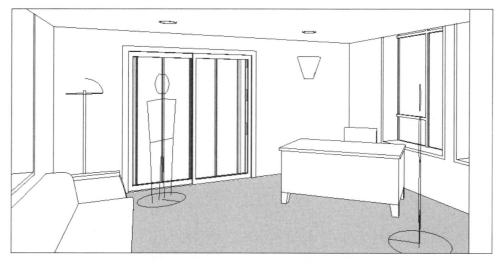

**Figure 9–68** *The field of view, adjusted—note the simplified appearance of the person components*

3.11 Choose Settings from the Render tab of the Design Bar. In the Scene Selection dialog, select New and Interior Scene. Enter the name **7th floor lobby**, as shown in Figure 9–69. Choose OK.

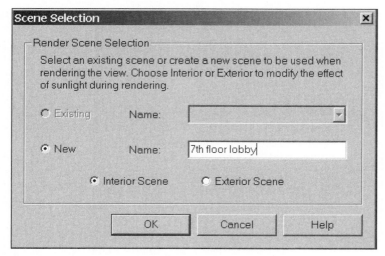

**Figure 9–69** *Name the new Scene*

3.12 The Render Scene Settings dialog will appear. Choose OK to close it for now.

## Lighting, Daylights, Radiosity

3.13   Select Lighting from the Rendering tab of the Design Bar, as shown in Figure 9–70. Note that a Light Group has been defined in this file—sconce lights on the first floor are all grayed out. Uncheck the On box for light group **1st Floor**, as shown in Figure 9–71. Choose OK.

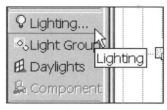

**Figure 9–70** *The Lighting tool on the Rendering tab*

| Mark / Name | Family type | On | Dimmer |
|---|---|---|---|
| Sun | Sun | ☑ | |
| 1st Floor | Light Group | ☐ | 1 |
| <EID: 58767> | Sconce : 100 watt Incandescen | ☑ | 1 |
| <EID: 58768> | Sconce : 100 watt Incandescen | ☑ | 1 |
| <EID: 58769> | Sconce : 100 watt Incandescen | ☑ | 1 |

**Figure 9–71** *Turn off the existing light group*

Light Groups function as component groups do in Revit, and are useful for managing groups of lights when you want to study or display alternate lighting conditions. We will not explore Light Groups in any detail in this exercise.

3.14   Choose Daylights from the Rendering tab of the Design Bar, as shown in Figure 9–72. There is no dialog associated with this tool. Select the two windows at the right side of the view. Select the double door to the elevator tower (see Figure 9–73). Choose Modify to terminate the selections.

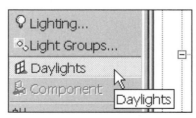

**Figure 9–72** *The Daylights tool*

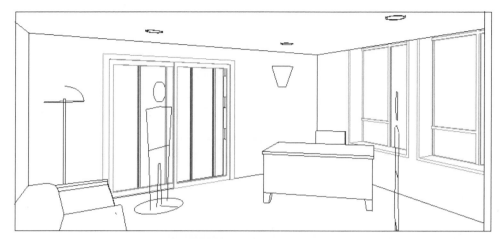

**Figure 9–73** *The windows and door selected to be Daylights*

3.15 Choose Settings from the Rendering tab of the Design Bar. Select Lighting in the Render Scene Settings dialog. Scroll to the end of the light list. Note that the new Daylights are now included. Choose Cancel to exit the Lighting dialog.

| | | | |
|---|---|---|---|
| 219 (daylight) | Fixed Multipane1 : 4'6 x 7' | ☑ | 1 |
| 220 (daylight) | Fixed Multipane1 : 4'6 x 7' | ☑ | 1 |
| 39 (daylight) | Sliding-2 panel : 72" x 84" | ☑ | 1 |
| <EID: 97293> | Sconce : 100 watt Incand | ☑ | 1 |
| <EID: 97310> | Sconce : 100 watt Incand | ☑ | 1 |
| <EID: 97699> | Floor Lamp - Hemispheric : | ☑ | 1 |
| <EID: 98192> | Downlight - Recessed Ca | ☑ | 1 |
| <EID: 98266> | Downlight - Recessed Ca | ☑ | 1 |
| <EID: 98369> | Downlight - Recessed Ca | ☑ | 1 |
| <EID: 98680> | Downlight - Recessed Ca | ☑ | 1 |
| <EID: 99007> | Downlight - Recessed Ca | ☑ | 1 |

[ OK ]   [ Cancel ]   [ Apply ]

**Figure 9–74** *The daylights now show in the lighting list, where they can be turned on and off*

3.16 Choose Environment. Select Clouds in the Advanced section on the Main tab of the Environment dialog. Choose OK.

3.17 Select Sun. Set the Place to Nashville, TN, USA, as before. Use the time slider to set the time to 8:15 pm, or just after sunset on the default day of June 30. Accept all the other defaults. Choose OK.

3.18 On the main page of the Render Scene Settings dialog, set the Use View's Section Box drop-down list to display **{3D}**. This applies the Section Box crop you performed earlier to this render setup in the exterior shaded view. Check Back Face Culling and View Culling to save some time in the rendering process (see Figure 9–75). Choose OK.

These options increase speed by rendering only a portion of the model. There may be undesireable artifacts in the image.

Use View's Section Box:

[{3D} ▼]   but NOT if rendered view has active Section Box.

☑ Back Face Culling     Ignore opaque surfaces facing away from the eyepoint

☑ View Culling          Ignore hidden objects and objects outside of the field of view.

Radiosity Settings

Quality:      Solution Goal:      Color Bleeding:

[Good ▼]   [25 ⬍] Steps    [0.25]

**Figure 9–75** *Limit the render carefully*

3.19 Select Radiate from the Rendering tab of the Design Bar, as shown in Figure 9–76. Revit will display a Radiosity Information panel, shown in Figure 9–77. Choose OK.

Raytrace...
Region Raytrace
Radiate...
Display M...
Radiate

**Figure 9–76** *The Radiate tool*

**Radiosity Information**                                                    ✕

Please read the following information about radiosity provided with AccuRender.

Radiosity is a powerful technique for calculating INDIRECT LIGHTING and performing LIGHTING ANALYSIS. When used properly, in conjunction with ray tracing, this technique can produce images which contain more subtle illumination effects than those produced using traditional ray tracing only.

Radiosity is generally most useful for SMALL ARCHITECTURAL INTERIORS. Use of this option WILL require more memory and processing time. These additional requirements will be sensitive to the size and complexity of the model.

Radiosity is usually NOT appropriate for the following situations:

* Non-architectural models
* Large or complex models
* Exterior renderings

☑ Don't show this message again.            [ OK ]      [ Cancel ]

**Figure 9–77** *The information box about Radiosity—note that it is not an all-purpose tool*

Revit will do a Radiosity calculation on the space and change the view gradually to a completely lighted rendering. This will take some time, so be patient. Note that the RPC persons will not display in the Radiosity (see Figure 9–78). They will display in the Raytrace.

**Figure 9–78** *The radiosity solution displayed graphically*

3.20 When the Radiosity is complete, choose Save Radiosity, as shown in Figure 9–79. Save the file in a convenient location. This creates a file with a *.rad* extension to hold the settings for this calculated Radiosity solution. This solution can be loaded later, to save creating another Radiosity solution for this view.

You can choose the Continue button on the Options Bar to cycle the Radiosity through more steps (25 is the default cycle). This will refine the calculations, but will not change the appearance of the radiated view.

**Figure 9–79** *The Save Radiosity tool*

3.21 Select Raytrace from the Rendering tab of the Design Bar. Set the Resolution field on the Options Bar to Medium (150 dpi). Note that the Image Size (grayed out to indicate that you cannot change it independently) changes as you change the Resolution value. Choose GO! from the Options Bar (see Figure 9–80).

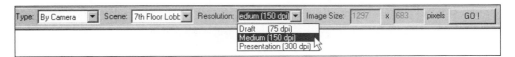

**Figure 9–80** *Options Bar controls for the Raytrace rendering*

3.22   If a warning appears, choose OK. This warning appears because you have saved the Radiosity, so Revit considers that it may now be out of date. Revit will take a few minutes to produce an image of the space (see Figure 9–81). Note that the window glass is both reflective and transparent—a few clouds are visible outside.

**Figure 9–81**  *The raytrace image with RPC people*

3.23   Capture or export the image as your instructor directs. Save the Revit file.

### Exercise 4. Create a Walkthrough

4.1   Open or continue working with the file *Chapter 9 render.rvt*. Open the Floor Plan Level 1. From the View menu, select View>New>Walkthrough, as shown in Figure 9–82.

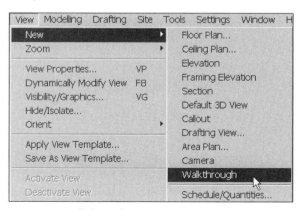

**Figure 9–82**  *Starting a new walkthrough*

4.2 The cursor will change to a pencil shape and the tooltip will display the message Click to place Walkthrough key frame. The Options Bar will display Walkthrough controls, as shown in Figure 9–83.

**Figure 9–83** *Walkthrough controls on the Options Bar*

4.3 Zoom in Region to the right side of the building. Place the first key frame inside the double door that opens north. Place key frames down the corridor past the curved desk, turn right down the hall past the escalator, and continue placing key frames between the columns and interior walls, moving towards the cafeteria seating (see Figures 9–84 and 9–85).

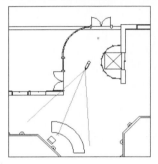

**Figure 9–84** *Start in the lobby*

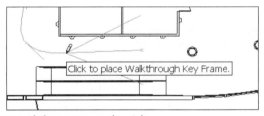

**Figure 9–85** *Move around the corner to the right*

4.4 Place 10–12 key frames in all. When the field of view indicator reaches the seating area, shown in Figure 9–86, choose Finish from the Options Bar.

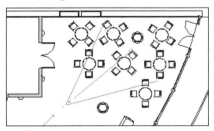

**Figure 9–86** *Finish in the seating area*

4.5 Select Edit Walkthrough from the Options Bar, as shown in Figure 9–87. A second set of controls appears on the Options Bar. Note that the Controls drop-down list, shown in Figure 9–88, lets you choose the type of walkthrough component you can edit.

**Figure 9–87** *Edit the walkthrough*

**Figure 9–88** *Walkthrough Editing controls*

4.6 Select the Frame settings button (it shows the number 300 as its caption). The Walkthrough Frames dialog opens, as shown in Figure 9–89. This allows you to set the number of frames and the distance between key frames if you do not want a uniform travel speed. Change the Total Frames value to **100**. Choose OK.

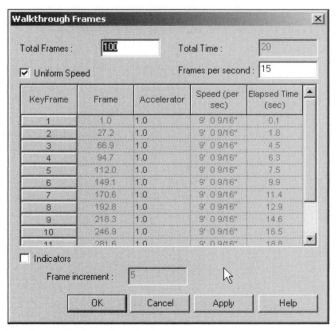

**Figure 9–89** *The Frame Settings dialog*

4.7 The camera is located at the last key frame. The Frame step controls on the Options Bar change the position of the camera along the path. You can step per Frame or key frame. Select the Previous key frame control one time. The Camera will move back one dot (key frame point) on the walkthrough path line. Select the camera Target Point control and pull the camera target to the right, pointed towards the couches, as shown in Figure 9–90.

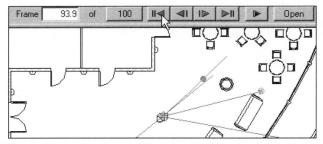

**Figure 9–90** *Change the camera direction by moving the target control point*

4.8 Step back two more key frames and point the camera target to the left, as shown in Figure 9–91.

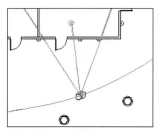

**Figure 9–91** *At this key frame, point the camera left*

4.9 Step back to a key frame in the middle of the entry corridor and point the camera target to the left of the path line, as shown in Figure 9–92.

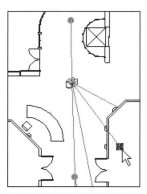

**Figure 9–92** *Point the camera left at this key frame*

4.10   Change the Controls value to Path and shift the grip on the 5th key frame from the beginning to the left of the original path line, slightly. Don't make too big a distortion of the path, or your animation will wobble.

If you accidentally select a point off of the walkthrough path, Revit will ask if you are finished editing the Walkthrough.

**Note:** The Reset Directions button at the right of the Options Bar will undo all the camera target edits, so that the same path can be set to different view directions and thus create different walkthroughs quickly.

4.11   With the Walkthrough highlighted, choose Open on the Options Bar. The current camera view will display. Its border will be highlighted and the grips available. You can edit the field of view—this setting will then hold for the entire walkthrough, not just the frame or key frame.

4.12   Use the key frame step controls to check the view at each key frame in your walkthrough. You can also type in the number of the frame you want to adjust (0 is the first).

If you need to change the camera location or direction for clarity, open the Level 1 Floor Plan and select Edit Walkthrough.

If the Walkthrough path is not visible in the Level 1 Floor Plan, expand the Walkthroughs section of the Project Browser, select Walkthrough 1, right-click, and choose Show Camera.

4.13   When you have finished editing the camera location and direction, choose Walkthrough 1 in the Project Browser, right-click, and choose Open. You must do this to make the next step possible.

## Export the walkthrough to an AVI file

4.14   From the File menu, select File>Export>AVI, as shown in Figure 9–93.

**Figure 9–93** *Export the AVI for this walkthrough*

4.15   Locate the new file in a suitable folder. Change the Frames per second value to **10**. This will make your walkthrough run for 10 seconds (see Figure 9–94).

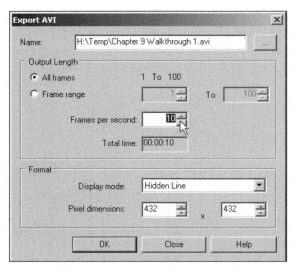

**Figure 9–94** *The Export AVI dialog*

Accept the Hidden Line display mode, but note that Wireframe, Shading, Shading with Edges, and AccuRender displays are available.

A rendered walkthrough in a complex model can take a very long time to create. Try experimenting with some settings and render your walkthrough overnight!

4.16 Choose Record in the Export AVI dialog. In the Video Compression dialog, choose a compression method, as shown in Figure 9–95.

Some compression routines are configurable. For our purposes, any one will do (see Figure 9–95). Using Full Frames (Uncompressed) creates a very big file. Choose OK. Revit will create the walkthrough frame by frame.

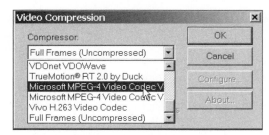

**Figure 9–95** *Select a compression method to keep file size down*

If the Export AVI dialog appears a second time, choose Close.

4.17 When the walkthrough has been created, find the new *.avi* file in your file manager and open it in your media player software application (see Figure 9–96). Play it through a few times; use the player's controls to move from view to view.

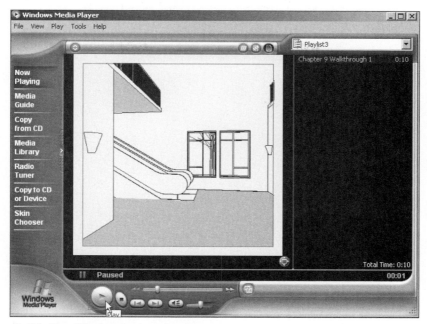

**Figure 9–96** *The AVI plays in a media player*

4.18   Save the Revit file.

## Exercise 5. Output formats

### Export Graphics via Plotting

Exciting graphics aside, Revit provides standard plotting and printing controls for putting images or sheets on paper. Revit includes a PDF writer so you can generate images or pages in that format, which is becoming a standard for exchanging documents.

5.1   Open or continue working in the file *Chapter 9 render.rvt*. Open the view {3D}. From the Window menu, select Window>Close Hidden Windows. Type **VP** to open the View Properties dialog.

5.2   Select the Print icon on the Toolbar, as shown in Figure 9–97.

5.3   Revit will open the Print dialog. Your list of available printers and output formats will look slightly different from what is shown in Figure 9–98. Check the Visible portion of current window radio button under Print Range. Choose Setup under Settings.

**Figure 9–97** *The Print icon*

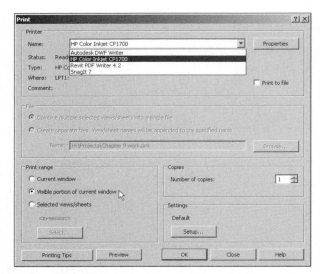

**Figure 9–98** *The Print dialog*

5.4 The Print Setup dialog opens. Note the available print settings, as shown in Figure 9–99. You can create different named setups for each printer and save them. Check the Fit to page radio button. Choose OK.

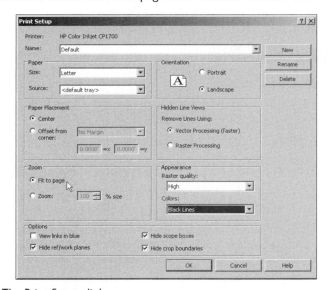

**Figure 9–99** *The Print Setup dialog*

5.5 Select Preview in the Print dialog. Revit will provide a print preview page of the active view based on the print settings (see Figure 9–100). Choose Close to exit the print dialog without generating a printed copy.

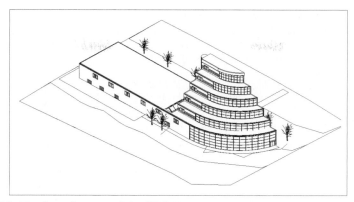

**Figure 9–100**  *The Print Preview of the {3D} View*

## Export via PDF

5.6    Revit ships with a module to create Adobe PDF output, which is also available as a free download. (This module is not automatically installed with Revit.) The following steps assume that you have installed Revit PDF Writer 4.2. Choose the Print icon on the Toolbar. Choose Revit PDF Writer 4.2 from the Name list. Select Properties. Note the two tabs available to establish settings in the PDF Writer 4.2 Document Properties dialog, as shown in Figure 9–101.

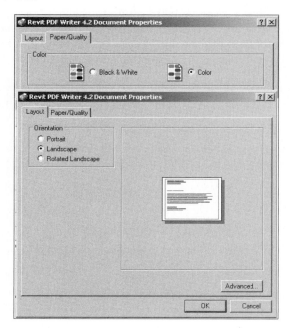

**Figure 9–101**  *The PDF Writer Properties dialog, with two tabs*

5.7 Select Advanced. Note the additional control settings for PDF output, as shown in Figure 9–102. Each of the underlined settings will open to an option drop-down list, if selected. Select the arrow for the Paper Size Letter (pti) setting to see the extensive list of available page sizes. Choose Cancel.

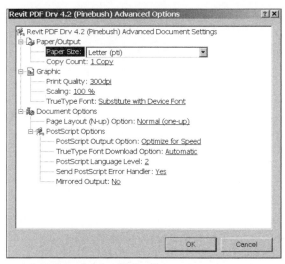

**Figure 9–102** *Advanced settings for the PDF writer*

5.8 Choose Cancel a second time to return to the Print dialog. In the File Name section, choose Browse. In the Specify a File dialog, give the output file a name and location you can find again. Choose Save. Choose OK to generate the PDF (see Figure 9–103). Revit will not display the new PDF.

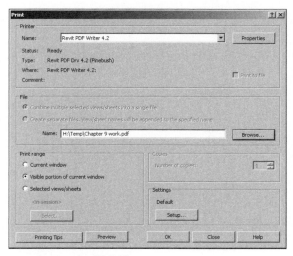

**Figure 9–103** *Prepare to print the PDF file*

5.9   Revit will generate a PDF file of the existing view, as you can see in Figure 9–104. Find the new file in your file manager and open it. If you are unable to open the file, Adobe Acrobat reader is a free download.

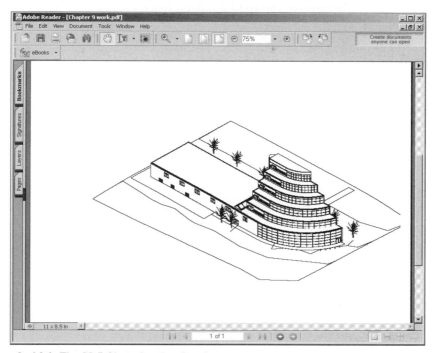

**Figure 9–104** *The PDF file in Acrobat Reader*

## Export via DWF

Autodesk is promoting the use of the DWF (Drawing Web Format) file output from its design products as a file format made for electronic sharing. DWF files are smaller than native (*.dwg, .rvt*) files, and can be embedded in Web pages for viewing with Autodesk's Express Viewer. The Express Viewer is a free download from Autodesk that, once installed, acts both as a standalone DWF viewer and a plug-in for Internet Explorer, so that DWF files can be viewed from a Web browser. Revit ships with a DWF output module on the product CD. This is also available as a free download. Like the PDF printer, it is not installed automatically with Revit. These following steps assume that you have installed the Autodesk DWF Writer.

5.10   Type **SD** to turn on the Shaded with Edges display in the current 3D view.

5.11   Select the print icon from the Toolbar. Choose **Autodesk DWF Writer** from the Name list. Choose Properties. Note the two tabs available to establish settings, as with the PDF Writer. On the Paper/Quality tab, select the Color option. Choose Advanced.

5.12 Explore the output settings in the Autodesk DWFWriter Advanced Options dialog. Set ICM Method to ICM Handled by Host System. Set Color Printing Mode to True Color (24bpp), as shown in Figure 9–105. Choose OK twice to return to the Print dialog.

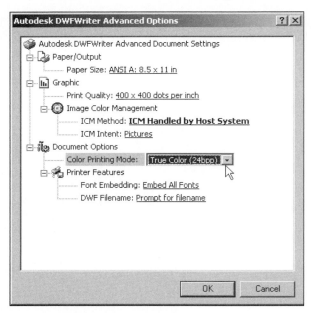

**Figure 9–105** *Advanced settings for DWF output*

5.13 In the Print range section, choose Selected views/sheets, and choose on Select, as shown in Figure 9–106.

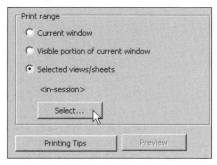

**Figure 9–106** *Print a selection*

5.14 In the View/Sheet Set dialog, check 3D View: 3D View 1, 3D View: 3D View 2, 3D View {3D}, and Floor Plan: Level 1. Choose New. Choose OK in the New name box to accept Set 1 as a named selection for DWF output, as shown in Figure 9–107. Choose OK to return to the Print dialog.

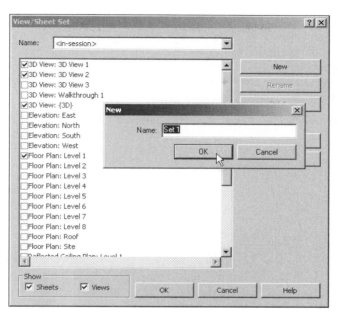

**Figure 9–107** *Check the views to show in the DWF file and supply a name to the selection*

5.15    In the File section of the Print dialog, select the option to Combine multiple selected views/sheets into a single file. Use the Name field and Browse button, as before, to locate the DWF file in a convenient place. Choose Save in the Specify prefix and extension dialog (see Figure 9–108). Choose OK to generate the DWF. It may take a couple of moments.

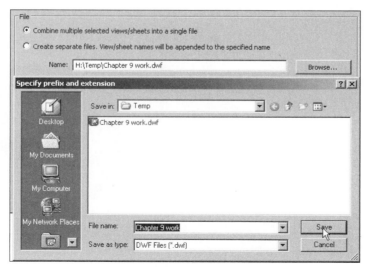

**Figure 9–108** *Specify a location for the output file*

5.16    Revit will not display the DWF. Launch the Autodesk Express Viewer, if it
        is installed on your machine, navigate to find the DWF file you just cre-
        ated, and open it (or navigate to the file location in your file browser and
        double-click on the file name). The Viewer will allow you to move from
        view to view, as you can move from page to page in a multi-page PDF file
        (see Figure 9–109).

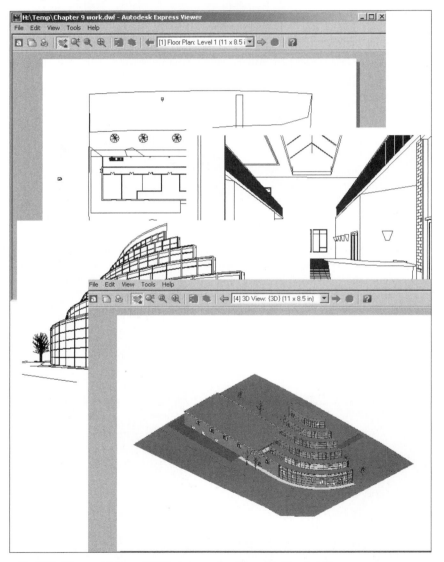

**Figure 9–109** *The multi-page DWF shows color from the Shaded View*

**Note:** Revit DWF output is designed for the Autodesk Express Viewer. Autodesk sells a DWF viewer with markup capability, called Volo View, to allow people without Autodesk CAD or modeling products to view and comment electronically on design files. DWF files from Revit can appear different in Volo View than Express Viewer.

## Export the model to an external renderer

For users who prefer to work with an external rendering application such as Autodesk VIZ or Architectural Desktop's VIZ Render, Revit's export to DWG capabilities will create files which can be imported into VIZ. A 2D Revit view (floor or ceiling plan, section elevation) will create a 2D DWG export. Make a 3D view current in Revit to generate a 3D DWG file.

To best receive DWG files into VIZ, all the model entities should be designated by either layer or color so that materials can be assigned. This preparatory work is done in Revit's Export Layer Settings.

5.17    Make the view {3D} current. From the File menu, select File>Import/ Export Settings>Export Layers DWG/DXF.

5.18    The Export Layers dialog will open. It uses a *.txt* file as its source. Revit has a number of default files to use as a basis for your own custom settings. The file shown in Figure 9–110 is *exportlayers-dwg-default.txt*. Note that no color IDs have been assigned. Choose Load.

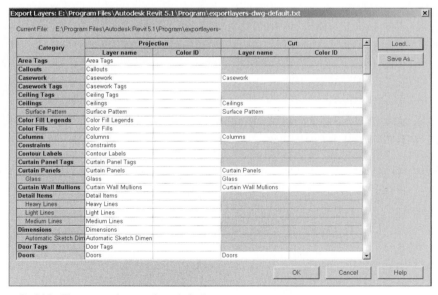

**Figure 9–110** *The exportlayers-dwg-default.txt export settings*

5.19    Navigate to the *Program Files/Revit 6.1/Data* folder. Choose the file *export-layers-dwg-AIA.txt*, as shown in Figure 9–111. Choose OK. Layer names and Color ID values in the Export Layers will change, as shown in Figure 9–112.

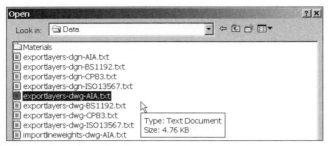

**Figure 9–111** *The exportlayers-DWG-AIA.txt export file location*

**Export Layers: E:\Program Files\Autodesk Revit 5.1\Data\exportlayers-dwg-AIA.txt**

Current File: E:\Program Files\Autodesk Revit 5.1\Data\exportlayers-dwg

| Category | Projection | | Cut | | |
|---|---|---|---|---|---|
| | Layer name | Color ID | Layer name | Color ID | |
| **Doors** | A-DOOR-SYMB | 1 | A-DOOR-SYMB | 1 | |
| Elevation Swing | {A-DOOR-SYMB} | 1 | {A-DOOR-SYMB} | 1 | |
| Frame/Mullion | A-DOOR-SYMB | 1 | A-DOOR-SYMB | 1 | |
| Glass | A-DOOR-SYMB | 1 | A-DOOR-SYMB | 1 | |
| Opening | A-ELEV-OTLN | 6 | A-ELEV-OTLN | 6 | |
| Panel | A-DOOR-SYMB | 1 | A-DOOR-SYMB | 1 | |
| Plan Swing | {A-DOOR-SYMB} | 1 | {A-DOOR-SYMB} | 1 | |
| **Electrical Equipment** | E-AUXL-EQPM | 4 | | | |
| **Electrical Equipment T** | E-ANNO-TEXT | 2 | | | |
| **Electrical Fixture Tags** | E-ANNO-TEXT | 2 | | | |
| **Electrical Fixtures** | E-DETL-LTFX | 2 | | | |
| **Elevations** | A-ANNO-SYMB | 6 | | | |
| **Entourage** | A-DETL-GENF | 3 | | | |
| **Existing** | -EXST | | -EXST | | |
| **Floors** | A-FLOR-OTLN | 6 | A-FLOR-OTLN | 6 | |
| Slab Edges | A-FLOR-OTLN | 6 | A-FLOR-OTLN | 6 | |
| Surface Pattern | A-FLOR-PATT | 8 | A-FLOR-PATT | 8 | |
| **Furniture** | I-DETL-FURN | 2 | | | |
| **Furniture System Tags** | I-SYST-IDEN | 1 | | | |
| **Furniture Systems** | I-SYST-FURN | 2 | | | |
| **Furniture Tags** | I-FURN-IDEN | 3 | | | |
| **Generic Annotations** | A-ANNO-NOTE | 2 | | | |
| Centerline | {A-ANNO-NOTE} | 2 | | | |
| Medium Lines | {A-ANNO-NOTE} | 2 | | | |

Load...  Save As...

OK  Cancel  Help

**Figure 9–112** *The exportlayers-DWG-AIA.txt export file settings*

5.20 Note that this file contains layer names based on the AIA CAD standards and color IDs. For export into VIZ, layers and Color IDs should be unique for each object that will receive a different material.

5.21 Enter new Door component Layer name and Color ID properties under Projection and Cut, based on the information shown in Figure 9–113.

| Export Layers: F:\Program Files\Autodesk Revit 6.0\Data\exportlayers-dwg-AIA.txt | | | | |
|---|---|---|---|---|
| Current: | F:\Program Files\Autodesk Revit 6.0\Data\exportlayers-dwg-AIA.txt | | | |
| Category | Projection | | Cut | |
| | Layer name | Color ID | Layer name | Color ID |
| **Doors** | A-DOOR | 1 | A-DOOR | 1 |
| Elevation Swing | {A-DOOR} | 1 | {A-DOOR} | 1 |
| Frame/Mullion | A-DOOR-FRAM | 12 | A-DOOR-FRAM | 12 |
| Glass | A-DOOR-GLAZ | 13 | A-DOOR-GLAZ | 13 |
| Opening | A-DOOR-OTLN | 6 | A-DOOR-OTLN | 6 |
| Panel | A-DOOR-panl | 14 | **A-DOOR-panl** | 14 |
| Plan Swing | {A-DOOR} | 1 | {A-DOOR} | 1 |

**Figure 9–113** *Change object layers and colors to provide unique identifiers*

5.22   Choose Save As to save your newly edited file for repeated use later. Give it a name to distinguish it from the default file you modified. The new file and any others you create can be accessed via the Load button.

5.23   Choose Cancel to exit the Export Layers dialog. To export a file as DWG, select File>Export>DWG, DXF, DGN.

5.24   The Export dialog allows you to select a location and name for the new file. You can also choose which Export Layers File to apply (see Figure 9–114).

**Figure 9–114** *Choose a name, location, and layer setting file for your exported DWG*

Note that Revit adds the name of the current view to the end of the file name. This is important to remember for file exchange purposes. As a Revit model develops you may export it numerous times. Be sure to keep the many files you can create from one model organized!

5.25    If you have AutoCAD, Architectural Desktop, or VIZ on your system, Save the {3D} view to a DWG file and open it in one of those applications. Otherwise, choose Cancel to exit the dialog.

5.26    Close the Revit file.

## SUMMARY

The lessons in this chapter have shown you how to make use of Revit's many capabilities for viewing your design within Revit itself, and for making your standard or enhanced views available as images, printed pages, or raw material for external graphics applications.

As with all the exercises in this book, you have seen only a few of the many possible alternative settings and combinations available. Particularly with rendered images, time spent working on the details and nuances of small portions of a building model can make a big difference in how people will see and understand the design intent behind the model. Practice with the render and walkthrough routines in Revit—you may surprise yourself and delight your clients!

# REVIEW QUESTIONS

## Multiple Choice

1. The Edit Walkthrough control allows you to

    a) change the Camera orientation at any frame point on the walkthrough path

    b) change the path location at any frame

    c) add or remove key frames at any point on the path

    d) all of the above

2. To make building components appear as desired in a rendering, you

    a) apply a Texture definition to a Material

    b) apply a Material definition to the building components

    c) a and b in that order only, or the definitions won't apply correctly

    d) a and b in any order

3. A Section Box is applied to a view to

    a) define custom spaces

    b) provide a restricted volume for rendering calculations

    c) make a link between floor plans and section views

    d) all of the above

4. You can change properties of a Camera

    a) using its Properties

    b) using View Properties of the Camera View

    c) using the Dynamic View control in the Camera View

    d) all of the above

5. To prepare a Revit file for export to rendering software, use the following menu selection:

    a) Settings>Materials

    b) View>New>Camera

    c) File>Import/Export Settings>Export Layers DWG/DXF (or Export Layers DGN)

    d) Render>Raytrace

## True/False

6. You must adjust the size of a camera view before you place any RPC people components in the model.

7. The Print Setup dialog lets you make adjustments to printer settings, but not PDF or DWF output.

8. You use Date, Time, and Place settings for an exterior rendering to adjust the angle of the sunlight.

9. DWF output from Revit is calibrated for the Autodesk Express Viewer.

10. Revit can export a Walkthrough to an animation (.avi) file in Hidden Line, Shaded, or Rendered view mode.

 Answers found on the accompanying CD.

# CHAPTER 10

# Augmenting the Design — Design Options and Logical Formulas

## INTRODUCTION

Building construction happens over time, as a result of many directed and realized choices. Revit provides design tools that take care of both space and time (phasing). As of release 6 Revit also allows explicitly for the process of choice. You can create, order, and display alternatives using Design Options. At any stage of the process you can resolve an option into the model and discard alternatives.

Revit uses parameters to drive any relationship the user wants to define. Parameters can hold formulas, which now include logical operators such as yes/no and if conditions. Families can now hold features or arrays that can be called or suppressed by family type or instance—cabinets with or without end panels, for example, or brackets that appear under a length of counter.

## OBJECTIVES

- Create, manage, and resolve Design Options
- Create two different stair styles with custom rails
- Create a custom wall style with a vertical structure
- Create and place a nested component family with family types based on size, utilizing parametric formulas

## REVIT COMMANDS AND SKILLS

| | |
|---|---|
| *Design Options* | *Family creation* |
| *Option sets* | *Family types* |
| *Options* | *Parametric formulas* |
| *Stairs* | *Conditional operators in formulas* |
| *Railings* | *Grid lines* |
| *Wall structure* | |

## DESIGN OPTIONS

In the previous exercises we developed various parts of building models. As most design projects progress, you will want, at some point, to explore multiple design schemes. Revit's Design Options allow you to develop alternate schemes, either simple concepts or detailed engineering solutions. Design Options coexist in the project file with the main model (all building elements that have not been assigned to a named option). You can study and develop each option independent of others, and easily create views to display option combinations. At any point you can resolve an option or option set into the main model to remove the alternates.

### Exercise 1: Design Option setup

1.1 Launch Revit. Open the file *Chapter 10 start.rvt*. The file will open to show the 3D View: Exterior Iso at the default southeast orientation.

This model is of the arts center for our hypothetical campus project. It contains performance halls, an art display area and cafe on the main floor, shop/storage space on the lower level, and classrooms/offices/rehearsal rooms on the upper levels. No seating has been created in the performance spaces, whose floors slope down to stages on the lower level.

1.2 Open the view Floor Plan: Lower Level. Zoom in to the upper half of the central hallway at the lower middle of the plan. Open the Visibility/Graphic Overrides dialogue for the view. Change the Underlay value to Level Above. Choose OK.

1.3 From the Tools menu, choose Tools>Design Options>Design Options (see Figure 10–1).

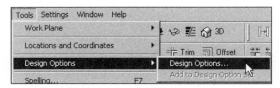

**Figure 10–1** *Open the Design Options dialogue*

1.4 Select New under Option Sets, as shown in Figure 10–2.

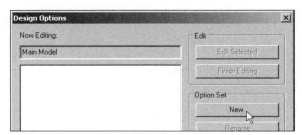

**Figure 10–2** *Create an Option Set*

Revit will open the Design Options dialogue and create a new Option Set named Option Set 1, with Option 1 (primary) as its default subset. You can create as many Option Sets as you desire, each with a tree of Options within it.

1.5 Select Option Set 1 in the tree pane. Select Rename under Option Set. In the Rename dialogue, type Stairs. Choose OK. Select Option 1 (primary). Select Rename under Option. In the Rename dialogue, enter Enclosed. Choose OK (see Figure 10–3).

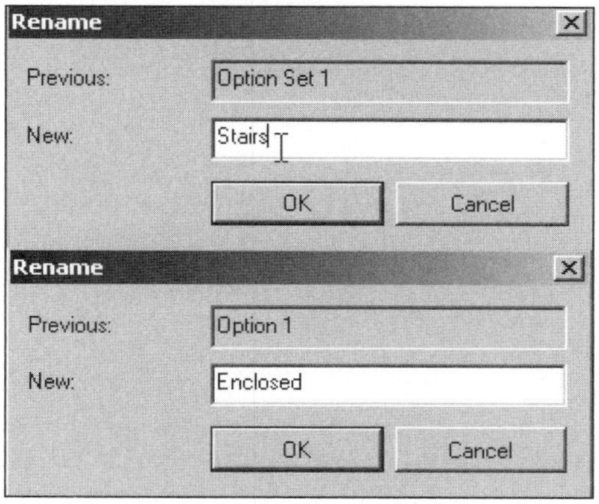

**Figure 10–3** *Name the new Option Set and Option as shown*

1.6 In the Design Options Dialogue, choose Edit Selected. The field text under Now Editing will change from Main Model to Option Set 1: Enclosed (primary) Stairs. Choose Close to leave the dialogue and work on the Design Option (see Figure 10–4).

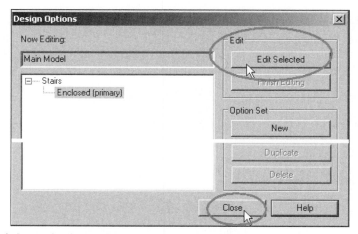

**Figure 10–4** *Begin the Option Editing process and close the dialogue*

## Create an Enclosed Stair, Landing, and Balcony Wall

1.7 From the Basics Tab of the Design Bar, select Ref Plane. Draw a horizontal reference place across the hallway space near the uppermost pair of sconce lights. Select Modify on the Design Bar. Copy the reference plane 7' up (at 90°—see Figure 10–5).

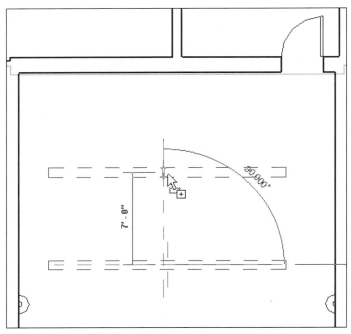

**Figure 10–5** *Create reference planes for stair creation*

1.8 From the Modeling tab of the Design Bar, select Stairs. Select Stairs Properties from the Design Bar. In the Instance Parameters Section of the Element Properties dialogue, change the Width to 6', and change the Top Level to Main Level, as shown in Figure 10–6. Choose OK.

Instance Parameters - Control selected or to-be-created instance

| Parameter | Value |
|---|---|
| Width | 6' 0" |
| Base Level | Lower Level |
| Base Offset | 0' 0" |
| Top Level | Main Level |
| Top Offset | 0' 0" |

**Figure 10–6** *Adjust Width and Top Level for the new stairs*

1.9 The cursor will show a pencil for Sketch mode. Click on a starting point near the right end of the lower reference plane. Pull the cursor left, until the temporary dimension reads 10'-1" and the light gray riser counter reads 12 RISERS CREATED, 11 REMAINING, and click to establish the first run of stairs.

1.10 Pull the cursor directly up to the upper reference plane. An alignment line will appear. Left-click and pull the mouse to the right (9'-2" minimum) until the sketch outline does not grow any more and the riser counter indicates that the layout is complete, then click again to establish the sketch (see Figure 10–7).

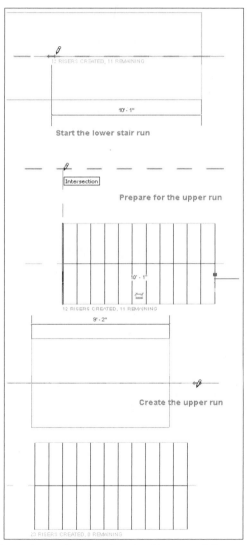

**Figure 10–7** *Sketch the stairs, with space for a landing—read from top to bottom*

1.11    Choose Finish Sketch. Select the reference planes and delete them. Select the stair's railing, not the stair itself. (Select both and use the filter, if necessary.) Select the Properties icon.

1.12    In the Element Properties dialogue, choose Edit/New. Choose Duplicate. In the Name box, type Single Wall Brick Cap. Choose OK. Choose Edit in the Rail Structure value field.

1.13    In the Edit Rails dialogue, there is a single rail defined for this style. For the Profile value, assign Brick Cap: Brick Cap, a custom profile that has been loaded into this project. For the Material value, click in the cell, then click the dialogue arrow that appears in the cell.

1.14    In the Materials dialogue, find Concrete-Precast Concrete in the Name pane, and select it. Click the arrow to the right of the AccuRender Texture field. In the Material Library dialogue, navigate to folder _ACCURENDER\ Concrete and select Exposed Aggregate, Tan, as shown in Figure 10–8. Choose OK three times to exit the Rail editor.

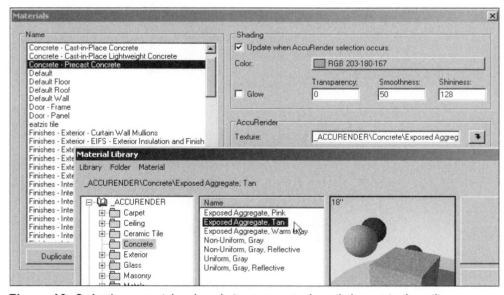

**Figure 10–8** *Apply a material and rendering texture to the rail element in the railing*

1.15    Choose Edit in the Baluster Placement field of the Element Properties dialogue. In the Edit Baluster Placement dialogue, change the value for Baluster Family to None in row 2 of the Main Pattern and all rows of the Posts, as shown in Figure 10–9. Choose OK three times to terminate the rail and stair editing.

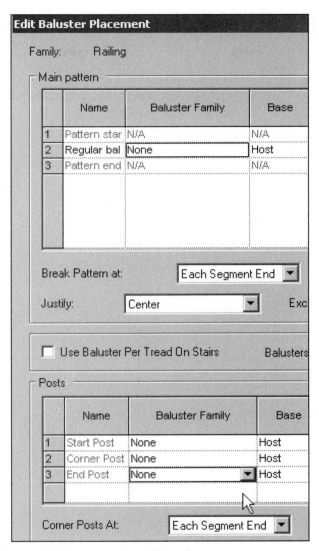

**Figure 10–9** *Remove all balusters in this railing style*

1.16    Select the Railing and Stair together. Choose the Move tool from the Toolbar. Select the upper-left corner of the railing as the starting point for the Move command, and select the intersection of the left side wall and the edge of the Main Level floor in the halftone underlay, as shown in Figure 10–10. Zoom out until you can see the whole stair to check the alignment.

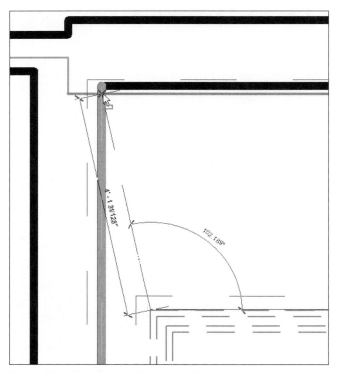

**Figure 10–10** *Move the new railing and stair together*

1.17   Open the Main Level view. Zoom in around the new stairs. Select the Match Type icon from the Toolbar. The cursor will change to an empty eyedropper shape. Select the outer stair rail to fill the eyedropper, and then select the inner rail to change its type from Rectangular to Single Wall Brick Cap (see Figure 10–11).

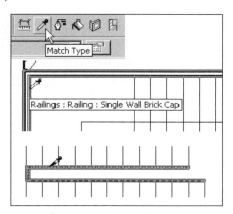

**Figure 10–11** *Match properties from one railing to the next*

1.18   Select the Floor tool from the Modeling tab of the Design Bar. Choose Floor Properties from the Design Bar. In the Element Properties dialogue, change the Type to Generic – 12" – Filled. In the Instance Parameters section, make sure the Level value is Main Level, and set the Height Offset From Level value to 0. (Revit will supply the units, as shown in Figure 10–12.) Choose OK.

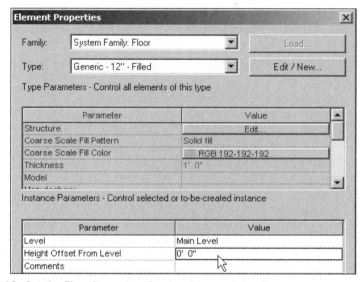

**Figure 10–12** *Set the Floor Properties for the upper stair landing*

1.19   On the Sketch Bar choose Lines, and check the Rectangle tool on the Toolbar. Select two points as shown in Figure 10–13: the lower end of the inner rail, and the intersection of the right-hand wall and the floor edge. Choose Finish Sketch to draw the upper landing.

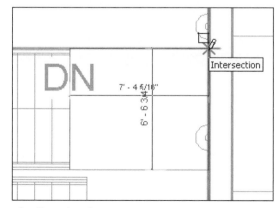

**Figure 10–13** *Draw the floor sketch*

1.20 Select Wall from the Modeling tab of the Design Bar. Select the Properties icon from the Options Bar. In the Element Properties dialogue, make the current type Generic – 4" brick. Select Edit/New.

1.21 Select Duplicate. In the Name dialogue, type Double Wall Brick with Cap. Choose OK. Select Edit in the Structure Value field. In the Edit Assembly dialogue, select Preview in the lower-left corner to open that pane. Change the view type to Section, as shown in Figure 10–14.

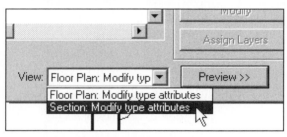

**Figure 10–14** *Select the Section preview of this wall type to facilitate editing*

1.22 Choose Row 2 in the Layers area. Choose Insert twice to create two new layers in the wall. Define the layers as shown in Figure 10–15: The Material for Row 2 should be set to Masonry – Brick and Thickness (typed in) **3 5/8"**, and Row 3 should be assigned Function Thermal/Air Layer, Material Air Barrier – Air Infiltration Barrier, and given Thickness **1 3/4"**. Set the Thickness for Row 4 to **3 5/8"**. Revit will show the Total thickness as 9".

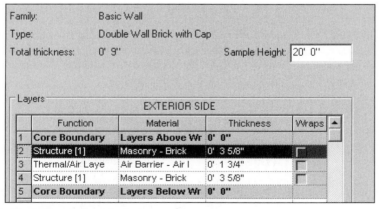

**Figure 10–15** *Layers defined in the new wall type*

Revit allows you to create walls of many layers. The Core Boundary layer defines the point at which finish meets structure, and determines how far floors penetrate into walls, as shown in drafting sections. Since this wall type is for decoration only, we will not alter the placement of the Core Boundary layers.

1.23   Select the Sweeps button at the lower right of the dialogue. In the Wall
       Sweeps dialogue, select Add. Assign the Profile Stone Cap: Stone Cap. This
       is a custom profile loaded into the project file. Set its Material to Concrete:
       Precast Concrete. Set the AccuRender material value for Precast Concrete
       to _ACCURENDER\Concrete\Exposed Aggregate, Tan, as you did before.

1.24   Choose OK twice to return to the Wall Sweeps dialogue. Set the From
       Value to Top and the Offset to **–0 4.5** (Revit will supply the units and frac-
       tion). Move the dialogue box off the preview if necessary and choose Apply
       to see the new sweep in place.

1.25   Select the Reveals button at the lower right of the Edit Assembly dialogue.
       In the Reveals dialogue, choose Add twice.

1.26   For both rows, make the Profile value Reveal-Brick Course: 1 Brick. Make
       the distance value **–0' 8"** for both rows. Make the From value read Top.
       Make the Side value Interior for Row 1 and Exterior for Row 2. Move the
       dialogue box off the preview if necessary and choose Apply to see the new
       reveals in place, as shown in Figure 10–16.

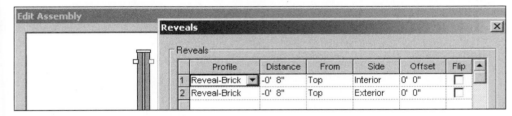

**Figure 10–16** *New Sweep and Reveals added to the wall type*

1.27   Choose OK twice to return to the Type Properties dialogue. Change the
       Wrapping at Ends value to Exterior. Choose OK. In the Element Proper-
       ties dialogue, set the Top Constraint to Explicit, the Unconnected Height
       to 3' 0", and the Location Line to Finish Face: Exterior, as shown in Figure
       10–17. Choose OK.

| Parameter | Value |
| --- | --- |
| Unconnected Height | 3' 0" |
| Comments | |
| Mark | |
| Location Line | Finish Face: Exterior |
| Wall top is attached | |
| Wall bottom is attached | |
| Structural Usage | Non-bearing |
| Base Constraint | Main Level |
| Base Offset | 0' 0" |
| Top Constraint | Explicit |
| Top Offset | 0' 0" |

**Figure 10–17** *Instance Properties of the wall you are about to place*

1.28 Place two instances of the new wall type at the edge of the landing and the floor, as shown in Figure 10–18.

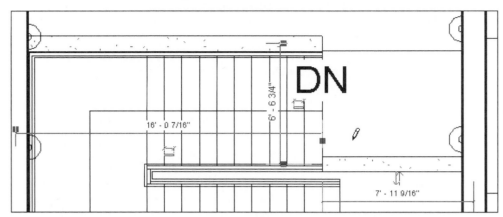

**Figure 10–18** *Landing walls in place for this Design Option*

1.29 Open the 3D View Atrium Staircase. Select the stair. Select the Properties icon on the Options Bar. Select Edit/New; and then select Duplicate. In the Name dialogue, type **Brick Sides**, as shown in Figure 10–19. Choose OK.

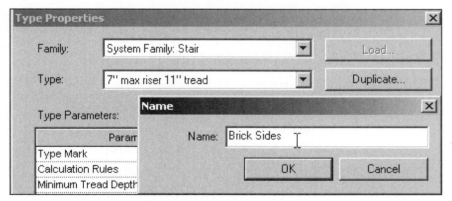

**Figure 10–19** *Define a new style for the stairs*

1.30 In the Type Properties dialogue, adjust the values as shown in Figure 10–20. Make the Stringer thickness **3 5/8"**; the Stringer Height **3' 5"**; the Stringer Carriage height **1"**; and the Landing Carriage Height **2"**. Make the material value for the Tread and Stringer Concrete – Cast In Place Concrete, and make the Stringer Material Masonry – Brick. Choose OK twice to finish the stair edits.

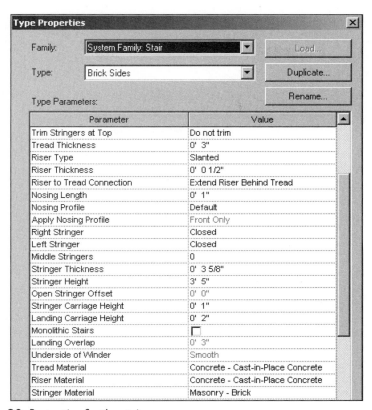

**Figure 10–20** *Properties for the stairs*

1.31 Select the Rail. Select the Properties icon from the Options Bar. Choose Edit/New. In the Type Properties dialogue, set the values for both Angled Joins and Tangent Joins to No Connector. Set Rail Connections to Trim (see Figure 10–21). Choose OK twice to finish the rail edits.

| Parameter | Value |
| --- | --- |
| Railing Height | 3' 0" |
| Rail Structure | Edit... |
| Baluster Placement | Edit... |
| Baluster Offset | -0' 1" |
| Use Landing Height Adjustment | No |
| Landing Height Adjustment | 0' 0" |
| Angled Joins | No Connector |
| Tangent Joins | No Connector |
| Rail Connections | Trim |
| Model | |
| Manufacturer | |

Type Parameters:    Rename...

**Figure 10–21** *Edit rail properties*

1.32　The first option for the stairs is now complete (see Figure 10–22).

**Figure 10–22** *The first stair option is ready*

1.33　Place the cursor anywhere over the Tool Bar and right-click. Choose Design Options from the list of available toolbars. In the Design Options toolbar, select the Design Options dialogue icon, shown in Figure 10–23.

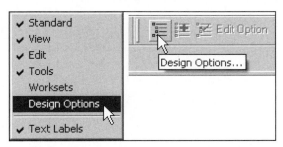

**Figure 10–23** *Make the Design Options toolbar visible and open the Design Options dialogue*

1.34   Choose Enclosed (primary) in the Stairs tree. Choose Finish Editing, as shown in Figure 10–24.

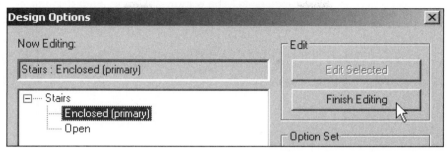

**Figure 10–24** *Finish editing the Design Option—you can open it for more edits later if necessary*

1.35   The 3D view will change appearance, as components of the model become available for editing. Select Open in the Stairs tree. Choose Edit Selected, as shown in Figure 10–25. The enclosed stairs disappear, and the model becomes gray again. Choose Close. Save the file.

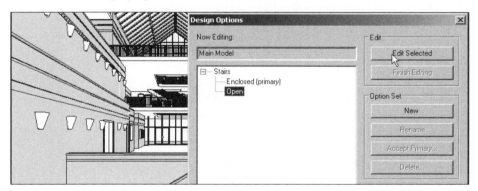

**Figure 10–25** *Start editing another named option*

## Exercise 2. Create an open stair with custom railings

2.1   Open or continue working with the file from the previous exercise step. Make view Lower Level current. Zoom in to the area you worked in before. Select Stairs from the Modeling tab of the Design Bar. Revit will go into Sketch mode.

2.2   Select Stairs Properties from the Sketch tab. Set the Stair properties as before: Width is 6', Base Level is Lower Level, Top Level is Main Level. Choose OK.

2.3   Start the run of stairs between the doors. Pull the cursor straight up (90°) until the temporary dimension reads 10'-1" and the riser counter shows 12 RISERS CREATED, 11 REMAINING. Left-click to establish the end of the run.

2.4 Pull the cursor straight up until a horizontal Reference Plane appears, along with the tooltip Midpoint, as shown in Figure 10–26. Left-click to start the second run, and continue pulling the cursor straight up until the stair outline is complete, as before. Click again to set the stair runs.

**Figure 10–26** *Starting the upper stair run*

2.5 The Stair sketch consists of three types of Model lines: green boundaries, black risers, and blue runs. Select the left-hand boundary of the upper run. Copy it 2' to the left.

2.6 Choose Boundary from the Sketch tab. Choose the 3-point arc tool from the Options Bar. Select the lower end of the new line for the first point of the arc; select the upper end of the original (copied) boundary for the end point of the arc. Pull the cursor to the right of the arc line until the arc bend temporary dimension reads approximately 20° and click to establish the arc sketch, as shown in Figure 10–27.

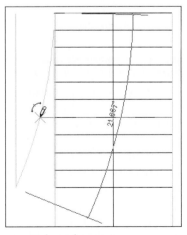

**Figure 10–27** *Drawing the new arc boundary*

2.7 The arc will display a radius dimension. Select the dimension and set it to 26, as shown in Figure 10–28. Revit will convert this to 26'-0".

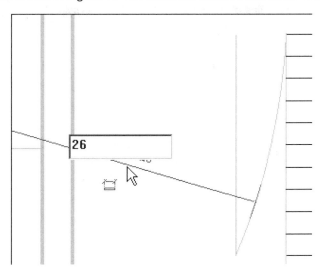

**Figure 10–28** *Set the arc radius*

2.8 Erase the two straight boundary lines for the upper run of the stairs and the copied sketch line. Select the new arc. Choose the Mirror tool form the Options Bar and mirror the arc to the right, using the center line of the stairs as the mirror line (see Figure 10–29).

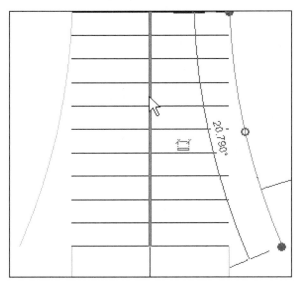

**Figure 10–29** *Mirror the arc*

2.9 Use the Trim tool with the multiple element option selected to extend the risers to the boundary lines, as shown in Figure 10–30.

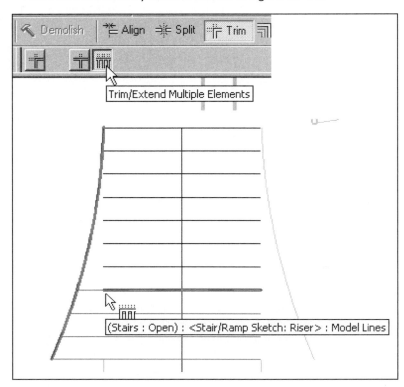

**Figure 10–30** *Extend the risers to the new boundaries*

2.10 Erase the boundaries of the landing. Move the boundaries of the lower stair run left and right 2' to line up with the wide ends of the upper runs.

2.11 Choose Boundary from the Sketch tab. Select the three-point arc tool. Select the lower end of the upper-left boundary and the upper end of the lower-left boundary as the start and end points, and pull the cursor to the left to define a 180° arc. Click to establish the arc.

2.12 Mirror the arc to the right to create a landing with rounded sides, as shown in Figure 10–31.

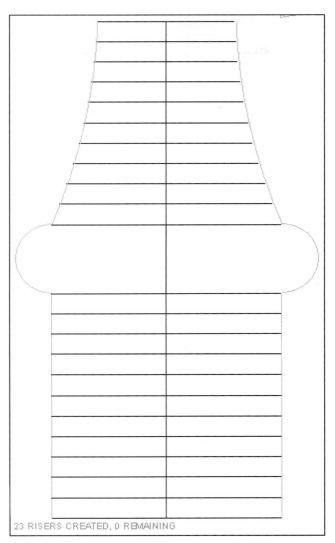

23 RISERS CREATED, 0 REMAINING

**Figure 10–31** *The completed stair sketch*

2.13   Select Stairs Properties on the Design Bar. Select Edit/New. Select Duplicate. In the Name box, type Open Riser Steel.

2.14   In the Type Properties dialogue, edit the following values: Tread Thickness 1"; Riser Type None; Stringer Thickness **2"**; Tread Material Metal – Steel; and Riser Material Metal – Steel, as shown in Figure 10–32. Choose OK twice.

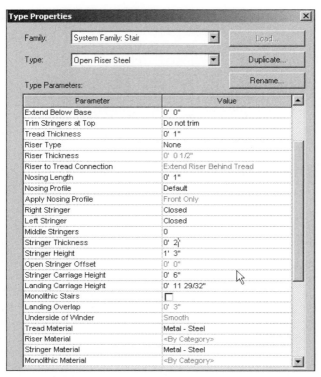

**Figure 10–32** *Stair Properties*

2.15 Choose Finish Sketch. Select the stair and both railings. Choose Move from the Toolbar. Select the middle of the top of the stairs for the first point, and select the middle of the Main Level floor as the second point (see Figure 10–33).

**Tip:** If you are having difficulty getting Revit to snap to the middle of the floor edge, type SM (for Snap Middle) to force the Midpoint snap override.

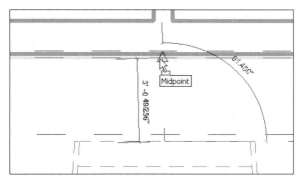

**Figure 10–33** *Move the middle of the stair to the midpoint of the floor edge*

2.16    Open the 3D view Atrium Staircase. Select one of the stair rails. Change the type in the Type Selector to Railing: Handrail – Pipe. Choose the Properties icon from the Options Bar. Select Edit/New. Select Duplicate. In the Name box, type Handrail – Pipe – Copper and Cherry. Choose OK.

2.17    In the Type Properties dialogue, select Edit in the Rail Structure value.

2.18    In the Edit Rails dialogue, change Rail 1 Material to Wood – Cherry. Change the Material for Rail 2 to Metal – Trim. While in the Materials dialogue, change the AccuRender Texture for Metal – Trim to _*ACCURENDER\ Metals\Copper\Polished, Plain*, as shown in Figure 10–34.

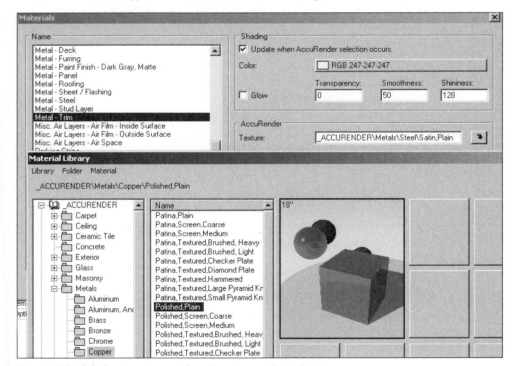

**Figure 10–34** *Assign a material and texture to the rails*

2.19    Choose OK twice to exit the Materials Library and Materials. Change the Material value for Rails 4 and 6 to match Rail 2.

2.20    Change the Profiles for Rails 3 and 5 to Fascia–Flat: 1" x 4". Make their Materials Wood – Cherry, as shown in Figure 10–35. Choose OK.

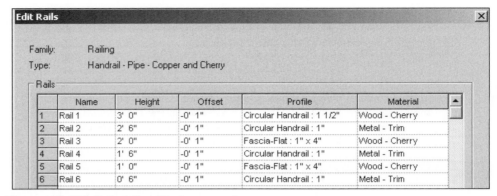

**Figure 10–35** *Rails defined in the new style*

2.21 Choose Edit in the Baluster Placement field. Change the Dist. from previous value in Row 2 of the Main Pattern to **3'**, as shown in Figure 10–36. Choose OK three times.

**Edit Baluster Placement**

Family: Railing                                  Type: Handrail - Pipe - Copper and Cherry

Main pattern

| | Name | Baluster Family | Base | Base offset | Top | Top offset | Dist. from previous | Offset |
|---|---|---|---|---|---|---|---|---|
| 1 | Pattern star | N/A | N/A | N/A | N/A | N/A | N/A | N/A |
| 2 | Regular bal | Baluster - Round : 1" | Host | 0' 0" | Rail 1 | 0' 0" | 3' 0" | 0' 0" |
| 3 | Pattern end | N/A | N/A | N/A | N/A | N/A | 2' 0" | N/A |

**Figure 10–36** *Edit the baluster spacing*

2.22 Select the rail you have not edited and change its type to Handrail – Pipe – Copper and Cherry. Right-click and select Zoom Out (2x).

2.23 Uncheck Active Option Only from the Options Bar. Select the bottom edge of the 3D View and drag it down to make more of the stair visible. Right-click and select Zoom to Fit.

2.24 From the Tools menu, select Tools>Design Options>Design Options. Choose Finish Editing. The 3D view will show the primary option, the enclosed stairs. Choose Close. Save the file.

## Exercise 3. Create balcony options

3.1 Open or continue working with the file from the previous exercise. Open the Design Options dialogue. Select Open in the Stairs Tree. Choose Make Primary under Option. The 3D view will change to show the new primary option. A warning box will appear that you can ignore.

3.2    Choose New under Option Set in the Design Options dialogue. Revit creates a new Option Set 1 and Option 1 (primary) under it. Choose New under Option to create a second Option in the new Option Set.

3.3    Select Option Set 1 and select Rename. In the Name box, type **Balconies**. Choose OK.

3.4    Select Option 1 and select Rename. In the Name box, type **Railings Only**. Choose OK.

3.5    Select Option 2 and select Rename. In the Name box, type **Twin Balconies**. Choose OK. Select the Railings Only Option and select Edit Selected. Choose Close.

3.6    Open the Floor Plan view Main Level. Select the Railing tool from the Modeling tab of the Design Bar. Select Railing Properties from the Sketch tab. Change the Type to Handrail – Pipe – Copper and Cherry.

3.7    On the Options Bar, change the Offset value to **0'-2"**. Draw a line along the edge of the floor left to right from the left wall to the end of the left stair rail, as shown in Figure 10–37. Choose Finish Sketch.

**Figure 10–37** *Sketch a rail at the left side*

3.8    Select the new rail. Use the Mirror tool to copy it to the right side.

3.9    Open the Design Options dialogue. Choose Finish Editing. Select the Twin Balconies Option, and choose Edit Selected (see Figure 10–38). Select Close.

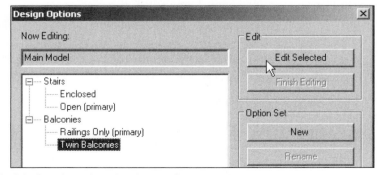

**Figure 10–38** *Switch to the other Design Option*

3.10　From the Modeling tab of the Design Bar, select Floor. Select Lines from the Sketch tab. Select the Rectangle tool from the Options Bar.

3.11　Draw a rectangle 7'-0" wide by 6'-0" deep starting at the left end of the main floor, as shown in Figure 10–39.

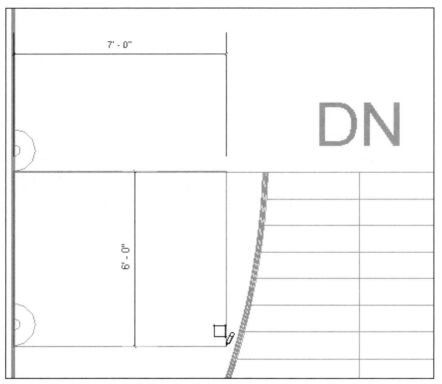

**Figure 10–39** *Draw a rectangular floor sketch*

3.12　Select Modify on the Sketch tab. Erase the bottom line of the floor sketch. Select Lines on the Sketch tab. Draw a 180° three-point arc.

3.13　Select the entire floor sketch. Mirror it to the right. Select Finish Sketch. Answer No in the question box.

3.14　Select Railing from the Modeling tab of the Design Bar. Select the Pick arrow from the Options bar. Set the Offset to **0'-2"** as before. Select the round edge of the left floor and the right side of that same floor, as shown in Figure 10–40.

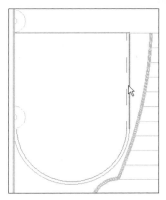

**Figure 10–40** *Use Pick and Offset to create these lines*

3.15   Select the Draw (pencil) icon from the Options Bar. Keep the 2" Offset. Draw a line from the end of the last line you created 1'-6" to the right. Revit will offset it from the floor edge. Use the Trim tool with the corner option to join the last two lines.

3.16   Select Finish Sketch. Mirror the new railing to the right (see Figure 10–41).

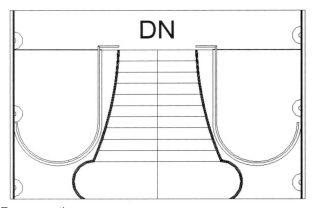

**Figure 10–41** *Two new rails*

3.17   From the Tools menu, choose Tools>Options>Design Options. Select Finish Editing. The balconies will disappear and the primary option will appear in the view.

3.18   Under Options, select New. Rename the new Option None. Choose Edit Selected. Select Finish Editing. Choose Close.

3.19   Select the 3D view Atrium Staircase in the Project Browser. Right-click and select Duplicate. Repeat to make two copies of the Atrium Staircase view.

3.20   Select View Atrium Staircase, right-click, and choose Rename. In the Rename View box, type Atrium Staircase, Enclosed Brick. Choose OK.

3.21     The view name will still be highlighted. Right-click and select Properties. Choose Edit in the Visibility field. The Visibility/Graphics Override dialogue will now have a tab for Design Options. Select that tab. In the Design Option field for Option Set Stairs, choose Enclosed. In the Design Option Field for Option Set Balconies, choose None. Choose OK twice to end the edits.

3.22     Select 3D view Copy of Atrium Staircase in the Project Browser. Right-click and select Rename. In the Rename View box, type Atrium Staircase Open, No Balconies. Choose OK.

3.23     Right-click and select Properties. Select Edit in the Visibility field. Select the Design Options tab. In the Design Option field for Option Set Stairs, choose Open (primary). In the Design Option Field for Option Set Balconies, choose Railings Only (primary). Choose OK twice to end the edits.

3.24     Select view Copy (2) of Atrium Staircase in the Project Browser. Right-click and select Rename. In the Rename View box, type Atrium Staircase Open, With Balconies. Choose OK.

3.25     Right-click and select Properties. Select Edit in the Visibility field. Select the Design Options tab. In the Design Option field for Option Set Stairs, choose Open (primary). In the Design Option Field for Option Set Balconies, choose Twin Balconies. Choose OK twice to end the edits.

3.26     Open each of the Atrium Staircase views in turn to check your work.

        If you wish to print your work, place the Atrium Staircase views on a sheet side by side. If you wish to keep the separate design options to experiment with, save a copy of the project file under a different name.

3.27     Open the Design Options dialogue. Select Stairs in the left pane. Select Accept Primary under Option Set on the right side. Answer Yes in the question box (see Figure 10–42).

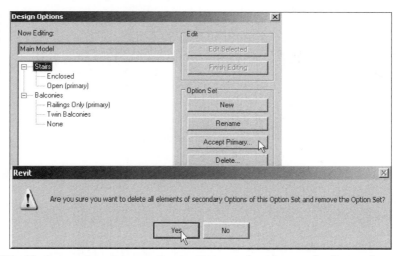

**Figure 10–42** *Accept the primary option to eliminate the others in the Option Set*

3.28    Select Twin Balconies in the left pane. Select Make Primary under Option. Select Accept Primary under Option Set. Select Yes in the question box. Choose Close. All Atrium 3D views now show the primary option. Delete the two views whose names are now incorrect. Save the file.

## Exercise 4. Create and place a nested family with conditional formulas

Revit's Families hold and organize definitions for nearly everything in Revit. Expand the Families section in the Project Browser, as shown in Figure 10–43, to see a list of content by Family. In this exercise we will create a nested column family using very simple drafting, and then apply Parameters, which we have already studied in previous exercises, to create design rules for types and instances within this family. The Parameters will use Formulas—mathematical, yes/no, and conditional (i.e. logical operations such as "if-then-else")—to create smart switches that control the appearance of the column as it is inserted.

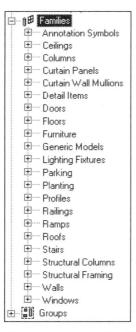

**Figure 10–43** *Even a blank Revit file contains many Families*

## Create some simple solids with materials

4.1    Open or continue working in the file from the previous exercise. Open view Floor Plans Main Level. From the Window menu, select Window>Close Hidden Windows.

4.2    From the File menu, select File>New>Family, as shown in Figure 10–44.

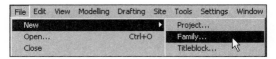

**Figure 10–44** *Start a new Family*

4.3  In the New dialogue box, select *Imperial Templates\Generic Model.rft*, as shown in Figure 10–45. Choose Open.

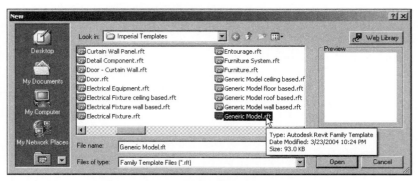

**Figure 10–45** *Use the Generic Model for this first Family*

4.4  In the new file, open view Elevations: Left. Zoom to Fit. Select Solid from the Family tab (the only one available) on the Design Bar. Select Revolve. Choose OK.

4.5  Sketch the profile shown in Figure 10–46. Do not create the dimensions.

The profile is basically a 2"-wide by 3"-tall rectangle set 4" to the right of the origin point, with 1" half-round stepped setbacks at 1/4". Use the Line and 3-point arc tool.

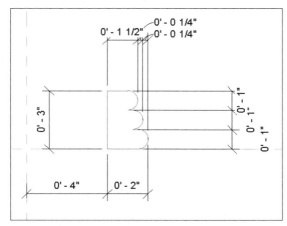

**Figure 10–46** *The profile for the trim ring family*

4.6    When the profile is complete, select Axis from the Design Bar. Draw a short line vertically from the origin, as shown in Figure 10–47. This profile will create a stepped ring when spun around the axis.

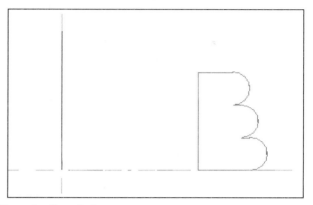

**Figure 10–47** *The axis of revolution*

4.7    Select Revolution Properties from the Design Bar. Verify that the End Angle is 360° and the Start angle 0°. Select the Material value field to open the Materials dialogue.

4.8    In the Materials dialogue, select Default in the Name panel, then select Duplicate. In the New Material name box, type Chrome. Select the arrow next to the AccuRender Texture field. Navigate to *_ACCURENDER\Metals\ Chrome\Satin, Plain*, as shown in Figure 10–48. Choose OK three times to set the Revolution Properties.

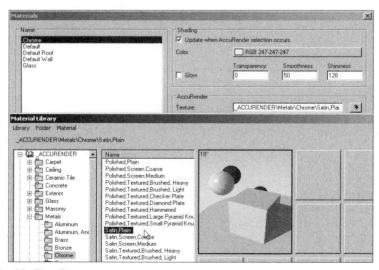

**Figure 10–48** *The Chrome material*

4.9 Select Finish Sketch. Select the Default 3D view from the Toolbar to see the finished trim ring (see Figure 10—49).

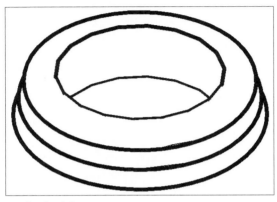

**Figure 10–49** *The revolved solid*

4.10 Save the file as **Trim Ring.rfa** in a place where you can find it later.

4.11 Return to the Left Elevation view. Choose the solid. Make a Mirror copy below the horizontal reference plane, and erase the original, as shown in Figure 10–50. Save the file as **Upper Trim Ring.rfa**.

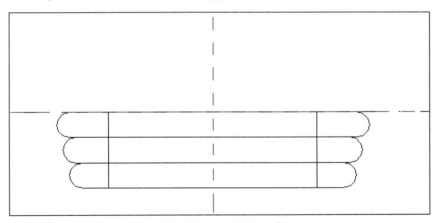

**Figure 10–50** *Mirror the solid and save it as a separate file*

4.12 Delete the solid. Select Solid from the Design Bar. Select Extrusion and choose OK. Draw the profile shown in Figure 10–51. Do not create the dimensions.

The profile is a bracket of 1"-thick material, 36" x 13", with a 2"-x-1" flange 1" from the right end and a 1/2"-thick arc brace. The exact length and placement of the arc are not critical. The vertical arm of the bracket is centered on the horizontal reference plane, and its left end is aligned with the vertical plane.

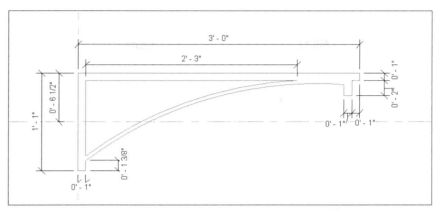

**Figure 10–51** *Draw another profile for extruding*

4.13   When the profile is complete, select Extrusion Properties from the Design Bar. Set the Extrusion End to **–0 1/2"** and the Extrusion Start to **0 1/2"**. Select the Material value field to open the Materials dialogue.

4.14   Select the Default material and choose Duplicate. In the New Material name box, type Black Iron. Click the arrow next to the AccuRender Texture field. Navigate to *AR2_ACCUREND\matte, black*, as shown in Figure 10–52. Choose OK three times to set the Extrusion Properties.

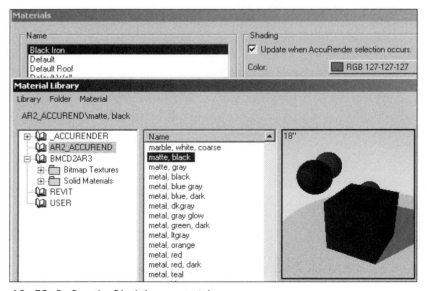

**Figure 10–52** *Define the Black Iron material*

4.15   Choose Finish Sketch. Open the default 3D view to see the results, as shown in Figure 10–53.

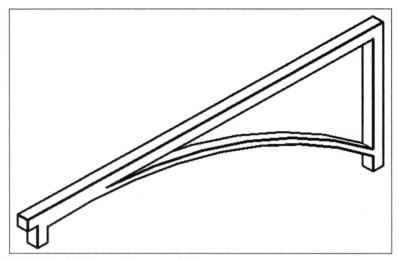

**Figure 10–53** *The bracket profile, extruded*

4.16 Return to the Left Elevation view. Select Solid from the Design Bar. Select Revolve and choose OK.

4.17 Draw the profile shown in Figure 10–54. Do not create the dimensions. The profile is a 90° arc from the midpoint of the flange bottom face approximately 1'-9" to the left by 12" down, then offset 1/4". Connect the endpoints of the arc to close the profile. Actual dimensions are not critical. The revolved solid will be a translucent light cover.

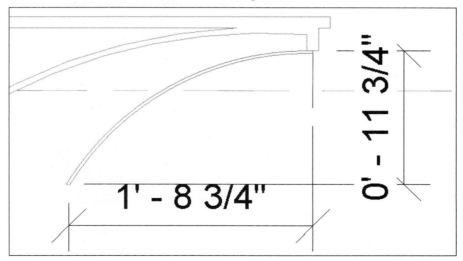

**Figure 10–54** *A profile for revolution—this will be a light cover*

4.18    Select Revolution Properties from the Design Bar. Verify that the End Angle is 360° and the Start angle 0°. Select the Material value field to open the Materials dialogue.

4.19    Select Glass from the Name list. Choose the arrow next to the AccuRender Texture field. Navigate to *_ACCURENDER\Glass\Tinted\White, Translucent, Frosted*, as shown in Figure 10–55. Choose OK three times to set the Extrusion Properties.

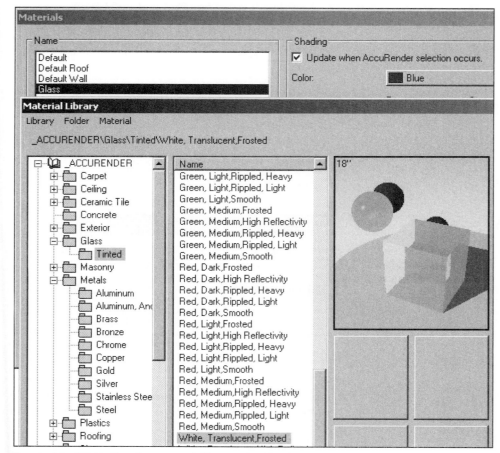

**Figure 10–55** *Define the material Glass in this family*

4.20    Open the default 3D view (see Figure 10–56). Save this file as **Bracket Light.rfa**. Close the file.

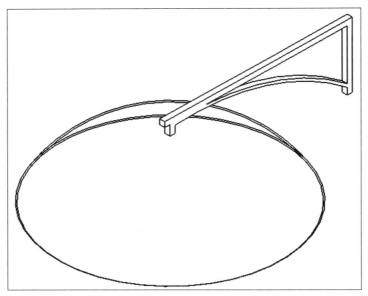

**Figure 10–56** *The bracket and light shade*

## Exercise 5. Create and apply conditional parameters in a family

5.1    Select File>New>Family from the File menu. Select *Imperial Templates\ Column.rft*, as shown in Figure 10–57, and choose OK.

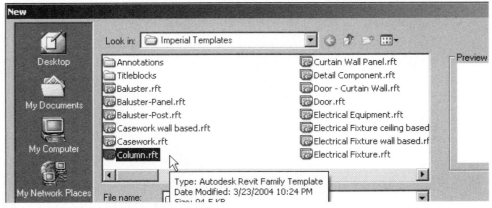

**Figure 10–57** *Open the Column template*

5.2    The new file will open to a Plan view. Select Solid from the Design Bar. Select Extrusion and choose OK. Sketch a 2'-x-2' square as shown in Figure 10–58, using the reference planes around the origin point.

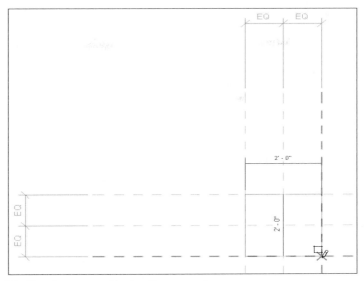

**Figure 10–58** *Sketch a square for the first extrusion*

**5.3**   When the profile is complete, select Extrusion Properties from the Design Bar. Accept the default Extrusion Start and End values for now. Select the Material value field to open the Materials dialogue.

**5.4**   In the Materials dialogue, choose Default in the Name panel, then choose Duplicate. In the New Material name box, type Brick. Select the arrow next to the AccuRender Texture field. Navigate to _ACCURENDER\Masonry\ Brown,_8",Running, as shown in Figure 10–59. Choose OK three times to set the Revolution Properties. Choose Finish Sketch.

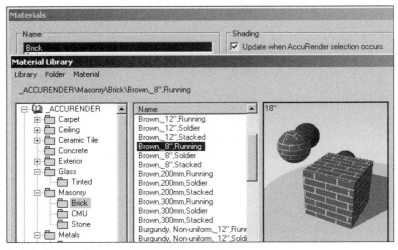

**Figure 10–59** *Define the material for this extrusion*

5.5 Choose Family Types from the Design Bar. Select Add in the Parameters section. In the Parameter Properties dialogue, verify that Family parameter is checked. Name the new parameter **Top height,** and select Length as the Type. Select Value stored by Type, as shown in Figure 10–60. Choose OK.

**Figure 10–60** *Define the Top height as a Length parameter*

5.6 Select Add in the Parameters section. Name the new parameter **Base height**, and select Length as the Type. Leave the other options set as for the previous parameter. Choose OK.

5.7 In the Formula field for the new Base height parameter, type **Top height/6**. Choose Apply. Revit will space the text according to its formula syntax. Revit will notify you if a formula is not correct. Formula text is case-sensitive.

 **Tip:** This dialogue box will resize. Stretch its borders left and right to give the Formula field plenty of space for typing. Adjust the width of the other columns as necessary.

5.8 Select New under Family Types. In the Name Box, type **12'** and choose OK. Repeat this step twice more, creating Family Types named **17'** and **23'**.

There are many ways to organize family types. For this column style we will differentiate the types by height. In this project, a 12' column will fit under the wing roofs at the front (North side) of the building file, and a 23' column will fit under the 3rd story overhang at the main entry. We will also create a 17' column, perhaps to use as a parking lot or walkway light. Columns of different heights will have different characteristics, which will not be drafted, but driven by formulas that use the column height as their basis.

5.9    Put the 23' column name in the Name list. In the Value field for Top height, type **23**. Choose Apply. Revit will set the units to 23' 0" and display a calculated value for the Base height, as shown in Figure 10–61.

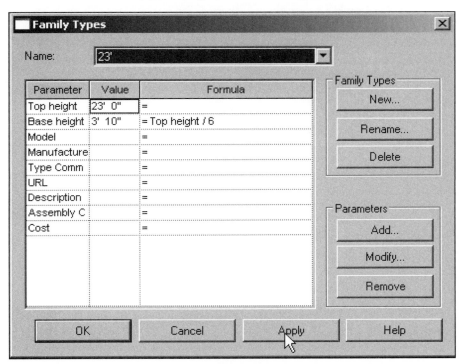

**Figure 10–61** *Create a Family Type and set its Top height—Revit will calculate the Base height*

5.10   Repeat this step for the 17' and 12' Family Types. Make the 23' Family Type the current Type. Choose OK.

5.11   Select the extrusion. Select Edit from the Options Bar. Select Extrusion Properties from the Design Bar. Click the small button to the right of the value field for Extrusion End (below the header panel with the equals (=) sign). In the Associate Family Parameter dialogue, select Base height and choose OK.

5.12   In the Element Properties dialogue, the Extrusion End value will display a different number, which will be grayed, indicating it is not editable by using this field. The small button to the right of the field will hold an equals (=) sign, indicating that a parameter is driving this value (see Figure 10–62). Choose OK. Choose Finish Sketch.

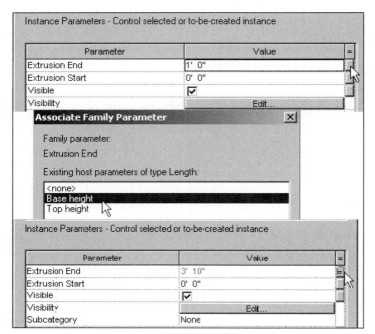

**Figure 10–62** *Associating a parameter with an Extrusion*

5.13 Choose Solid in the Design Bar. Choose Extrusion. Choose OK. Select the Circle tool from the Options Bar, and draw a circle of 4" radius at the origin, as shown in Figure 10–63.

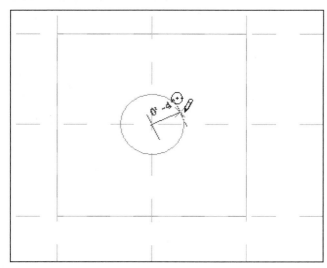

**Figure 10–63** *Draw the next extrusion using the Circle tool*

5.14  Select Extrusion Properties from the Design Bar. Select the Associate Parameter button as before. Choose Top height and choose OK.

5.15  Repeat this step for the Extrusion Start value, using Base height as the Associated Parameter. Choose OK.

5.16  Select the Material value field. Select Brick and choose Duplicate. Name the new material **Aluminum**. Use the AccuRender Texture field to set the material to _ACCURENDER\Metals\Aluminum\Satin, Plain, as shown in Figure 10–64. Choose OK three times to exit the dialogues. Select Finish Sketch. Except for placing components, drafting for this family has now finished, and the rest of the work will be with parameters and formulas.

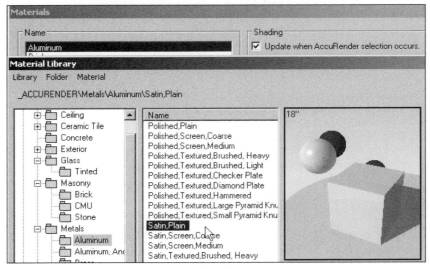

**Figure 10–64** *Define the column's Aluminum material*

5.17  Open the default 3D view. Choose Family Types from the Design Bar. Cycle the Family Types as you did when you created them, using the Apply button for each Type even though you did not make changes. Move the Family Types dialogue box so you can see the column shaft and base change height.

 **Note:** Checking family types is an important part of creating them effectively. Check your work often when adding parameters, formulas, or reference planes with dimensions, to make sure that your families work as expected.

5.18  Move the Family Types dialogue to the center of the screen. Stretch its sides left and right to make typing easier.

5.19  Select Add in the Parameter section of the dialogue. Create a Family parameter named **Light Height** as a Length value stored by Type. Choose OK.

5.20  Add a Family Parameter named **Trim Height** as a Length value stored by Type. Choose OK.

5.21  Repeat the Add step to create a Family parameter named **Lights 1** as a Yes/No value stored by Type (see Figure 10–65). Choose OK.

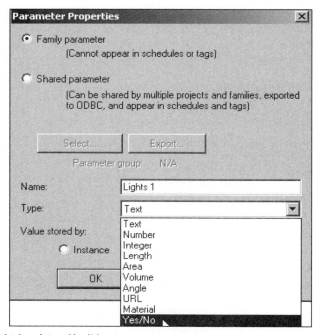

**Figure 10–65** *Lights 1 is a Yes/No–type parameter*

5.22  Add a Family parameter named **Lights 2** as a Yes/No value stored by Type. Choose OK.

5.23  Add a Family parameter named **Middle Trim** as a Yes/No value stored by Type. Choose OK. The Family Types dialogue will display the five new parameters you have created. They will apply to all the Family Types, not just the 23' type that was active when we started adding parameters (see Figure 10–66).

**Family Types**

Name:  23'

| Parameter | Value | |
|---|---|---|
| Trim Height | 0' 0" | = |
| Top height | 23' 0" | = |
| Middle Trim | ☑ | = |
| Lights 1 | ☑ | = |
| Light height | 0' 0" | = |
| Lights 2 | ☑ | = |
| Base height | 3' 10" | = Top height / 6 |

**Figure 10–66** *The new parameters defined – compare this list with Figure 10-61*

5.24    Enter formulas into the appropriate fields as shown in Figure 10–67. Choose Apply after entering each one to check your formula.

| Parameter | Value | |
|---|---|---|
| Trim Height | 11' 6" | = Top height / 2 |
| Top height | 23' 0" | = |
| Middle Trim | ☑ | = Top height > 16' |
| Lights 1 | ☑ | = Top height > 8' |
| Light height | 0' 0" | = |
| Lights 2 | ☑ | = Top height > 16' |
| Base height | 3' 10" | = Top height / 6 |

**Family Types**

Name:  23'

**Figure 10–67** *Formulas for the parameters*

The formulas mean the following: The Trim height will always be 1/2 the height of the column. The Middle Trim value will be Yes when the column height is greater than (>) 16'. This condition is satisfied in the family types 17' and 23' tall, but not in the 12' type, so any element keyed to the Middle Trim parameter will not appear in that type. The same condition applies to Lights 2. Lights 1 will appear in all column types taller than 8'. The three column types we have defined all meet this condition—a shorter column style could be defined, maybe for interior use, that would not have any elements keyed to Lights appear.

5.25    Carefully enter the following text into the formula field for Light height:

**If (Top height > 20', Top height –4', If (Top height < 15', Top height –2', Top height –3'))**

Choose Apply to check your formula.

 **Note:** Revit will supply spaces in formulas, as we have seen before, but be very careful with spelling, case, and punctuation. Open and closing parentheses must match.

This formula means: If the column is taller than 20', the Light height will be 4' down from the top. If the column height is less than 15', the Light height will be 2' down from the top. For all heights between 20' and 15', the Light height will be 3' down from the top. A complete study of conditional syntax is beyond the scope of this book.

5.26    Cycle through the Family Types to see changes. In the 12' type, the Middle Trim and Lights 2 value check boxes are empty, since this height is below the threshold value defined by the formula (see Figure 10–68). Choose OK to finish the parameter and formula setup.

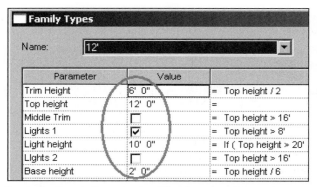

Figure 10–68 *Check the Family Types—the 12' column will display empty checkboxes for Middle Trim and Lights 2*

5.27    Save the file as **decor column.rfa**. From the File menu, select File>Load From Library>Load Family. Navigate to the folder(s) where you saved the *bracket light* and *trim ring* files, select them, and choose Open.

**Tip:** If the files are in the same folder, you can hold down the CTRL key and select multiple file names.

5.28    Open the Lower Reference Level plan view. Select Component from the Design Bar. Put Trim Ring in the type selector. Locate an instance of the trim ring at the origin point, as shown in Figure 10–69.

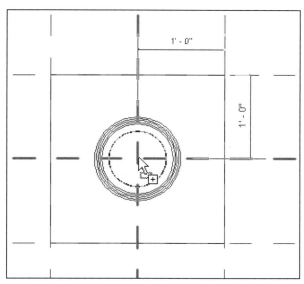

Figure 10–69 *Locate the trim ring at the origin*

5.29  Select Modify to terminate the insert. Open the Left Elevation view. Select the new trim ring. Select the Properties icon from the Options Bar.

5.30  In the Element Properties dialogue, select the Associate Parameter button next to the Elevation value field. Select Base height and choose OK (see Figure 10–70).

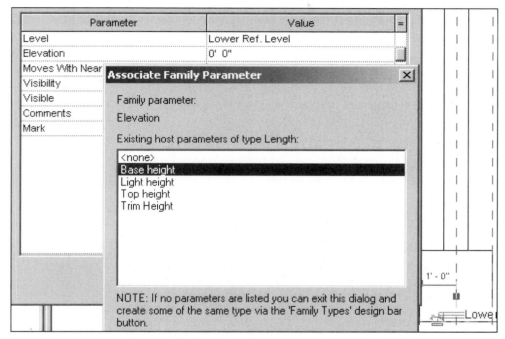

**Figure 10–70** *Associate the new component Elevation with the parameter Base height*

5.31  The trim ring component will still be highlighted. Copy it some distance straight up (the actual distance does not matter.) Choose the Properties icon from the Options Bar.

5.32  In the Element Properties dialogue for the new instance of the trim ring, associate its Elevation value with the parameter Trim height. Choose OK. Associate the Visible value for this instance with the parameter Middle Trim. Choose OK twice.

5.33  Open the plan view again. Select Component from the Design Bar. Put the upper trim ring in the type selector. Locate an instance as the origin, as before. A warning will appear, which you can ignore. Choose Modify.

5.34  Open the Left Elevation view. Select the new upper trim ring instance. Select the Properties icon from the Options Bar. In the Element Properties for this component, select the Associate Parameter button for its Elevation and associate its value with Top height. Choose OK twice.

5.35 Copy the upper trim ring some distance down, as before. Repeat step 5.32 to associate the new copy's elevation with the Trim height parameter and associate its Visible value with the Middle Trim parameter (see Figure 10–71).

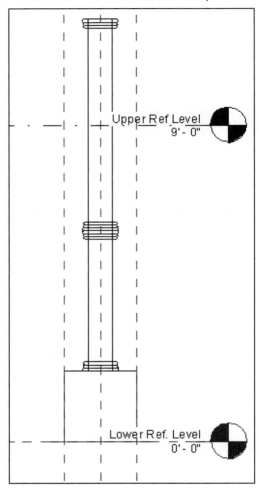

**Figure 10–71** *Trim rings in place*

5.36 Open the Lower Ref. Level plan view. Select Component from the Design Bar and choose bracket light in the Type Selector. It will appear with the bracket facing up (plan north). Place the instance at the face of the column extrusion, as shown in Figure 10–72. Select Modify.

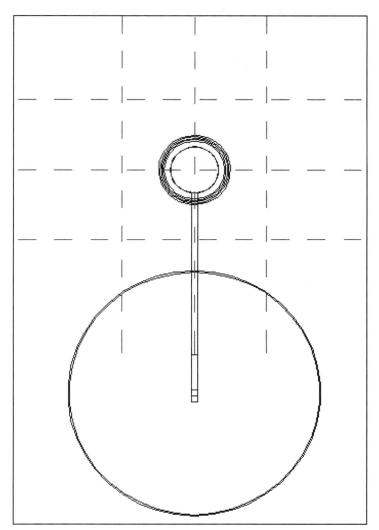

**Figure 10–72** *The first light component placed in the plan*

5.37    Select the light. Use the Mirror tool and the horizontal reference plane to make a copy at the other side of the column.

5.38    Copy the new instance to the left, rotate it 90°, and place the new instance at the column face, as shown in Figure 10–73.

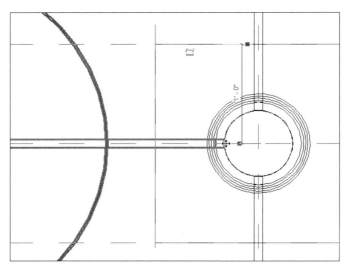

**Figure 10–73** *Place the bracket endpoint at the intersection of the column and reference plane*

5.39 Mirror the new light as before, for a total of four lights in place around the column (see Figure 10–74).

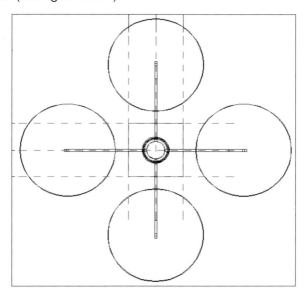

**Figure 10–74** *Four lights placed in the plan*

5.40 Open the Left Elevation view. Select the four lights and choose the Properties icon from the Design Bar. In the Element Properties dialogue, associate the Elevation value of the lights with the Parameter Light height. Choose OK twice. The lights will move up.

5.41   Open the default 3D view. Select the front and rear lights, as shown in Figure 10–75. Choose the Properties icon. Associate the Visible parameter of those components with Lights 1. Choose OK twice.

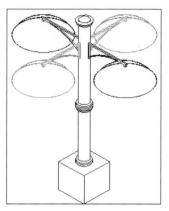

**Figure 10–75** *Associate these lights with parameter Lights 1*

5.42   Select the other two lights. Associate their Visible parameter with Lights 2. Choose OK twice.

5.43   Cycle through the Family Types to make sure all components obey the elevation formulas. The Visible condition does not show in the family model, so lights and trim will not turn on and off. Save the file. Close the file.

## Exercise 6. Place families in a project

6.1   In the Main Level view of the Chapter 10 project file, make the Basics tab of the Design Bar active. Select Grid, as shown in Figure 10–76.

**Figure 10–76** *The Grid tool from the Basics tab*

We will not provide a complete examination of grids and their uses in this text. Columns snap to grids, so we will quickly place grid lines to locate columns, and then delete the grid lines. Therefore the location or numbering of grid-line bubbles does not matter.

6.2 Select the 3-point tool on the Options Bar. Set the Offset value to **2' 0"**. Select the endpoints right to left and then the middle of the curved front plaza, as shown in Figure 10–77.

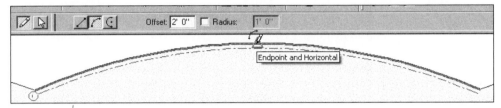

**Figure 10–77** *The first curved grid line follows the front walkway*

6.3 Select the Line tool on the Options Bar. Type **SC** to force a Center snap. Select the curved curtain wall at the main foyer, as shown in Figure 10–78.

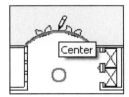

**Figure 10–78** *Start a grid line at the center of the front entrance curved wall*

6.4 Pull the cursor straight up at 90° until it crosses the first grid line, as shown in Figure 10–79. Left click to create the grid line. Choose Modify.

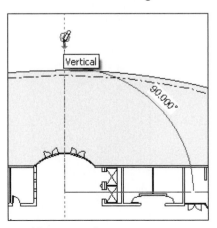

**Figure 10–79** *Pull the new grid line vertically*

6.5 Array the new vertical grid line four times left and right at a distance of 24', as shown in Figure 10–80.

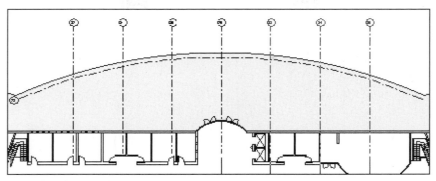

**Figure 10–80** *Array grid lines*

6.6 Right-click to the View Properties dialogue. Set the Underlay value as Level Above. Choose OK. Make the Modeling tab active on the Design Bar.

6.7 Choose the Column tool on the Design Bar. Select Load on the Options Bar. Navigate to the folder where you saved the *decor column.rfa* file, and choose Open.

6.8 In the Type selector, make decor column: 23' the active type. Place instances at the intersections of the grid lines along the rim of the front plaza, as shown in Figure 10–81.

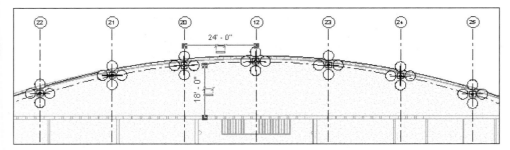

**Figure 10–81** *Columns placed along the front*

6.9 Change the column type to decor column: 12' and check Rotate after placement on the Options Bar. Place four instances, rotated 110° or –70°, at the corners and middle of the roof on the front right of the building. Precise placement is not critical for this exercise (see Figure 10–82). Note that these columns only have two lights.

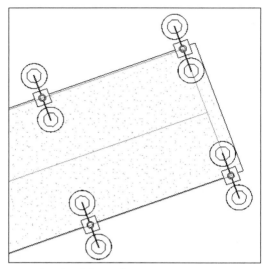

**Figure 10–82** *Columns under the right roof*

6.10 Change the column type to decor column: 17' and place two instances about 24' apart near the front entrance, as shown in Figure 10–83. Precise placement is not critical.

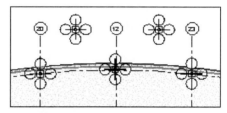

**Figure 10–83** *17' columns near the front*

6.11 Mirror the four 12' (rotated, two-light) columns to the left of the building using the middle grid line. Delete the grid lines. Open the View Properties dialogue and return the Underlay value to None (see Figure 10–84).

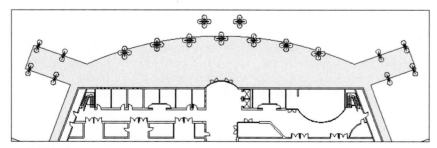

**Figure 10–84** *Columns in place*

6.12    Open the 3D view Exterior Iso. From the View menu, select View>Orient>Northeast. Zoom in a little to study the three types of columns. The 12' columns under the wing roof have two lights and no middle trim, as you defined in the family type formulas (see Figure 10–85).

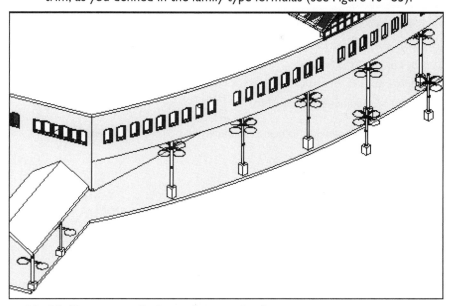

**Figure 10–85** *Three types of columns from one family*

6.13    There are other 3D views set up in this file to examine the columns and other parts of the model. Save the file. Close the file.

## SUMMARY

This final chapter has showed you some of the very latest features in Revit, and some of its most useful ones, particularly where busy designers are concerned. Design Options allow you to create many alternatives inside any project on the fly, and then view them, change them, and ultimately resolve them into your preferred options.

Family creation in Revit is an extremely powerful toolset—practice creating your own component families to suit your purposes. Experiment with writing simple formulas that will take advantage of the design rules inherent in any component so that you can make many versions of the same basic family structure without extensive drafting.

# REVIEW QUESTIONS

## Multiple Choice

1. Family parameter types include

   a) Text, Number, Integer

   b) Length, Area, Volume

   c) Angle, URL, Material

   d) all of the above, plus Yes/No

2. Which of the following is not available when editing a Wall Structure?

   a) Sweeps

   b) Reveals

   c) Balusters

   d) Layers

3. A Design Option Set holds

   a) Family Parameters

   b) as many Design Options as you define, one of which will always be primary

   c) only the Design Option that is currently being edited

   d) the key to a happy life

4. Stair Parameters that you cannot edit include

   a) Nosing, Thickness, and Profile

   b) Riser Type, Thickness, Material, and Tread Connection

   c) Stringer Thickness, Height, Material, and Trim at Top value

   d) all of the above can be edited

5. Revit does not provide family template files (*.rft) for

   a) Doors, windows, furniture

   b) Roofs, floors, walls

   c) Lights, columns, electrical fixtures

   d) Profiles, plumbing, mechanical equipment

## True/False

6. Revit keeps Design Options as separate files from the main project model.

7. Extrusion Start and End values can be set by parameters.

8. Length parameters cannot use formulas.

9. You can edit a Railing baluster placement value only once if the railing is a Design Option.

10. Stair boundaries can be edited to change the shape of a stair.

 Answers will be found on the CD.

Congratulations! You have studied and experimented with the major features of a powerful, efficient, new building-design application, and you have joined the growing ranks of those able to use the latest technology that the market has to offer. Practice and experience with Revit will provide speed in your work and the confidence to explore its numerous capabilities.

It's now up to you and your organization to determine how best to implement what Revit has to offer. Take a look at the Appendix for a discussion of factors that affect the implementation process. You can blend Revit into your workflow by combining it with standard CAD applications, or start out with a project planned and executed completely in Revit. Either way, you can refer back to the concepts and exercises in this text to help you grasp and make effective use of Revit's massing, modeling, drafting, data, and illustration capabilities in your own work process. It's important to match your expectations with your resources. It will be hard to ask too much of Revit, but it can be easy to underestimate the time and effort necessary to make it do exactly what you think you are asking of it.

Those of you who develop expertise with Revit can and should lead the way in providing guidance, support, and training for others in your team or company. Patience and sympathy can be key—not everyone learns at the same pace, hits the same stumbling blocks, or finds the same solutions to procedural problems. Revit is very well designed in that there is more than one way to do almost anything. This can be confusing for both instructors and their students at first, but practice will help everyone develop flexible working methods and the knack for finding answers to questions unaided. Don't forget online resources such as user groups and their ongoing discussions. You can engage enthusiastic, experienced, articulate Revit users worldwide with very little effort.

We wish you goodspeed moving on from this point. It is the authors' sincere hope that you will soon feel that, just like the fabled Mr. Natural, you can Keep On Trucking through your design work with your hands firmly on The Right Tool For The Job.

# What It Really Takes to Implement Revit

 **Author's Note:** This appendix is taken from a highly-rated presentation that co-author Jim Balding gave at Autodesk University 2003. His audience consisted of executives, managers, and engineers from firms seeking to understand what they would be facing while migrating to Revit. The term Building Information Modeling (BIM) refers to database systems such as Revit, rather than vector-based CAD applications.

Readers from educational institutions or other organizations with in-house design departments may readily identify similar concerns that their programs or departments will have when facing upgrade, deployment, and training issues.

## WHY CHANGE?

There are many reasons to change to the Building Information Model (BIM). In the beginning, the most compelling is the three-dimensional aspect of designing and documenting a project. In addition to that, many find the constant, complete coordination of the single model to be a compelling reason. If that doesn't convince you, the fact that firms using Revit are consistently providing better service and deliverables either in less time or with less staff and occasionally both, may sway you. The bottom line is that a totally efficient design process has always been the "Holy Grail" of design and documentation, and only recently have the hardware and software caught up and become able to deliver on this elusive promise.

## INTRODUCTION—POINTERS, NOT PRESCRIPTION

We have been asked countless times, "How do I implement Revit at my company?" The answer to this question is very simple—we have absolutely no idea. That may seem a little odd coming from two people writing a book about the subject, but it is very true. All firms and institutions are different in so many ways that there is no way to create a one-size-fits-all, step-by-step instruction book on how to implement a software such as Revit in any given firm. With that in mind we have identified seven factors that will influence the implementation process in just about any office, and have outlined the issues to look for while implementing Revit.

## THE SEVEN FACTORS OF IMPLEMENTATION

While there are many different factors that will affect the success of implementation efforts, most can be grouped into one of seven categories. The following list is in order of importance.

Firm size

Project size

Architectural style

Project type

Scope of services

Firm locations

Firm culture and age

## FIRM SIZE—WHAT'S YOUR SCORE?

The size of the firm is the number-one issue when it comes to implementing the Building Information Model. While there are many factors that firm size will affect during implementation, the key factors are project size, technology, training/support/ R&D, personnel, decision making, and standardization. Below are general observations regarding key factors related to company size. Each category has been rated as holding an advantage towards either larger or smaller firms. This is by no means an absolute system, and the advantage could be great or slight—use your own experience when considering these issues and evaluating your own situation.

### PROJECT SIZE—ADVANTAGE: SMALLER FIRMS

Large firms generally have larger projects, and when it comes to Revit, project size can be viewed as an obstacle with regards to implementation. If you consider that you are adding all of the building data to one file, that file can certainly grow, and become larger than the referenced files you may be presently using. BIM products including Revit have strategies to alleviate the issue of large projects; however, many firms feel that they need more time to get to a comfort level using full Revit implementation on such large-scale projects. It is important to note that this does not count out larger projects, they will simply require a little more planning and more experience.

This is where the smaller firms tend to shine. They typically have smaller projects that are ideal for Revit. There is not as much information to be communicated in a smaller project, which translates to smaller files and project teams.

One thing to keep in mind when selecting projects based on project size: It is advisable not to base the decision entirely on the floor plan area but the "spatial size," including both the amount of mass and detail to be built. For example, designing and documenting a 750,000–square foot (s.f.) warehouse using Revit can be far easier than creating a 20,000–s.f. gothic cathedral.

## TECHNOLOGY—ADVANTAGE: LARGER FIRMS

Due to the fact that Revit models are larger and more complex than ordinary 2D drafting files, the software and models require fast machines with plenty of RAM and good video cards. When it comes to technology, the larger firms usually have the upper hand. They typically have the latest hardware, fast Local Area and Wide Area Networks, and perhaps even a dedicated Information Systems team to support the systems.

The small firm generally finds themselves behind the eight-ball in this arena. They may have made a recent, significant investment in connectivity, hardware, or software, and can't afford to budget large-ticket items such as these every year. A small firm that has made a substantial investment in technology could find themselves in a very advantageous position when implementing Revit.

Firms of any size should pay close attention to the hardware requirements of BIM software, as well as performance on typical firm project types. Maintaining up-to-date hardware and software often pays dividends not only in the speed of deliverables, but also in perceivable pride of ownership, which can provide incentives to learn and maintain knowledge, thus reinforcing a company's commitment to technology.

## TRAINING/SUPPORT/R&D—ADVANTAGE: LARGER FIRMS

The larger the firm, the larger the operating expenses. With this in mind, the ability to create and maintain in-house expertise to provide training and support, as well as perform research and development roles for firm-specific tasks, becomes significant. This phenomenon often runs parallel with a firm's CAD manager role. If the firm has a full time CAD manager, typically it will have a full time Revit manager; who are often one and the same person.

The smaller firm will tend to have a part-time CAD manager, often an architect who knows the ins and outs of computers better than the rest of the firm. Firms can expect additional overhead and responsibilities when it comes to implementing and steadily using BIM. The return on investment over time is proving to be much larger proportionately when comparing results from Revit to 2D CAD.

Training and support are two of the keys to success, discussed in greater detail later, and they should be given serious thought when implementing Revit.

## PERSONNEL—ADVANTAGE: LARGER FIRMS

When planning the implementation of BIM, one of the first things you need is a "champion," the person responsible for heading up the change. However, one person cannot do everything required to make the move to Revit. You will need talented people to carry out the plan and use the software effectively enough to deliver on the promise.

Because larger firms have larger talent pools to select from, they generally have the luxury of hand-selecting the users that best suit the chosen tool. One thing to note, however—the proportion of talent and roles within a firm will tend to be about the same for both larger and smaller firms.

### DECISION MAKING—ADVANTAGE: SMALLER FIRMS

When it comes to making the decision to change, it seems to be better to have a single entity make that decision rather than a committee. Nowhere does the saying "too many chefs spoil the broth" apply more appropriately than in implementing change. Larger firms with multiple locations will have more challenges in the way of purchasing, planning, and organizing an implementation plan.

Smaller firms, generally based in one location, have just one "chef" who makes the decision to implement BIM, and then it is time to move on. When changes in the plan occur, there isn't another round of negotiations, planning, etc., to slow progress down.

### STANDARDIZATION—ADVANTAGE: SMALLER FIRMS

Here again, the larger firm will tend to have many "chefs," from the top to the bottom. Somewhere in between might be a local CAD manager or two with individual CAD standards and techniques. If your firm is international you have, potentially, different languages, building techniques, documentation techniques, cultures, and infinitely more issues of diversity to face.

Smaller firms get the nod in this arena. With one location, standards can be established, and even voted on, in one location, perhaps at one time. When changes to the standard occur, communicating that change is simple, and most users will understand the reasoning behind it as it undoubtedly has arisen from issues within the local firm.

### FIRM SIZE—CONCLUSION

While there are many factors listed here, there are many more that will be firm-specific. As stated above, firm size is the number-one consideration when deciding how to implement Revit at a firm. And while each subject has been labeled with a score giving an advantage to one size firm or the other, the intention of this section is to make firms aware of the issues regarding firm size and to assist firms in their implementation process.

Firm Size Final Score:   Larger firms – 3

Smaller firms – 3

## PROJECT SIZE

While project size is a major factor in Revit implementation, it also is influenced more generally by firm size. Refer to the Project Size section under the Firm Size discussion above.

## ARCHITECTURAL STYLE

When referring to architectural style and the implementation of Revit, the foundation of the distinction is the size, shape, and amount of detail involved. Revit does not recognize the literal difference, for instance, between Romanesque, Renaissance, and Post-Modern architecture.

Common sense, while not always common, does, in fact, make sense here. When it comes to modeling a building, straight, square forms are easier to model/build than the organic styles of Mr. Frank Gehry. This is not to say that it is impossible to build a sweeping tilted wave of a facade, just that more time and technique are required. If you can imagine the effort required to physically build the building, there is not a great difference to building it virtually using Revit. Simply put, more detail, shapes, and mass equals more time to develop and build.

While modeling all of the components in a project has desirable effects, keep in mind that it is not always necessary. During the planning stages of a project it should be determined what will be modeled and what can be "represented" with lines, arcs, and circles. A simple example: if you have raised wood paneling on a door, there is no harm in drawing lines on the surface of the door to represent the paneling and moving on with the project. Some might argue that it would not render properly or that there is a different material in the panel. The point is that there needs to be an understanding, on a per project basis, of what is important enough to model and what isn't. Three-dimensional modeling in Revit is not an all or nothing proposition.

## PROJECT TYPE

Project types often affect different firms in different ways. Generally speaking, however, some project types are better suited for Revit than others. Again, this is not to say that a certain project or project type cannot be done effectively. Some of the aspects of project types that dictate the ease of implementation include elements that are repetitive, component-driven, area-driven, or those that make use of—and find great value in—the different visualization and/or scheduling advantages of Revit.

Below is a general outline of the advantages and challenges typically presented by various project types.

### ENTERTAINMENT/RETAIL

Some of the advantages of Revit in the entertainment and retail industries might be the presentation value gained in color fills for area or room types, or the coordination of area calculations. Perhaps there is a need to schedule and color the tenant spaces by lease expiration, retail type, or rent-per-foot. Bi-directional, live scheduling may also have advantages, whether that includes display racks, flooring, or parking stalls.

Some of the challenges facing the entertainment/retail architect might include custom furniture and fixtures, or project size and scale. It has also been noted that due to the nature of a retail mall, individual retailers often have architects of their own, so there can be many different designs and ideas flying about at any given time.

## HEALTHCARE/HOSPITALITY

Healthcare and hospitality designs, of course, are going to have significant gains using the repetitive, modular, or component-driven aspects of Revit. It could be argued that this is also the case with CAD drawings, where xrefs, blocks, or cells are used, but Revit goes beyond that. When duplicating the data along with graphic representation of that data, you have the benefit of scheduling, coordination, and visualization. When a single change to one guest room sink and vanity can affect the plans, interior elevations, sections, scheduling, and specification information for 500 rooms, there is great efficiency in place.

Generally, healthcare and hospitality projects are large, very complex designs with large project teams. Incorporating and coordinating the necessary data set can become a significant burden. Planning a project of this nature is possible, but requires advanced planning and experienced Revit-capable staff.

## RESIDENTIAL

Residential projects gain tremendously from using Revit since they are generally smaller in size, as discussed above. Beyond that there is also the fact that residential clientele may not read or understand floor plans, sections, and elevations, but certainly would be able to understand a perspective or a rendering.

Challenges facing a residential architect might include the smaller scale of the project as well, surprisingly enough. Residential work tends to be shown at a larger scale, and therefore includes more detailing in each drawing. This tendency can give rise to the notion that all of the information must therefore be modeled. The residential architect would be wise to evaluate and plan the model simply.

## PUBLIC/RELIGIOUS

This is a very broad spectrum, and includes projects large and small, simple and complex. What they have in common is that there is a large body of critics to please. Whether it is the congregation of a church or the citizens of a city, communication is the key ingredient here. This brings the visualization aspect of Revit to the forefront. City councils can be shown the proposed building model within the proposed setting. The public can readily understand the aesthetics of a rendering and the color-fill diagrams showing the location of the services to be provided.

Challenges facing an architect specializing in these areas might include the fact that many of these building types can be rather large and ornate in nature, requiring additional time spent modeling. The need to provide many different presentation-level images requires additional time working out materials, and possibly significant computer rendering time.

## SCOPE OF SERVICES

Due to the nature of Revit—building a virtual model of the project rather that representing one with lines—it has been noted that more of the effort is front-loaded in the schematic design and design development phases. In other words, you are working out constructability issues earlier in the design process. This generally provides great gains in efficiency in the construction-documentation phase. Architects need to be cognizant of this issue, and perhaps make adjustments in the area. When the scope of services ends at what could be described as the traditional design-development phase, there is a greater level of information within the model than before. Architects are beginning to understand this, and respond in many ways, such as marketing additional services, adjusting fees (front-loading) to maintain projects through their entirety, and using a project information base as a sales tool for current and future work.

## FIRM LOCATIONS

It may go without saying, but the more office locations there are in a firm, the more complex the implementation process gets. There are many things to consider, as noted above. Training, support, and standardization also become more important (and slightly more difficult) when there are multiple office locations. It is not impossible to accomplish widespread implementation efficiently; it will, however, require a little more planning. The planning will need to take into account the different office cultures, personnel, and so on.

## FIRM CULTURE AND AGE

Firm culture and age may not seem important in the planning stages of implementation; however, it will become readily apparent during the implementation itself. How well does the firm accept and anticipate change? Is change part of the corporate culture? How do the lines of communication operate and what is the management organization? Who will need to back and support the initiative?

When age is considered, generally speaking, younger firms are more open to change, while older firms might tend to be more set in their ways and resistant to a change of this nature.

## FOUR KEYS TO SUCCESS

In addition to the seven factors of implementation outlined above, we have developed four keys to success in implementation, listed below. We have found that when focusing on these steps, the potential for success increases greatly.

Plan

Communicate

Train

Support

## PLAN

The first stage of implementation is planning. Of course, each firm is as individual as the employees within the firm. Careful planning is the foundation to a successful implementation. Carefully considering the seven factors of implementation outlined above can supply a firm with guidelines, but each firm should consider its own unique characteristics at every step. Planning the implementation of any strategy or technology should take into account all of the individuals and groups that the change will affect. It is advisable to discuss these issues with key individuals in these groups.

## COMMUNICATE

Once the plan is in place, it will become imperative that everyone knows what the plan is. It is very important to share the plan with the entire office. This does not mean inviting the office as a whole to a meeting to show them the latest tool. There are many different roles within any firm; any new tool will affect users in different ways, and therefore users will be interested in different aspects of the tool.

The principals or senior leadership will be interested in how the new technology affects the bottom line—what is the net gain? They will also be interested in what kind of investment in capital and staff that it will require. This group will also be interested in why the firm would make the change from the process currently in place. Generally speaking, their question will fall along the lines of "what does this mean to the business of architecture"?

The project managers will want to know how it affects their work flow. Will it really allow them to get work out sooner or with less staff? Revit's coordination of drawings, details, and grid bubbles is greatly appreciated by those who experience it. Managers are also interested in the coordination between consultants and how a team might share the data. What does this mean to the process of architecture?

The architects and designers that will be using the tool will generally be concerned with the user interface, toolsets, and learning curve. They will also be asking questions about how to create specific detailed models. What does this mean to the design and documentation of architecture?

## TRAIN

When implementing a technology like Revit, it is advisable to have dedicated training. While on-the-job training can be a great learning experience, it can also be detrimental. When there are deadlines and revisions flying about, trainees tend to want to jump back to their comfort zone, 2D CAD.

The training should be organized and tailored to cover a firm's specific topics and issues. Having general training to cover the basics of the tool can be helpful initially; however, having an expert in the firm covering firm-specific issues will pay dividends later.

## SUPPORT

Once the plan is complete and disseminated to your staff, and when training has been finished, the implementation is not yet complete. Firms should plan on maintaining and updating information through ongoing support. It is advisable to have regular meetings to discuss issues as they arise, from topics including new techniques to changes in structure and beyond.

## WHAT HAS WORKED

- **Planning**—There is no substitute for good planning. Take the time to consider the issues within your firm and plan on addressing each and every one. It is a good idea to plan on re-planning too, as firm-specific issues arise and need to be addressed.

- **Formal training**—While on-the-job training can be the best-quality training for the real world, real deadlines, revisions, and owners can often frustrate users and prompt them to want to give up and go back to good old CAD.

- **On-the-job training**—Occasionally, with the right personnel, project, schedule, team, and support, a user can get up to speed while working on an active project. Great care needs to be taken if this is the path you choose to take.

- **Baby steps**—Take a few of the overall benefits of Revit and focus on those issues on a particular project, mastering those, and then take on a few more on the next project. Some firms use pilot projects to take these steps.

- **Partial projects**—Consider the "horizontal" approach; use Revit to design and document the plan view and scheduling while using CAD to document the vertical drawings, like the sections, elevations, and perhaps the details.

- **Small projects with experienced users**—It may seem obvious, but overloading the first few small projects with experienced users works well and serves purpose down the road. The theory here is that these users will gain real-world experience and confidence. These users can then go on to work with others and spread the knowledge.

- **Management buy-in**—It will be important for the entire company to have the management and upper management understand the issues and lend support to the projects and users.

- **Rewards and recognition**—The early adopters of these technologies often have to endure the nay-sayers, critics, and the frustration of learning a new software and design process. With rewards and recognition they will, at the very least, feel that it is worth the effort. Some firms reward new users by upgrading their computers first, or doling out preferred seating in meetings or recognition in newsletters.

- **On-going training**—Many firms, recognizing the fact that trainees cannot learn a large application overnight, organize and support weekly mini-training sessions. These sessions generally last one hour and cover a single topic.

- **Exit Strategy**—On the rare occasion that the BIM is not working out on a particular project, it may be necessary to exit and return to AutoCAD. It will be helpful if that exit is prepared ahead of time.

## WHAT TO AVOID

- **Indifferent communication**—As noted above, users need to know not only how the tools work, but also the concepts of Revit and why the firm is going in this direction. Understanding what BIM is, how it is to be implemented, and why, can be just as important as learning how to use it.

- **On-the-job training without full support**—If it is impossible to get formalized, non-project-related training, it will become essential that the new users have a support mechanism available as close to full time as possible. When it comes down to crunch time with new functionality and there's no one in-house to assist, a deadline can become a breakdown point in implementation. The support team could be an in-house expert or a reseller/training center.

- **Attempting to hit a home run**—Some parts of Revit can be easier to learn than 2D CAD. It is, however, incredibly comprehensive. Take baby steps into the software. After all, most people did not learn CAD on the first project. Generally speaking, users become very comfortable and highly productive on their 2nd or 3rd project.

- **Isolating users**—This goes back to training and support. Users that are set free on a project without consistent support will tend to lose interest and become increasingly frustrated while they attempt to figure everything out themselves.

- **The oversell**—Avoid selling all of the benefits of Revit to clients prior to actually having a few projects under your belt. Be certain that the firm can deliver on promises. Avoid repeating vendor marketing promises without in-house verification.

## WRAP UP

This appendix is, just that, a little bit of extra information at the end of this book. There are any number of combinations and additional factors that can, and very well will, affect the implementation of Revit at any given firm. The intent of this appendix is to give general guidance to those on the implementation path. Good luck, and good hunting.

# INDEX

# D

# K

# L

# R

## S